*Advances in
Underwater Technology,
Ocean Science and
Offshore Engineering*

Volume 27

SUBTECH '91
Back to the Future

ADVANCES IN UNDERWATER TECHNOLOGY
OCEAN SCIENCE AND OFFSHORE ENGINEERING

CONFERENCE PLANNING COMMITTEE

Cyril Lafferty, *British Gas plc* (Co-Chairman); Tom Hollobone, *AODC* (Co-Chairman); O. C. Andersen, *Stolt Nielsen Seaway (UK) Ltd*; Steve Bibby, *Texaco North Sea UK Ltd*; Ian Brown, *The Marine Technology Directorate Ltd*; Paul Henderson, *Total Oil Marine plc*; Ken Hulls, *Comex UK Ltd*; Bil Loth, *W. D. Loth & Co Ltd*; Barry Moore, *Stena Offshore Ltd*; Jean Pritchard, *SUT*; Duncan Sharp, *Shell UK Exploration and Production*; John Smith, *Chevron UK Ltd*; Stuart Smith, *Amerada Hess Ltd*; Trevor Smith, *BP Exploration*

Advances in
Underwater Technology,
Ocean Science and
Offshore Engineering

Volume 27

SUBTECH '91
Back to the Future

Papers presented at a conference
organized by the Society for Underwater Technology
and held in Aberdeen, UK, November 12–14, 1991.

SPRINGER SCIENCE+BUSINESS MEDIA, B.V.

ISBN 978-94-010-5569-7   ISBN 978-94-011-3544-3 (eBook)
DOI 10.1007/978-94-011-3544-3

Printed on acid-free paper

Contents

Society for Underwater Technology

The Society was founded in 1966 to promote the further understanding of the underwater environment. It is a multi-disciplinary body with a worldwide membership of scientists and engineers who are active or have a common interest in underwater technology, ocean science and offshore engineering.

Committees

The Society has a number of Committees to study such topics as:

Diving and Submersibles
Offshore Site Investigation and Geotechnics
Environmental Forces and Physical Oceanography
Ocean Resources
Subsea Engineering and Operations
Education and Training

Conference and Seminars

An extensive programme is organized to cater for the diverse interests and needs of the membership. An annual programme usually comprises four conferences and a much greater number of one-day seminars plus evening meetings and an occasional visit to a place of technical interest. The Society has organized over 100 seminars in London, Aberdeen and other appropriate centres during the past decade. Attendance at these events is available at significantly reduced levels of registration fees for Members or staff of Corporate Members.

Publications

Proceedings of the more recent conferences have been published in this series of *Advances in Underwater Technology, Ocean Science and Offshore Engineering*. These and other publications produced separately by the Society are available through the Society to members at a reduced cost. A careers pack 'Oceans of Opportunity' has been produced by the Society in response to the growing demand by students schools and colleges for up-to-date information.

Journal

The Society's quarterly journal *Underwater Technology* caters for the whole spectrum of the inter-disciplinary interests and professional involvement of its readership. It includes papers from authoritative international sources on such subjects as:

Diving Technology and Physiology
Civil Engineering
Submersible Design and Operation
Geology and Geophysics

Subsea Systems
Naval Architecture
Marine Biology and Pollution
Oceanography
Petroleum Exploration and Production
Environmental Data

An Editorial Board has responsibility for ensuring that a high standard of quality and presentation of papers reflects a coherent and balanced coverage of the Society's diverse subject interests; through the Editorial Board, a procedure for assessment of papers is conducted.

Endowment fund

A separate fund has been established to provide tangible incentives to students to acquire knowledge and skills in underwater technology or related aspects of ocean science and offshore engineering. Postgraduate students have been sponsored to study to MSc level and subject to the growth of the fund it is hoped to extend this activity.

Awards

An annual President's Award is presented for a major achievement in underwater technology. In addition there is a series of sponsored annual awards by some Corporate Members for the best contribution to diving operations and oceanography, and for the best technical paper in the Journal

FURTHER INFORMATION

If you would like to receive further details, please contact
Society for Underwater Technology, The Memorial Building, 76 Mark Lane,
London EC3R 7JN.
Telephone: 071-481 0750; Telex: 886481 I Mar E G; Fax: 071-481 4001.

Part 1
Back to the Future

PRODUCTION TECHNOLOGY FOR SUBSEA DEVELOPMENT WELLS

S.D. GOMERSALL
Marathon Oil UK Ltd
Rubislaw Hill
Anderson Drive
Aberdeen   AB2 4AZ

ABSTRACT.   This paper provides a discussion of the many points
which should be considered when planning the engineering aspects
of subsea development wells.   The multi-discipline issues in a
subsea project are presented together in order to emphasise the
requirement for effective project management and close teamwork.
Drilling activities including well design and rig selection are
considered along with downhole completion design.   This includes
the differing requirements of exploration wells and development
wells with respect to subsequent well utilisation.   The
requirements of the subsea completion, installation equipment and
subsea facilities and their impact on other well design issues are
presented.   Recent   industry   efforts   on   subsea   equipment
standardisation through API 17D and other North Sea initiatives
and   the   potential   for   further   industry   standardisation   are
discussed.   The paper should form a useful reference for those
involved in subsea projects by way of highlighting the issues
which must be considered, without always presenting solutions.
Information has been drawn from many references reflecting
primarily North Sea operations in addition to the author's own
experience.

INTRODUCTION.   The issues presented are arranged under the
following seven headings; Exploration and development well design
issues, rig selection, downhole completion, subsea completion,
subsea facilities, standardisation and project management.   Subsea
well production technology is the theme for the paper and it is
intended to compliment Reference 1 which concentrates on drilling
related subsea development well issues.   The nature of a subsea
development implicitly involves many disciplines and it is
essential that from the earliest planning phase the potentially
conflicting requirements of all the disciplines are taken into
account.   This paper presents these issues in order that they can
be considered by personnel involved with subsea projects and
together can form a coordinated approach to production technology
for subsea development wells.

3

Volume 27: Subtech '91, 3–28.
© 1991 *Society for Underwater Technology.*

1.   EXPLORATION AND DEVELOPMENT WELL DESIGN ISSUES

The suitability of a subsea well for completion is dependent on its nature as an exploration or development well and more specifically on the way the well was planned. This section highlights various key points which, if considered early in the well design, can help to make the well more suitable for re-entry and completion should it be required.

Seabed Location

One of the earliest design decisions for any well is the seabed location. In the case of an exploration well the seabed location is often chosen to be vertically above the target. This is usually no problem, unless other wells have been drilled or are planned close by, in which case future operations may be significantly complicated by overlapping mooring patterns or seabed wellhead obstructions. (Reference 1).

As wells become more appraisal in nature then the seabed location becomes more of an issue. A decision has to be taken on an optimum location or locations to group wells together in order to avoid the problems mentioned above. If this decision is not taken and wells are drilled individually then the potential savings from being able to reuse existing wells for either a subsea development or a platform development are reduced. The economics of a subsea project make grouped wells much more attractive through the use of combined facilities, such as flowlines and umbilicals, and the ease of future access between wells without needing a rig move. The chosen seabed location should take into account potential flowline and umbilical lengths and routings in addition to directional drilling considerations to target locations.

Well Spacing

Having decided to group wells together at one seabed location the next decision is how close they should be and how they should be spaced out and arranged. The four primary options available are; separate without accurate spacing, separate with accurate spacing, modular close spaced and template drilled. Since the cost of having wells accurately spaced but mechanically separate is only marginally more than having wells loosely spaced it is generally preferable to use some form of accurate spacing. Knowing the exact subsea location of a group of wells with respect to each other allows several future options; platform tiebacks, installation of a common subsea protection frame and more rapid installation of subsea tie-ins.

Wells can be accurately spaced at low cost using slings installed over the adjacent well posts. (Reference 1). In this case the well arrangement is usually linear with as many wells as are required in a straight line and up to 30' (9m) between centres (Figure 1). Accuracies within 6" (152mm) of separation and 3 degrees of rotation have been achieved over a three well cluster. The primary advantages of this method of well spacing are its low cost and its operational flexibility allowing additional wells to be included as necessary. Another key advantage from the drilling point of view is that the wide spacing makes directional drilling in the top hole interval less critical than with closer well spacings.

Closer spaced wells can be similarly clustered as required using proprietary modular guidebases. These guidebases are cantilevered from a primary guidebase and share two guideposts (Figure 2). The distance between well centres of 8'6" (2.6m) is extremely close to allow subsea completion. Although adjacent wells have been completed on such guidebases such spacing is not recommended for future subsea development wells. (Reference 14)

The final option for grouped wells is a template, designed either for drilling or for drilling and production. The cost of a template of either kind cannot usually be justified during early exploration or appraisal drilling and may only be a realistic option if such drilling has been successful. A production template can be used for the final part of a drilling programme which was begun using any of the previously described well spacing options.

Guidebases

Normal exploration wells utilise temporary and a permanent guidebases with the temporary guidebase (TGB) providing a foundation upon which the permanent guidebase (PGB) is set. In hard seabed conditions this means that any subsea tree set on the PGB is several feet above the mudline allowing the necessary clearances for subsequent flowline and umbilical tie-ins. In soft seabed conditions the TGB may sink along with the PGB making completion an impossible task. With these conditions it may be preferable to miss out the TGB and to cement the 30" conductor and PGB well above seabed. Alternatively using a 48" diameter stub on the base of the TGB may prevent hole washouts during drilling of the 36" hole which in turn may prevent the TGB sinking. Yet another alternative is to use bolt on boxes to give the temporary guidebase additional surface area to prevent it sinking.

The height of the guidebases also affects the ability to dispose of top hole cuttings and cement returns. An ROV should be used to ensure the build up of debris on or around the guidebase is not excessive. Close spaced development wells are particularly prone

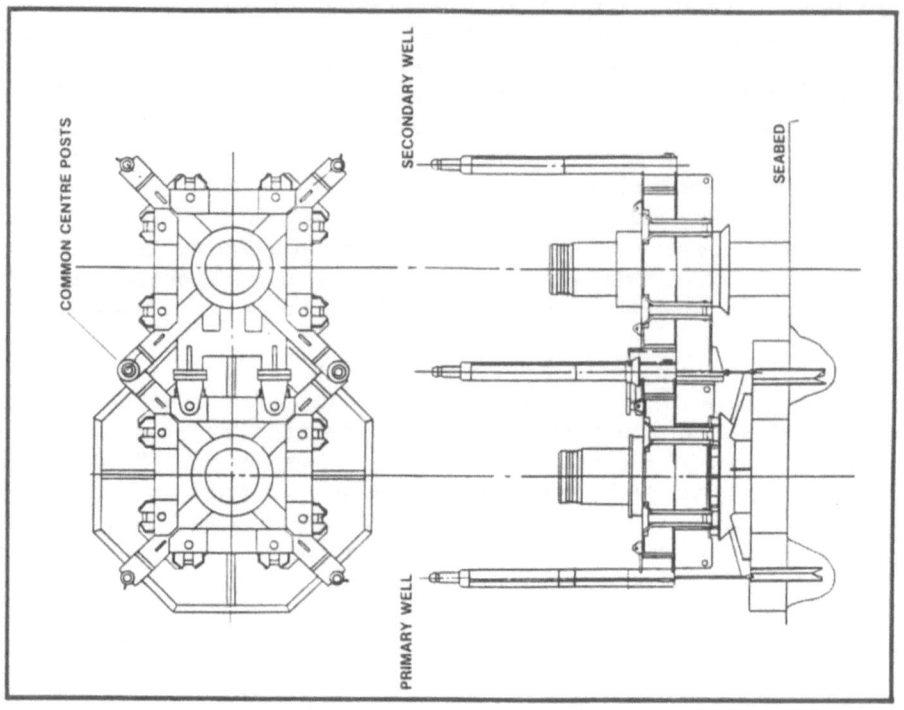

Fig. 2: Modular Guidebase and Cantilever

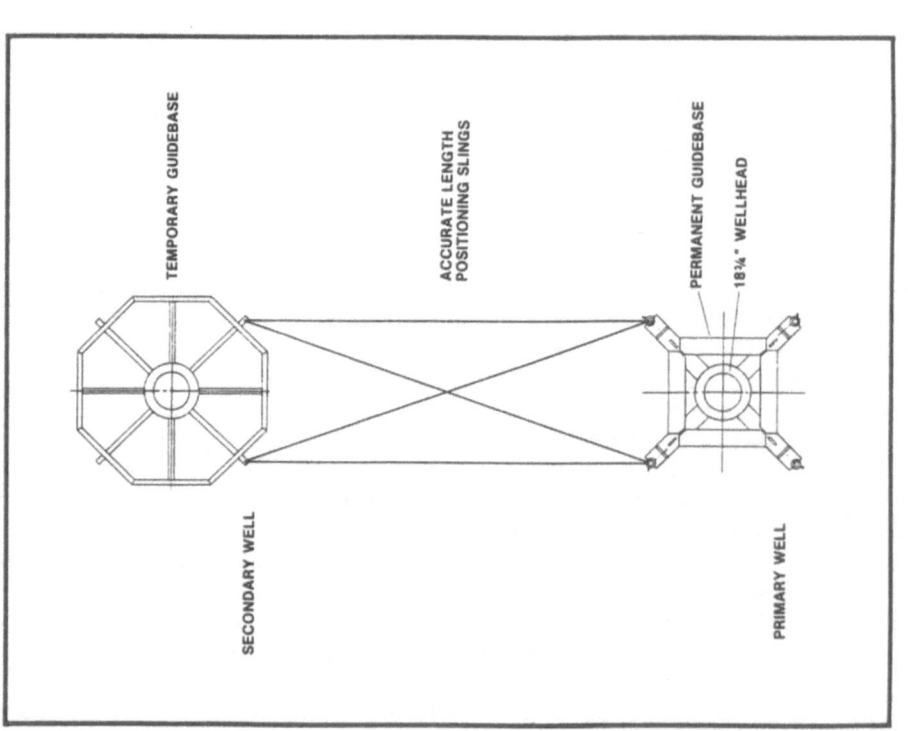

Fig. 1: Accurately Spaced Clustered Wells

to problems with debris and care should be taken to remove it.

Consideration should be given to preparing a permanent guidebase for production, prior to its installation on any well. The following areas should be considered; orientation to allow additional wells at the same location, mudline clearance, near horizontal levelling, preparation for flowbends, flowline tie-ins or controls tie-ins as required and removal of any unnecessary padeyes etc. which may subsequently cause fouling.

As part of a more general guideline, an accurate record of all guidebase equipment must be kept to allow planning at the completion phase. This must include the manufacturers part numbers of all components and details of any non standard items, especially any extra fixtures and fittings.

Wellheads

Subsea wellheads have been developed primarily for their drilling function. Their use as a key part of the completion system has been an afterthought by manufacturers who have designed tubing hanger/tree to wellhead interfaces around existing casing hangers and wellhead profiles.

In general subsea wellheads used in the North Sea are 18-3/4" one stack systems with either 10,000 psi or 15,000 psi pressure ratings. Traditional casing sizes are most often used; these being 30", 20" 13-3/8", 9-5/8" and 7" liner. Although not always possible it is generally preferable to be able to run 7" casing (either as a liner tieback or a full casing string) and still land a tubing hanger in the wellhead. This contingency allows for long term use of a well in various production scenarios where the integrity of the production casing is essential, such as gas lift or high pressure gas injection. Platform experience indicates that over the producing life of a well the possibility exists of the 9-5/8" production casing losing its integrity thus requiring the use of a 7" cemented tieback.

The most obvious requirements of a subsea wellhead for production are the ability to land, seal and lockdown the tubing hanger internally and to land, seal and lock down the xmas tree externally. Tubing hanger lock down mechanisms can be either to the wellhead or the previous casing hanger with a preference for the improved integrity of a wellhead lockdown. The xmas tree locking and sealing profiles in the wellhead are generally the same as those used for the BOP stack and are therefore usually reasonably simple and reliable. Care should be taken however during the drilling phase not to damage the seal areas during BOP installation or retrieval.

During the drilling of an exploration well, lock down rings available with each casing hanger to lock the casing into the wellhead are rarely run. Their use prevents the rapid recovery of the wellhead system during abandonment. In a subsea development well their use is almost essential in order to prevent upward movement of each casing string with thermal expansion. Their use in an appraisal well which may be reused should be evaluated in advance, the alternative being to rely on a single lock down assembly installed in the wellhead during completion. One area which has previously caused operators problems is the loss of lock down rings when a lock down assembly is retrieved. This can be overcome by careful design of the assembly body to retain the ring on retrieval.

The selection of a subsea wellhead system which may be used for production should take into account several specific issues;

Wellhead Seals: These must be suitable for long term production and therefore be rated correctly for temperature and material. The mechanism may include a metal to metal seal or may be limited to an elastomer which is compatible with produced fluids and annulus completion fluids. If a wellhead is to be used which gives the option of platform tieback then it is likely to have an 'annulus shut off insert' which, when removed, allows tieback access to the 9-5/8" x 13-3/8" annulus. This insert includes an elastomer seal which must be upgraded for production service.

Wellhead Metallurgy: The short term exploration requirements for subsea wellhead metallurgy are unlikely to be the same as those required for long term production service. The composition of produced fluids and annulus completion fluid must be considered when specifying wellhead metallurgy. Unlike a platform wellhead, most subsea wellhead components are impossible to inspect and replace.

Fatigue: The fatigue life required from a subsea wellhead used for production is significantly greater than that required just to drill the well. Precautions must therefore be taken to ensure a subsea development wellhead has the maximum fatigue life possible. Other authors have addressed this subject (References 1, 13) and suggested; a wellhead with two point reaction load transfer between wellhead and conductor using at least 80 ft (24.4 metres) of 30 inch (762 mm) x 1.5 inch (38 mm) wall thickness X52 conductor with pre-loaded connectors and ensuring a good cement job by cementing the conductor using a stinger until returns appear at the seabed.

Standardisation: Common sense indicates that within any one company a standard wellhead system should be used on all wells which may be completed subsea. This will allow the use of common subsea completion components and minimise any redesign costs.

Depending on operational activity, availability and cost, using a standard wellhead may or may not be preferred for any particular well.

Casing Design

From a production well point of view the casing should be designed to house the required completion. This includes being of acceptable internal diameter, having acceptable pressure ratings (compatible with the worst design case of the well on production or injection) and being of suitable metallurgy. Often these are not the primary design considerations on exploration or appraisal wells but their early consideration will improve the flexibility of the wells' future for minimum incremental expense.

The casing internal diameter required should be suitable for the required tubing size along with its completion components in particular large diameter items such as gas lift mandrels and tubing retrievable safety valves. At surface the ID must be suitable to house the production bore offset if a dual bore tubing hanger and tree are used. All production casing connections should be of the premium threaded type to minimise the possibility of casing leaks.

A key issue of well design for subsea production is the build up of annulus pressure causing casing and tubing collapse. Of particular concern is the 9-5/8" x 13-3/8" annulus, the pressure in which is not normally monitored on subsea wells. As a well warms up on production, the fluid in this annulus expands and as a result the pressure increases. Sufficient pressure can be generated, especially with the high coefficient of thermal expansion of oil based muds, to collapse the 9-5/8" casing and tubing within (References 1,3, 7). A solution to the problem is to cement the 9-5/8" casing short of the 13-3/8" casing shoe and subsequently allow annulus pressure to bleed off to the formation. In this case the formation leak off pressure must be low enough to accept fluid before any other mechanical problems may occur. The design of the liner must take into account the proposed completion design. Usually the liner forms a part of the produced fluid flow path and must therefore be considered as part of the completion. If the completion is a liner top tie back then polished bores and anchor points may be needed in the liner hanger. Provision should be made for a liner top packer in the event of a leaking liner lap and the size, length and metallurgy of the liner must all be considered.

Well Testing and Suspension

Any well testing and well suspension activities must take into account that the well may be re-entered for long term production.

Well test aspects including test interval, water tests, injectivity tests and tubing conveyed gun debris must be considered. Any of these could compromise the primary objective of long term production and they must be weighed against the requirement for well test data.

The way in which zones are suspended/isolated may impact subsequent production and the overall well suspension plan must be compatible with a simple and reliable re-entry. Cement plugs set across perforated intervals should be avoided and the use of bridge plugs or cement retainers immediately above the liner top should be avoided. Suspension fluids should be chosen to prevent corrosion and minimise handling problems when drilled out, in addition to being of reliable kill weight and minimising cost.

2.0 RIG SELECTION

The selection of a semi submersible vessel for drilling and completion activities on a subsea project must be a compromise between drilling requirements, completion requirements, cost and availability.

Suitability for Completion and Drilling Activities

The amount of time to be spent in each phase of the operation, drilling and completion must be carefully considered. Because the rig specification and selection is often carried out by a drilling department, emphasis towards drilling equipment and performance is usually given the highest priority. However during a subsea development as much as two thirds of all rig time is taken up with non-drilling activity (References 1,7). Therefore the efficiency of the rig during the completion phase is critical to the operation and completion requirements should be given significant attention during rig selection. Furthermore whereas semi-submersible vessels are generally designed and equipped for subsea drilling, they are not always designed and equipped for subsea completion activities.

Rig Facilities

The necessary equipment and services to complete a well subsea must be available and adequate capacity. Probably the most important aspect is the handling of BOPs and subsea trees on deck and in the moonpool areas.

Sufficient space should exist to store both the rig BOP and a subsea tree in the moonpool area simultaneously. The height available in the moonpool should be sufficient to allow stack-up of a lower riser package (LRP) and emergency disconnect package (EDP) on the tree or an alternative stack-up procedure must be developed during rig selection. Skid beams should be provided to allow acceptable deck load distribution and the easy movement of subsea equipment into and out of the moonpool. The launching method should be carefully evaluated on an individual rig basis considering the use of the BOP 'fork lift transporter' equipment or movable spider beams in the moonpool. Using an additional set of cross beams may overcome limitations of existing spider beams.

During well completion it is often necessary to backflow wells to the rig for clean-up. In this case it may be necessary to use all or part of a standard well test hook-up with; choke manifold, test separator, tanks, pumps, booms and burners. If a full well test is to be performed then a significant amount of space will be occupied by this equipment. A rig's suitability for well testing must be considered, including issues such as; rig test piping (length, route, certification, pressure rating, metallurgy and temperature rating), boom and burner availability, and equipment positioning.

The other primary rig facilities which must be considered during rig selection for subsea completion are; Crane capacity, deck load capacity, electrical power tie-in for control systems, air points and vessel stability and heave characteristics.

Space

A large amount of equipment of varying shapes and sizes must be onboard during completion activities. Of necessity the rig deck space must therefore be considered during the early stages of rig selection. Space must be available on skid beams for the subsea tree, LRP, EDP, access stand, tree test base, LRP test base and any other shipping skids. The workover control system must be housed either on the rig floor or in the moonpool area. The pipe deck has to contain the rig's drillpipe and drill collars, a string of tubing, several completion baskets, drilling riser and completion riser (each with 50' joints) as well as other containers and possibly items of well test equipment. Dive facilities including saturation vessels, dive bell, rescue craft, gas storage, dive workshop and dive control must all be fitted in. During the final stages of rig selection all the major items of equipment must be identified and their location on the rig decided in order to ensure sufficient space exists. These considerations are especially relevant during the first completion, with a set of equipment which may never have been gathered together before.

The other major space concern is that of accommodation. During completion activities the large number of contract personnel can easily reach the accommodation limits of normal semi-submersible vessels. Usually careful control of personnel movements can avoid problems but if the rig facilities are to be used for multiple operations such as simultaneous subsea wireline work, accommodation can be limiting (References 12 and 17).

Rig Crew

As important as the rig facilities and equipment is the experience level of the rig crew. A crew which has experience with previous subsea completions is a major asset, since personnel will be familiar with types of equipment and general procedures. If this is the case, discussing at length proposed procedures with the crew can quickly highlight problem areas and solutions, especially in the area of tree handling and deployment. These discussions should take place as soon as possible, preferably initiated during a rig visit shortly after rig selection. The principle can be effectively extended further, by having the contractor's rig engineer work with project staff in the operators office during the pre-completion phase. The importance of involving those personnel who will be involved during the offshore installation should not be overlooked.

Dive Facilities

An important consideration, if saturation divers are to be used on a project, is the location and space available for dive facilities. Often during rig selection no dive facilities will be installed, and a standard location will exist on the rig where these facilities have been located in the past. In either case discussions should be arranged with the diving contractor to agree the dive spread layout, dive moonpool access and emergency evacuation methods. The United Kingdom Health and Safety Executive currently requires two independent means of evacuation for all the divers on the rig in saturation. This can cause significant problems with dive spread layout and facilities.

3.0   DOWNHOLE COMPLETION

During various stages of a subsea development the downhole completion must be considered. Previous mention has been made of the requirements for the well design, at the drilling stage, to take into account tubing size and liner metallurgy amongst other issues. The downhole completion must be considered as an integral part of the subsea production system and all aspects of completion design must be considered in parallel with tree and subsea equipment design. This requirement is to ensure that the downhole issues are adequately reflected in subsea equipment.

In general downhole completion design and equipment for a subsea well is no different to that for a platform well. However certain aspects require significantly more attention. Because of the cost to re-enter wells the following requirements should be addressed;

(i)   Minimise workovers for mechanical reasons.

(ii)   Minimise re-entries for reservoir monitoring.

(iii) Minimise workovers for reservoir initiated reasons.

In order to avoid unnecessary workovers for mechanical reasons, only the most reliable equipment should be used. Additional expense, if it can be shown to increase reliability, can be easily justified if it helps avoid future workovers. The use of corrosion resistant alloy completion equipment in the case of marginally corrosive produced fluids, being a good example. Using equipment with the minimum complexity and with a proven track record will help improve reliability. The downhole safety valve is often the least reliable piece of equipment in a well and significant attention should be given to the type of valve used and its previous failures (Reference 7, 18).

Reservoir monitoring requirements can be minimised in two ways. Firstly using permanently installed monitoring devices such as downhole pressure and temperature gauges, so long as these do not compromise the mechanical reliability of the system. The reliability of the gauge system itself has often been relatively poor, but more recent installations have shown much improvement (Reference 16). Secondly the education of reservoir engineering staff and management that reservoir monitoring on a subsea project will be significantly different to that available on a platform. Decisions will be required with less data and reservoir modelling will have to take place with fewer history match points.

Workovers which are required for reservoir initiated reasons include the need for water shut-off, gas lift, smaller tubing and well conversion. Where possible workovers for these reasons should be minimised or avoided. It is often feasible to install completion and subsea equipment to allow for either production or injection at some stage during the life of the completion (Reference 4). The use of monobore type completions, with the minimum internal diameter across the reservoir interval, will allow through tubing wireline isolation of water zones. Sizing tubing for the long term production from a well and not for short term high rates and the use of pre-installed gas lift strings will help avoid premature workovers. Artificial lift of subsea wells has most often been carried out using gas lift (Ref. 5, 7, 9). This requires consideration of gas supply, annulus access, tubing and casing design, flowline multi-phase flow at high GOR's as well as detailed analysis of the system pressures. A detailed review of all artificial lift options and the selection of the preferred system should be carried out early in the project design phase.

4.0 SUBSEA COMPLETION

Design Considerations

The primary design considerations for subsea tree and tubing
hanger are obviously size and pressure. Typically 5" x 2" 5,000
psi or 4" x 2" 10,000 psi dual bore systems and 5" 5,000 psi
concentric systems have been used in the North Sea. Very few
developments have required subsea equipment pressure ratings in
excess of 5,000 psi and no subsea production equipment has been
built with a 15,000 psi rating.

The issues in choosing between concentric and dual bore systems
are complex with both options offering certain advantages and
disadvantages. Most North Sea equipment is dual bore but several
concentric projects exist. The primary argument against dual bore
systems is the problem of orientation for the tubing hanger and
tree. Several different orientation systems exist and have proved
successful in use (References 4, 6, 10). The primary argument
against concentric systems is the lack of positive annulus shut
off in the tubing hanger when the tree is removed. A spring
loaded sleeve or valve, which cannot usually be tested prior to
removing the tree, contains annulus pressure during tree removal.
The positive setting and testing of wireline type plugs in both
production and annulus bores of dual bore tubing hangers is
generally preferred.

Overall tree configuration is becoming more standard with usually
double master valves, wing valve and swab valve on the production
bore. Similar valving is typical on the annulus bore, of dual
bore systems, although usually with only one master valve. The
tie-in between annulus side and production side is either inboard
or outboard of the wing valves and may include a crossover valve
or pressure relief system.

The issues which give rise to a variety of additional features
are; ability to test individual wells, remote well kill facility,
production start up/shut down philosophy, downhole safety valve
and tree valve integrity testing, downhole safety valve
equalisation, annulus pressure monitoring and bleed down
(References 4, 7), injection of inhibitors and pressure and
temperature monitoring facilities. The way in which each of these
requirements is handled is reflected in the final tree
configuration.

Other overall design requirements should be to utilise proven and
reliable subsea equipment where possible, and to incorporate
built-in redundancy to key systems (Reference 5). By example this
may include a secondary release mechanism for the tubing hanger
running tool and a mechanical release for the subsea tree.

Appraisal Well Re-entry

The re-entry and completion of an exploration or appraisal well may require the use of an existing wellhead from one vendor when the tree and tubing hanger have subsequently been sourced elsewhere. Different operators have solved this problem of manufacturer interface in two ways. The first is to use a Lock Down Seal Assembly (LDSA) which can be landed in the wellhead and which forms the landing profile for the tubing hanger (Reference 1, 10). In this case the wellhead manufacturer has to provide information regarding the wellhead internal profile to allow the LDSA to be designed. The second method is to use a separate 'wellhead spool' consisting of a short wellhead housing (Reference 14). This is mounted above the existing wellhead and connected using a standard 18-3/4" AX profile. The primary advantages of this method are; the reliance only on the AX gasket profile and not the wellhead internals , no requirement for wellhead vendor information, optional 7" casing hang off point, new sealing surfaces for the tubing hanger and the ability to mount flowbends on the associated wellhead spool guideframe. This method is however more expensive.

Along with the tree design and order, several miscellaneous areas must not be neglected. Amongst the most important of these is the requirement for testing and shipping frames/baskets for all equipment since damage of equipment in transit can cause significant lost time and problems offshore. The purchase of a scale model although often relatively expensive can prove invaluable during periods of troubleshooting. Adequate drawings available offshore and onshore, especially complete stack ups, are essential in understanding and solving problems.

Installation Equipment

The design of installation equipment must identify key operational requirements such as; recoverable valving configuration, cutting facilities and disconnection facilities. Most recent developments make use of a Lower Riser Package and an Emergency Disconnect Package with which the tree is run. The LRP typically contains a production bore valve or set of BOP's, an annulus valve and a crossover valve allowing access from the production bore to the annulus bore. This is used to circulate the riser clean of hydrocarbons, for pressure testing and for placing hydrate inhibitors. Many recent developments have utilised a dual bore completion riser to run both the tubing hanger and subsequently the subsea tree. These risers have proved efficient in operation and rapid to deploy.

Diver and ROV Access

For installations which use divers for routine subsea operations
it is essential that equipment is diver friendly (References 5,
14). Divers should be consulted early in the design of a project
in order to advise on their requirements. As indicated
previously, with the rig crew on a drilling vessel, the divers
will have extensive experience of the good and bad points from
previous projects, which can be usefully incorporated into a
project.

Access, ease of operation and clear identification are probably
the three subjects requiring most attention for divers. Equipment
should be designed to be easy to see, easy to reach and with
plenty of space to manoeuvre. Items which are intended for diver
manipulation should be especially accessible, clearly identified
and with positive indicators of correct operation. All valving
and components should be labelled with anti-fouling markers and
non degradable attachments. Everything should be painted matt
white or yellow and silver high reflective surfaces should be
avoided.

The use of ROV's within saturation diving depths, for operations
other than remote television, is rarely efficient. Specially
designed ROV interfaces and ROV tooling requires extensive
development and once installed can cause major access problems for
divers. Systems should be designed either to be diver friendly or
diverless, most middle options cause operational difficulties.
With over 96% of all subsea completions currently installed being
within saturation diving range, experience of diverless systems
remains limited (Reference 16).

Projects to design generic diverless systems for as yet
undiscovered fields have major limitations. Systems have a
tendency to over complexity due to their attempt to meet too many
design criteria (Reference 11). The current generation of
diverless systems will have to remain extremely simple in design
with specific equipment to meet specific requirements.

5.0 SUBSEA FACILITIES

Subsea facilities in addition to the subsea completion make up the
subsea production system. Since this paper is intended primarily
to address well related issues the points on subsea facilities
discussed in this section are those necessary to ensure successful
well operations. These include template or manifolding systems,
protection, and certain aspects of control systems and flowlines.

Manifolding and Templates

The use of templates as a drilling guidebase has been discussed earlier.   In general a template is only used in this mode for planned production wells rather than exploration or appraisal wells.   If a template is chosen for the development phase of a project it can usefully serve many functions.   These include; drilling guidebases, protection, control system tie-ins, flowline tie-ins, annulus line tie-ins, diver work platform, flowline manifolds and pigging loops.   The template can serve as a useful way of spacing wells whilst incorporating many other features which are of long term benefit keeping the subsea layout logical and tidy.   High cost   along with an uncertain development programme and timing are the main reasons templates are not always chosen.

One alternative is a well cluster spaced close or widely, the flowlines from which are manifolded together at another nearby location.   (Reference 15).   Drawbacks of this option are; separate protection is required possibly on a well by well basis, tie-ins for both flowlines and controls are long and more awkward, dropped object protection is no better than template spacing, flowline tie-ins are unprotected, diver access to wells is more awkward, and simultaneous subsea wireline whilst drilling is much more difficult (References 2, 12, 17).   The advantages of a clustered system are a minimal capital expenditure and a flexible development with expenditure as required.

The remaining option is to utilise single satellite wells tied back on a well by well basis.   For a one or two well development this option is probably to be preferred.   The drawback with more wells is the expense of control system umbilicals and flowlines, the requirement for single well protection, increased subsea congestion with wellheads causing interference in anchoring, and a requirement to move the rig to get from one well to another.

Maximum flexibility should always be designed such that any well can be tied in as an injector or a producer and facilities such as gas lift lines should be available for each well.   This allows flexibility in well utilisation as reservoir requirements develop (Reference 5).

Protection

Most subsea wells in the North Sea utilise some form of protective structure with more significance being given to the subject in the Norwegian Sector than in the UK Sector (Reference 16).   As discussed previously this can form part of a larger template structure or be a specific structure in itself.   Most wells outside the North Sea utilise no specific protective structures.

Protection is primarily against trawling equipment but also to give some protection against anchor chains, and dropped objects. Cathodic protection is usually given by way of aluminium anodes on subsea trees and other hardware. It is essential that these are adequately sized and take into account at least part of the 30" conductor surface area when performing corrosion rate calculations.

Flowlines

The number of flowlines required for a development is dependent on the number of services required and the way in which bulk production is handled. Three options exist for flowing production back to the process facilities; commingled production with individual well subsea chokes, commingled production with no chokes, and separate flowlines for each well. All three systems have been operated successfully in several projects. The use of chokes, although offering several operational advantages, such as gradual well opening and production allocation, can lead to well down time as a result of choke failures. Subsea chokes, if used, should be easily recoverable for maintenance and repair.

Whilst separate flowlines for each well is a high cost approach, the use of commingled wells without subsea chokes represents an operational compromise. The lack of chokes means wells tied into a common manifold have identical tubing head pressures and the rate from each well cannot be controlled independently. Using downhole pressure gauges in these wells helps to produce them within the optimum range of reservoir conditions.

The routing of flowlines and umbilicals must be considered very early in the project when appraisal well locations are being chosen. Necessary flowline corridors must be set aside outside the anchor pattern and avoiding other constraints. The cost of flowlines is often a significant percentage of the project cost and attempts should be made to review the types of flowline available and the methods used for installation. Flexible lines can be easily installed and have been used for most types of flowline. Lines manufactured from polymer have been used as gas lift lines and tubing strings installed from a semi-submersible drilling vessel have been successfully laid as flowlines (Reference 7).

Consideration must be given to corrosion of all subsea lines and the use of subsea injected corrosion inhibitor or the specification of superior metallurgy may be often justified. Where seabed conditions are not perfect and if a drilling vessel is planned to anchor repeatedly over a period of years the use of permanently installed piled moorings may be required. These avoid the major pitfalls of normal anchors such as slippage and heavy

Standardisation Potential   Table 1

Area	Status and Scope	Potential Benefits/Drawbacks
Minimum System configuration and requirements (API 17D)	Already agreed industry wide	Simplified equipment specifications
One Specific interface (Top of LRP connector)	Discussions well advanced and general industry agreement (at least in North Sea) likely	Allows interchange of riser systems
Several key interfaces	Potential exists for industry standardisation	Allows flexibility exchange of key pieces of equipment
All relevant interfaces and certain key items of equipment	Unlikely to be standardised other than within an individual operating company	Total flexibility
All equipment components and interfaces	Beyond standardisation	Restrictive

Table 2

Shell/Texaco/BP Interface Proposal

LRP Connector (tree running tool) to LRP safety valve block (LRP BOP)

Production Bore	5-1/8"	(130.2mm)
Annulus Bore	2-1/16"	(52.4 mm)
Pressure Rating	10,000 psi	
Production bore offset	1.875"	(47.6 mm)
Production/Annulus bore spacing	5.375"	(136.5 mm)
Main flange interface	13-5/8", 10,000 psi, API 6BX BX159 ring gasket	

lifts during installation, and allow higher anchor tensions to be used if required.

Control System

An extensive discussion of subsea control system is outside the scope of this paper. However since the control systems represent the most vulnerable and unreliable part of the subsea system several point should be discussed. Control system umbilicals and subsea hoses have frequently caused problems (Reference 4). Various leak detection fluids have been used to identify the location of hydraulic leaks. Fluorescent dyes and dark contrast dyes have both been used, but probably the most efficient dye for subsea use is a light contrast dye such as one which become bright milky white in contact with seawater. It is essential that all chemicals introduced into the control system, including dyes, are chemically compatible with the control fluid and control system hardware.

An often neglected area in a subsea project is the workover or completion control system. Often this is manufactured by the same manufacturer as the primary control system requiring an additional design interface with the subsea tree vendor. Purchasing a workover control system from the tree manufacturer would negate this problem. Workover controls are frequently over complex, awkward to operate, insufficiently robust and poorly laid out.

6.0 STANDARDISATION

'Standardisation' of subsea well completion equipment is a topical but ill defined area of development. The objective within this paper is to present what is meant by standardisation and discuss some of the current proposals and their benefits. By standardisation of subsea equipment we mean designing equipment for different projects (possibly for different operators or by different manufacturers) in such a way that a part or all of the equipment could be used on either project. Under this definition standardisation does not include specifications such as API 17D, which does not and should not address specific equipment sizes or interface details. Additional complexity is introduced to the subject when considering who is standardising with whom ? Within each company and within the industry several potential levels of standardisation exist. It could be within one regional office of an operating company, within the whole company or within the whole industry on a regional or international basis. The status and scope for standardisation has been summarised in Table 1 which indicates that although potential for standard equipment is relatively limited, major benefits do exist from agreeing to certain key interfaces and equipment details. The motivation for manufacturers to standardise, has to come from operating companies since they are the ones who will benefit from the change. Specifically it is the oil companies most active subsea who must direct the thrust of standardisation because it is they who stand to gain most and to whom the manufacturers are prepared to listen.

API Specification 17D 'Subsea Wellheads and Trees' is currently being finalised and should be issued for use during 1991. The specification includes subsea trees, wellheads and mudline suspension equipment but does not cover completion riser, control, or subsea manifolding systems. In many ways it is very similar to Specification 6A, the major modification is the requirement for Performance Verification Testing which is within the body of the text rather than in an Appendix. This makes the testing obligatory if the equipment is to be API monogrammable. The other difference between 6A and 17D is that 17D contains details of 'minimum standard valving requirements' which details the required number of valves in the production and annulus flow paths and other block penetrations along with requirements for fail safe valving. (Figure 3). The information in 17D is a significant step forward given that no specifications previously existed for subsea equipment. It will simplify the equipment specification required from the operator and will help provide the operator with more consistent equipment between different manufacturers. However it will not progress the issue of equipment standardisation which must be pursued separately and outwith the API.

Regional efforts are ongoing in the North Sea to begin serious standardisation by agreeing a common interface. Shell, BP and Texaco have produced a proposal which was circulated for industry comment in October 1990.

The interface proposed is between the LRP Connector (tree running tool) and the LRP safety valve block (LRP BOP) (Figure 4). This interface has been selected because it is made up on surface and not subsea, the interface performs no special function, an API connection is adequate and it allows the use of proprietary tree and tree cap running tools. The proposal is for 5" by 2" dual bore system, pressure rated at 10,000 psi. Complete interface details are given in Table 2. It seems likely that this proposal will be accepted and will form the basis for more standard riser systems and possibly for additional key interface standardisation. The proposal does not necessarily define the bore spacing or centres for all subsea trees using the standard LRP interface since a transition spool could be used to change spacing and centres above the tree. Similarly no problems are envisaged using small bore sizes for the tree or tubing string if that is required by the completion design.

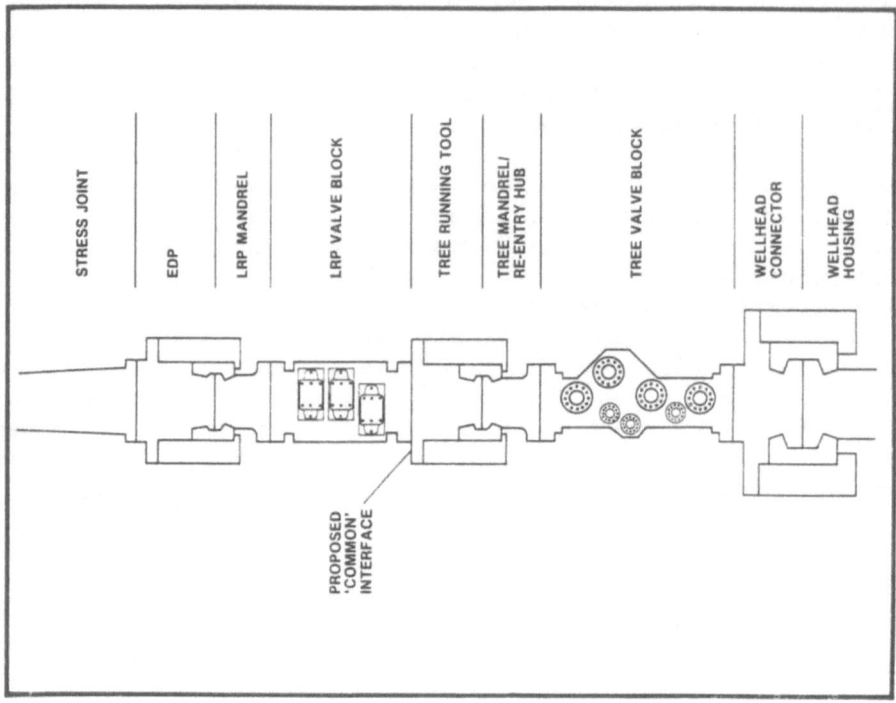

Fig. 4: Proposed LRP Common Interface

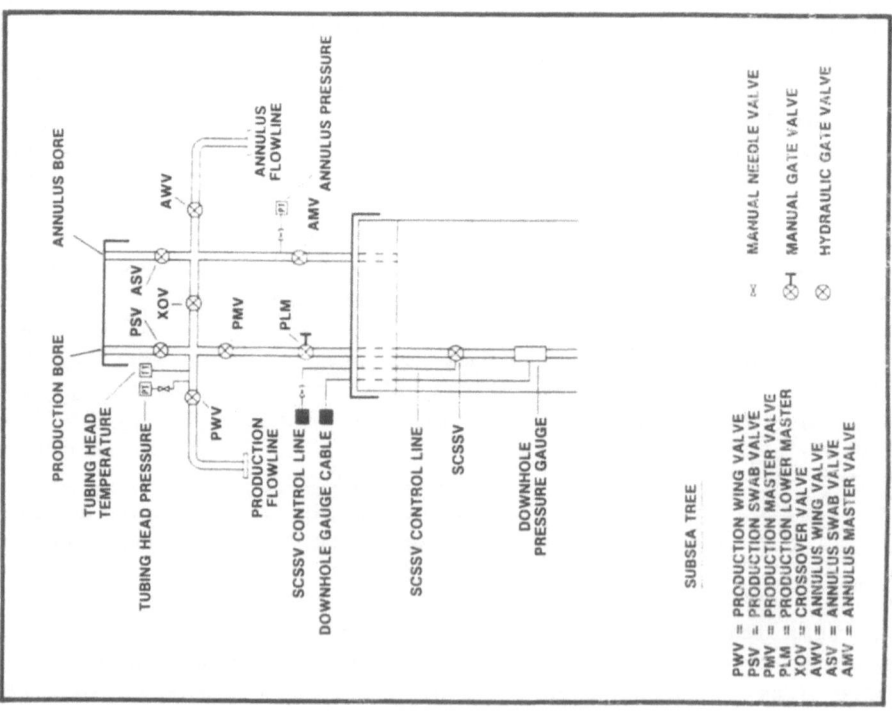

Fig. 3: Typical Subsea Tree Valve Schematic to API 17D

The Shell, Texaco, BP proposal is intended to allow an increase in the utilisation flexibility of completion risers with different subsea trees. Within each operating company this will allow; flexibility in equipment mobilisation after initial well commissioning, reduction in the number of riser systems to be purchased and maintained and a reduction in the inventory of spares and the amount of expertise to be maintained. Within the industry manufacturers will be in a position to supply rental completion riser equipment on the basis that more than one potential client exists and more equipment will be available in order to fulfil planned or unplanned operational requirements for simultaneous operations on one field. Typically the capital cost for a Lower Riser Package, Emergency Disconnect Package, completion riser, running and handling equipment will be approximately the same as the cost of one subsea tree. Completion riser systems are already being used on a rental basis for one or two well developments and the adoption of the common interface will make this option more attractive.

Independently within several companies, internal standardisation is taking place which should allow relatively standard system configurations, and specifications to be drawn up. This should allow individual companies to go at least one better than industry standardisation by specifying standard interfaces and certain standard components. In order that these efforts are of general benefit to the industry it is essential that they are made public as soon as possible. This will allow other companies to adopt them as their own, if they see fit, and prevent the growth of a number of separate internal standards.

In general standardisation within an operating company will lead to the following significant advantages;

Shorter Deliveries: Since a major component of equipment delivery is the specification and design, any standardisation will reduce the delivery up to the point where ordering a standard tree requires only the manufacturing lead time.

Standard Procedures: The use of standard equipment would allow the use of common rig manuals, operations manuals and offshore procedures.

Flexibility: The use of standard equipment allows increased flexibility through the use of common installation equipment, common spares, common tooling and the option of using a tree from one field directly on another without modification.

Shared Expertise: Manpower within the operating and manufacturing companies normally required for designing, monitoring and progressing a totally new tree design can be substantially redeployed.   Operational expertise from a previous project is directly relevant for the next project.   Preparation and planning for subsea operations can become routine rather than one off.

Reduced Costs: Each of the above items separately contributes to a reduction of costs.   However they are not all immediately quantifiable and only the cost of equipment without separate riser systems and tooling may be apparent.   A general reduction in rig times for subsea operations, as experience increases, should be expected and therefore the use of standard equipment eliminates a new learning curve and will save rig time.

Obviously the way in which a particular company handles standardisation depends on both management and individual attitudes.   A strong and operationally aware management is required in order to standardise at all, because without guidance each project will design, specify and build its own subsea tree, without reference to what has gone before.   The benefits of standardisation are only appreciated on a multi-project basis and therefore requires management support to succeed.

7.0   PROJECT MANAGEMENT

As a result of the number of different disciplines involved in a subsea project it is essential that effective project management is in place to bring together the relevant issues.   The most successful subsea projects have been run with small, fully integrated project teams which have full responsibility for all technical decisions and only occasionally require external approvals.   Clear objectives must be established at all levels within the project and each member of the project team must have a clearly defined but flexible role.   Previous experience and continuity through a project minimises the necessary size of the team.   Clearly defined responsibilities, and the ability to make decisions which goes with that, allows clear and rapid communication between vendor and operator.   This minimises any confusion and avoids any delays caused by the operator.   An earlier work (Reference 6) has highlighted the requirement for a certain minimum management organisation at the suppliers office. This includes a project manager and a separate technical manager allowing the necessary emphasis to be placed on both areas of work simultaneously.

Staffing levels within the project team must be carefully considered.   A project group which is too large is not only expensive but become inefficient since a large amount of time is occupied in communicating internally and not progressing the project.   A smaller project group is usually to be preferred but flexibility should exist to co-opt other staff into the group as required for either specific pieces of work or to help during a busy period. The team of personnel working on a subsea project necessarily changes over the project life as a project moves from design through installation and into production. However, it is essential that several key members remain in the post-production operations group in order to ensure efficient trouble shooting as problems arise.

Several specific project management recommendations have been made by previous authors. An operator using as many internal operations staff as possible has been shown to improve two aspects of a subsea project (References 7 and 8). Firstly during equipment specification and design, operations staff can provide essential input which will ensure that the final product is 'user friendly'. This results from both the necessary technical expertise and being familiar with existing facilities, company objectives, practices and prejudices. Secondly any expertise gained during the project should be kept within house to allow subsequent improvements for the next project and maximise the knowledge base. The use of in shop expediting at the vendor's facility has previously proved effective in ensuring project delays are brought to operator's attention rapidly and solved easily (References 5 and 6). A resident QA inspector has also proved effective to improve expediting.

As has been presented earlier, detailed input to the subsea project should be sought from all disciplines very early in the project. Only by involving all the relevant groups in the design and procurement phase will a subsea project be successful.

CONCLUSIONS

1. The potentially conflicting requirements of the various disciplines involved in a subsea project must be taken into account from the earliest design phase.

2. Exploration and appraisal wells can be routinely drilled for subsequent use as development wells at minimum incremental costs, if early consideration is given to the required design.

3. Subsea appraisal wells should be drilled from a clustered seabed location, where feasible, with wells accurately but widely spaced.

4. Rig selection must take into account specific completion requirements as well as drilling considerations.

5. The expertise of external groups such as the drilling
    contractors staff and dive contractors staff should be
    incorporated into the project.

6. The reliability of all completion equipment, both downhole
    and subsea must be given the highest priority during
    equipment design and selection.

7. Subsea wells can be routinely completed using a tree and
    wellhead from different manufacturers if this proves cost
    effective.

8. Subsea systems in water depths less than the saturation
    diving limit should be designed for efficient diver
    intervention. Only in deeper water should ROV interfaces be
    incorporated.

9. Maximum flexibility must be maintained in subsea facilities
    to allow for a variety of development scenarios, equipment
    failures and operational problems.

10. The efficient design of a workover control system should be
    considered a high priority.

11. Standardisation efforts, at an industry level, are extremely
    valuable but are unlikely to proceed beyond agreement to
    several key interfaces.

12. Standardisation of subsea equipment within individual
    operating companies could allow large financial savings.

13. A small, integrated, multi-discipline team with the necessary
    expertise and effective project management is a near
    guarantee for a successful subsea project.

References

1. King G., 'The Drilling Engineering for Subsea Development Wells'. Paper SPE/IADC 18687 presented at the 1989 SPE/IADC Drilling Conference, New Orleans, Louisiana, February 28-March 3.

2. Stoddard B., Campbell J., 'Don – A Cost Effective Approach to Subsea Design'. Journal of Petroleum Technology (April 1991) pp 386-391.

3. Pujol L., Hoyland S., Mortensen H., 'Predicting casing stresses caused by annulus pressure'. Oil and Gas Journal (October 1987) pp 79-83.

4. Dawson A.P., Murray M., 'Magnus Subsea Wells: Design, Installation and Early Operational Experience'. SPE Production Engineering (November 1987) pp 305-312.

5. Smith G.D., Batcheler G.H. 'Overview of the Highlander Field Development'. SPE Production Engineering (November 1987) pp 313-318.

6. Davies P.J.R., Pond R.J., 'Development of the Subsea Completion System for the Highlander Field', Journal of Petroleum Technology (April 1986) pp 453-460.

7. Huber D.S., Burnett R.C., 'Subsea Systems of the Argyll Area Fields'. SPE Production Engineering (May 1990) pp 107-112.

8. Workman D.M., Methven J.O., Kearns J., 'Extending the Innes Field life by Cost Effective Subsea Technology'. Journal of Petroleum Technology (September 1988) pp 1197-1202.

9. Horne J.S., Berkoz A., Wood L.R., 'Development of a Marginal Property: Petronella Field'. Journal of Petroleum Technology (June 1988) pp 723-728.

10. Emptage R.J., 'A Review of the Satellite Production System (SPS) Ness Development'. Paper OTC 5723 presented at the 1988 Offshore Technology Conference, Houston, Texas, May 2-5.

11. O'Brien E.J., Hetland T., 'The Underwater Production System'. SPE Production Engineering (February 1991) pp 33-39.

12. Hopper C.T., 'Simultaneous Wireline Operations from a Floating Vessel using a Subsea Lubricator'. Paper SPE 18239 presented at the 1988 SPE Annual Technical Conference, Houston, October 2-5.

13. Hopper C.T., 'Vortex Induced Oscillations of Long Marine Drilling Risers'. Paper presented at 1983 Deep Offshore Technology Conference, Valletta, Malta, October 17-19.

14. Gomersall S.D., Hopkinson J.M., 'The Conversion of a Two-Well Modular Guidebase with Shared Post to Subsea Production'. Paper OTC 6389 presented at the 1990 Offshore Technology Conference, Houston, Texas, May 7-10.

15. Howes J., 'Conoco to develop Lyell 15 years on'. The Oilman (March 1991) pp 97-98.

16. Hansen R.L. 'Subsea Production: The decade past and a look ahead'. December 1990.

17. Lorimer D., 'New techniques applied to concurrent drilling, subsea wireline logging'. Offshore (March 1991) pp 20-27.

18. Molnes E., Holand P., Sindet I., Lindqvist B., 'Reliability of Surface Controlled Subsurface Safety Valves, Phase III - Main Report'. October 1989, SINTEF.

FLEXIBLE PIPE TECHNOLOGY – A DECADE OF CHANGE

J.M. NEFFGEN
Stena Offshore Limited
Westhill Industrial Estate
ABERDEEN   AB32 6TQ

ABSTRACT

The utilisation of flexible pipe has gained ever increasing prominence over the past decade. Improvements in pipe design, construction quality, strength, and overall reliability have resulted in an increased industry confidence. Increasing confidence has permitted flexible pipe technology to become a key factor enabling the safe and reliable exploitation of marginal and deepwater hydrocarbon fields. Founded largely on early flowline and riser experience gained in shallow waters offshore Brazil, flexible pipe has evolved to become the composite material technology for the 1980s. However during this same period due to increased competition among manufacturers, reduced industry activity levels, and low margins there has been a reduction in both the number of manufacturers and overall production capacity. This change may have been somewhat mitigated by greater R&D activities during this decade resulting in advances in analysing pipe stresses and loads, and in determining fatigue lifetime. Greater awareness and levels of offshore industry activity have lead to increased applications of flexibles particularly in dynamic riser service for floating production. Some lessons have of course been learned from the use of simple catenary and multi-riser configurations. This paper reviews both the early flexible pipe experience with particular emphasis on the Brazilian experience and it charts the course of the main applications throughout the 1990s giving particular emphasis to several pertinent North Sea examples. The paper identifies important considerations for future designs of riser systems and outlines the trends the flexible pipe industry is likely to pursue during the 1990s.

Volume 27: Subtech '91, 29–62.

Introduction

The concept of using flexible, reelable pipe to transport liquids, gases, and vapours is not a new one. As early as the 1940s a steel braided elastomeric pipeline was developed for the Allied Forces in order to transport fuels to support the Normandy Beacheads. In fact, the longest flexible pipeline ever constructed is likely to be that laid across the English Channel as part of 'Operation Pluto'. The methodology used to handle and instal such pipe is also not new. Ellis (1943, London) in an early patent specification identifies three basic objectives for a flexible pipelining method. These are: prefabrication of the pipe onshore; coiling of the pipe on suitable drums or reels; and using such reels to lay pipe from anchored or motorised barges.

The design concept for flexible pipe is also not a new invention given that flexible hoses and umbilicals have been in service for more than sixty years. A break-through was however achieved by the French Institute of Petroleum in the early 1970s when they developed an improved steel reinforced pipe structure having a high axial loading capacity which utilised corrosion and hydrocarbon resistant polymers to extend pipe service lifetime. This early pipe design utilised established cable making techniques to apply steel armour and axially and radially reinforce alternating layers of polymer sheaths. The pipe was primarily developed as a flowline for use in static seabed applications. Such flexible pipe had obvious advantages over rigid pipe laid at that time by conventional lay barges primarily in the manufacturer's ability to prefabricate and test the structure as a whole onshore and to assemble it onto readily transportable reels. The necessary installation equipment and offshore vessel sizes consequently were reduced and significant savings in overall installation time and costs were realised. To date more than 1700 km of flexible pipe have been similarly installed.

FLEXIBLE PIPE CONSTRUCTIONS

Flexible pipe has been defined by the American Petroleum
Institute (1988, Washington) as being ".....a composite of
layered materials which form a pressure containing conduit.
The pipe structure allows large deflections without a
significant increase in bending stresses". Two fundamental
constructions have evolved, the "Bonded" and "Non-Bonded"
types. The selection of the method of pipe construction by a
particular manufacturer was largely determined by his prior
industrial experience in the manufacture of high pressure
elastomeric hoses or in the production of submarine power
cables. Fundamental differences exist between the two types
with regards to manufacturing, armouring techniques, and
design philosophy. Examples of both constructions are shown
in figures 1a and 1b.

In the "Bonded" construction alternating layers of elastomeric
or thermoplastic elastomers and fabrics are radially applied
around a steel mandrel or are extruded over an internal
former. Steel armouring generally consists of applying cross
wound steel chords at approximately 55° to the neutral angle
to provide axial reinforcement. Helically wound steel spiral
or an interlocked steel carcass provide radial reinforcement
and resistance to hydrostatic collapse. All layers are bonded
to each other using adhesives and a curing method known as
'vulcanisation' is used. Vulcanisation involves applying heat
and pressure either electrically, in an autoclave, or using
microwave ovens to ensure adequate polymer molecular cross-
linking results. Such a construction method can offer
significant benefits with regards to mechanical behaviour in
that there is virtually no internal differential movement or
slip of layers thus reducing internal friction, and
consequently wear.

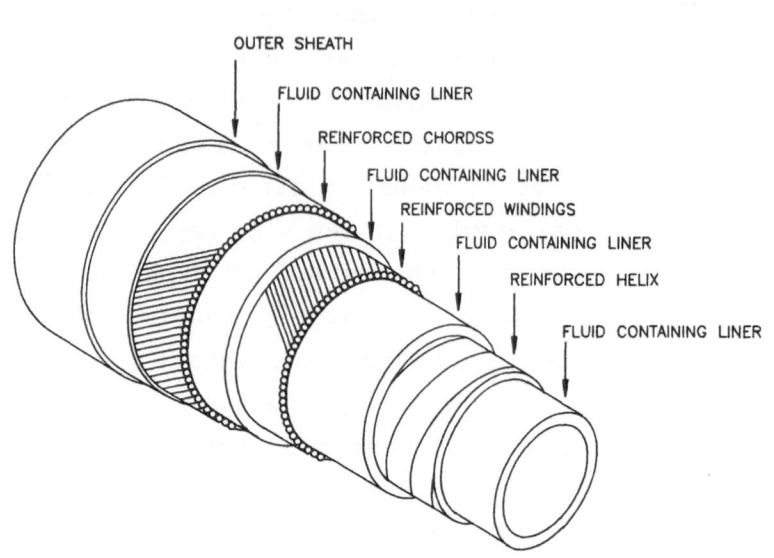

Figure 1a : **EXAMPLE OF A BONDED FLEXIBLE PIPE**

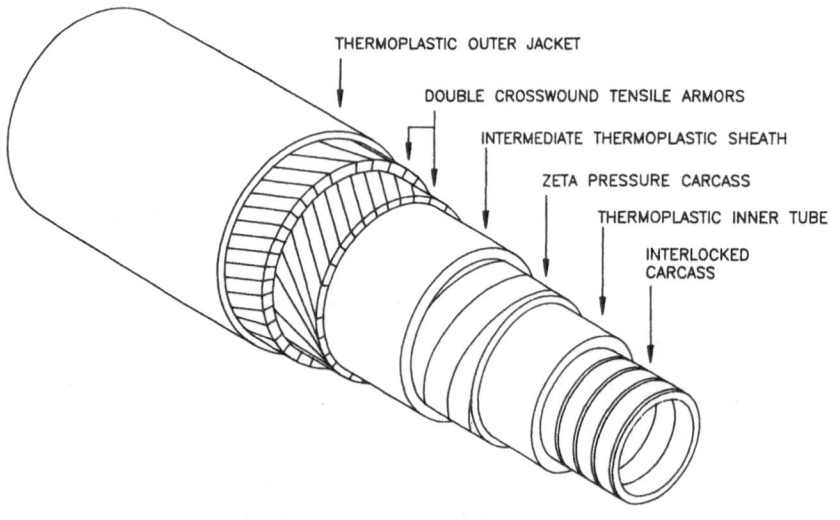

Figure 1b : **EXAMPLE OF AN UNBONDED PIPE**

Lotveit and Often (1990) reported that largely due to the presence of synthetic rubber materials, such as NBR (Nitrile Butadiene Rubber), CR (Chloroprene), or HNBR (Hydrogenated Nitrile Butadiene Rubber), low material stiffness properties gives the elastomer a tendency to act more like a fluid than a stiff membrane. For this reason, individual layer stress levels can largely be ignored. In fact, Perzborn et. Al (1987) reported that after long-term fatigue testing of some 24 off 6 inch pipe samples failure was likely to be synonymous with failure of the steel reinforcement, not the elastomer matrix. Manufacturers of bonded pipes, current or past, are: Pag-O-Flex, Dunlop, Taurus and Uniroyal Manuli. Today, most of the above manufacturers primarily produce relatively short length products or those for dynamic jumper applications.

"Non-Bonded" pipes are those whereby alternating layers of steel, polymers, and fabrics (tapes) are applied without the use of bonding agents or vulcanisation techniques to affect curing. Pipe construction primarily centres around the internal conduit which can either be made from a thermoplastic material (i.e. substance which becomes plastic on heating, hardens on cooling and is able to repeat these processes) such as Nylon 11 or Polyethylene; or it can be made from steel strips. Where live crude or gas are to be transported an internal conduit in the form of an interlocked steel carcass is used. According, to Calquhoun (1990) the other purposes of the carcass are twofold. The first is to provide support to the fluid barrier layer against the collapsing influence of external hydrostatic pressure. The second is to provide support against the squeezing action of the helical armour were layers under the influence of axial tension. The carcass strip is made from plain carbon steel or stainless, generally 316L material and is formed into an S-section.

The carcass is surrounded by barrier layers which are designed to contain the fluid. The barrier layers are restrained against collapse by the carcass beneath and against burst by the armour above. The selection of the barrier layer(s) is critical as it is these layers which are exposed to primary effects of the aggressive constituents of the fluid as well as being subject to diffusion forces of migrating gases. In the case of high fluid temperatures, the number of barrier layers may be increased to include a PVDF or polyamide material to extend the normal operating temperature range. The carcass material composition may also need to be altered for high temperature service (90°C or greater) to an Austenitic or Duplex material. Some development in this area has taken place in 1980s with long-term testing to be addressed in the 1990s. For high temperature service a figure of 100°C is generally considered today to be a critical cut-off point for extended service life with peak temperature resistance able to rise for short periods to 130°C. A short synopsis as to the blistering resistance of thermoplastic material is given in Table 1.

Table 1 : Blistering Resistance for Thermoplastic Materials

	Behaviour According to Number of Cycles (200h)				
Material	1	5	10	20	50
High Density Polyethylene	Good	Good	Fair	Poor	–
Polyamide 11	Good	Good	Good	Good	Good
Fluorinated Polymer	Good	Good	Good	Good	Good
Polyamide 12	Good	Good	Good	Good	Good

Note: Table taken from Moore, F. (1989) "Materials for Flexible Riser Systems: Problems and Solutions", Eng. & Structures Journal, Vol. 11. No. 4 p.211, Guildford Surrey.

The barrier layer(s) are supported by the circumferential layer(s) to resist burst pressure. The steel layers are separated by thermoplastic wear layers to reduce internal heat build-up and friction. Two helical armour layers are applied in the form of flat wires to minimise residual stresses and prevent 'birdcaging' when the wires are cut.

The outer helical layers can be wrapped using a tape layer to provide a bed for outer sheath plastic extrusion. An outer sheath is applied to protect the armour layers from seawater and/or mechanical damage. When additional insulation is required, this can be achieved by using an extra layer between the helical armour and outer sheath, or by applying an external syntactic foam insulating material.

Manufacturers of "Non-Bonded" Pipes are Coflexip, Wellstream, Furakawa/NKT, with Pirelli remaining in the background. Of this group, only two suppliers are currently active in the market.

MARKET FORCES AND THE MANUFACTURER

Flexible pipe can in some cases offer an alternative pipeline methodology to that of rigid pipe especially where small pipe diameters (< 10-inch i.d.) are called for. Examples of such applications are tie-in jumpers, intrafield flowlines, or risers. Flexible pipes can be used in subsea and floating production systems to connect subsea wells to PLEMS (pipeline end manifolds) or directly to a surface termination flange. Risers can be made from flexible pipe and a limited manufacturing range currently extends to 18.75 inch i.d. with one supplier. Static flexible risers can be made and are generally required when retrofitting existing facilities. Such risers are most often pulled through available J-tubes or vertical steel caissons. They are utilised where low minimum bend radii or other constraints prohibit the use of steel pipe.

Some applications of flexible pipes are shown in Figure 2. Dynamic risers represent the most considerable challenge to the pipe manufacturer. Such risers represent the primary link between the subsea well or riser base and the surface facility. Because risers pass through the splash zone, they are subjected to high inertial and cyclic loads together with high environmental forces. The design and construction of such dynamic risers, therefore, requires specialist knowledge not only of the material properties of the pipe, but also of its mechanical behaviour under varying loads.

Perhaps, the ultimate market but not necessarily the most profitable one which beacons the pipe manufacturer is that involving the supply of dynamic risers. The products concerned offer higher gross margins but this is coupled with higher technical risks and lower product volumes than for flowline supply. Flowline purchase orders are generally measured in tens of kilometres, whereas riser or jumper orders often extend only to several hundred metres in length.

Most pipe production costs are reflected in the initial plant set-up not in the mass production costs, therefore low volume orders are often synonymous with unprofitable ones. Despite the relatively high unit price for flexible pipe, the manufacturer could not survive on revenues from a single product line. This statement is modified when purchase orders are coupled with high volume flowline orders or value added contributions from associated services, such as is the case in developing marginal fields. Generally, few pipe manufacturers have sufficient experience or track record to tender for the supply of flowlines and risers together as a turn-key field development contract. Pipe manufactured cost consequently is therefore influenced by a number of relevant factors, such as: total material and labour costs; investment costs; production output and efficiency; and development costs.

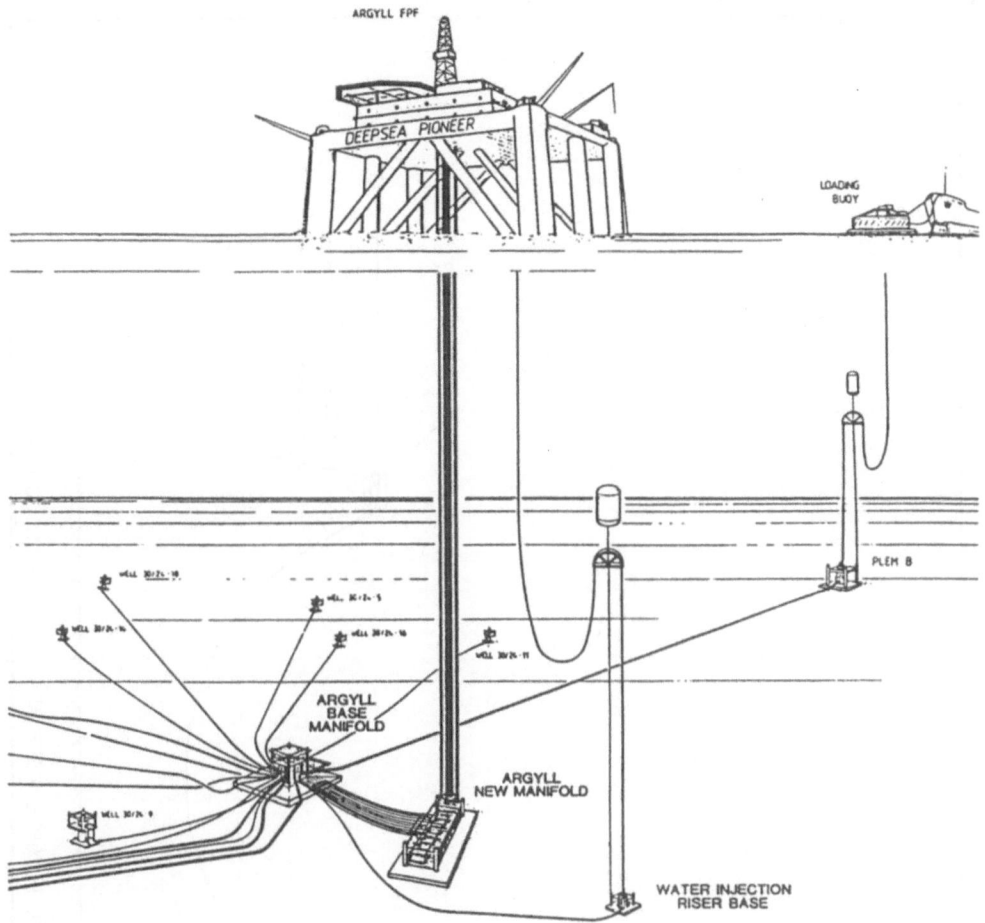

Fig. 2·A   Argyll Field

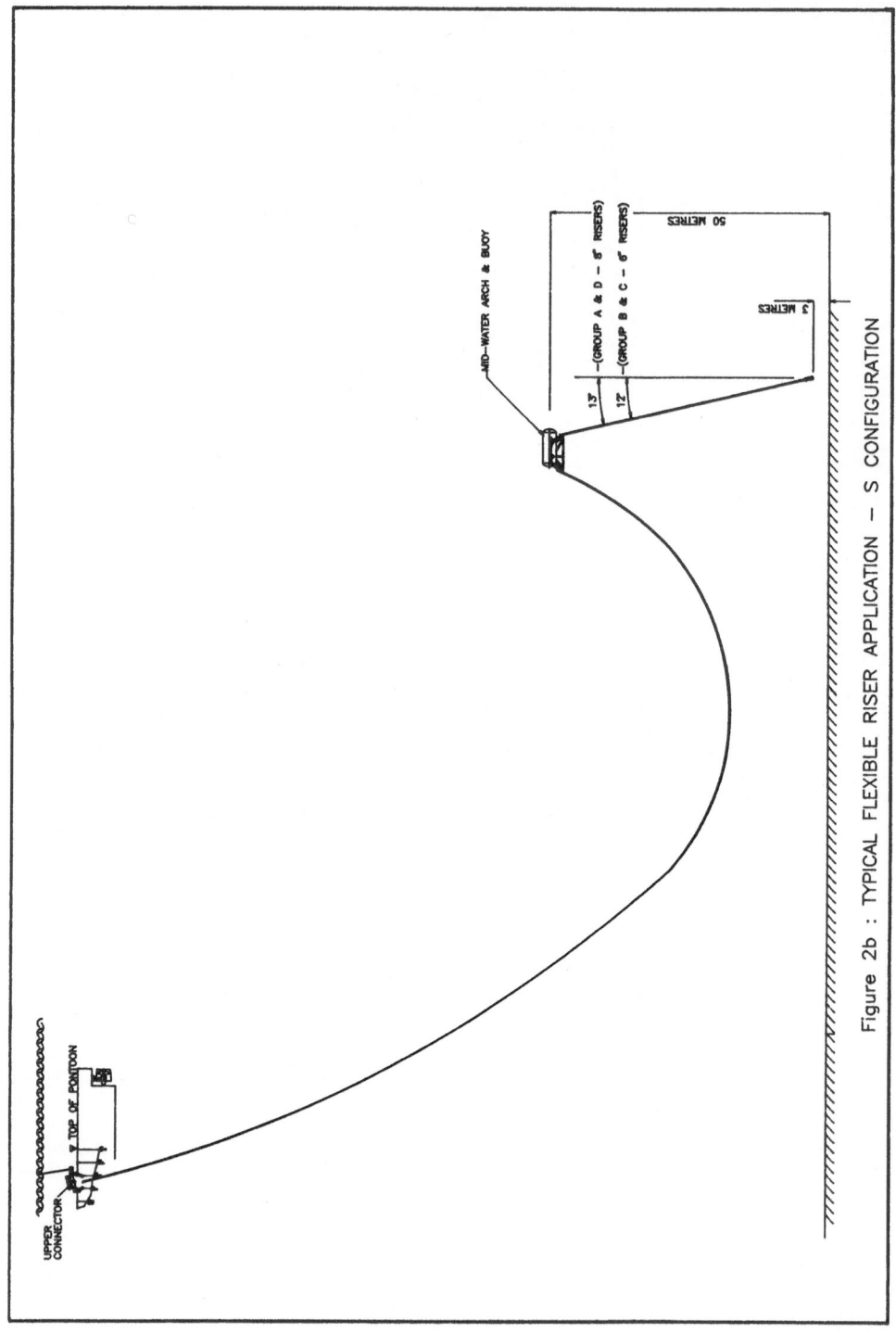

Figure 2b : TYPICAL FLEXIBLE RISER APPLICATION – S CONFIGURATION

Pipe selling price is however influenced not only by the foregoing, but also by volumetric considerations, market forces, and the level of competition. It is in this area that the suppliers have suffered most in the past decade and consequently the supply side has reduced.

Table 2a compares the number of manufacturers engaged in the design, manufacture and supply of flexible pipes in 1980s and 1990s. It can be seen that the early successes with flexible pipe in the 1970s lead to a proliferation in the number of suppliers in the 1980s. By 1990 due largely to the effects of reduced market demand, lack of R&D funding, reduced investor confidence, and greatly increased competition the market today is principally left with two primary suppliers of the full product range with a limited number of specialist suppliers for short-length products. As can be seen there are a number of serious gaps in the market if sufficient capacity and corresponding quality of supply are to be realised to meet 1990s demand.

Table 2b taken from a confidential market report (Seaflex A.S., 1990) illustrates the estimated annual demand for flexible pipes in the main offshore regions of the world based on typical requirements for 4 inch i.d. pipe. As can be seen demand could easily reach 350 km per year by the end of the 1990s.

TABLE 2a : Comparison of Pipe Suppliers 1980s & 1990s

Manufacturer	Country	1980s			As at 1991		
		Riser	F/line	Other	Riser	F/line	Other
Coflexip	France	Yes	Yes	Yes	Yes	Yes	Yes
Brasfex (Coflexip)	Brazil	No	Yes	Yes	No	Yes	Yes
Simplex Inc.	USA	No	Yes	Yes	No	No	No
Dunlop Inc.	U.K.	Yes	Yes	Yes	No	No	Yes
Pag-O-Flex	Germany	Yes	No	Yes	No	No	No
Furakawa/NKT	Japan	No	Yes	Yes	?	Yes ?	Yes
Pirelli	Italy	No	Yes	Yes	No	No ?	No
Uniroyal Manuli	Italy	No	No	Yes	No ?	No	Yes
Wellstream	USA	No	Yes	Yes	Yes	Yes	No
Taurus	Hungary	No	No	Yes	No	No	Yes

NOTE :   List is not exclusive and includes primary product suppliers with a
          track record published for such products.

TABLE 2b : Annual Demand for Flexible Pipes in Principal
           Offshore Regions 1991-1998

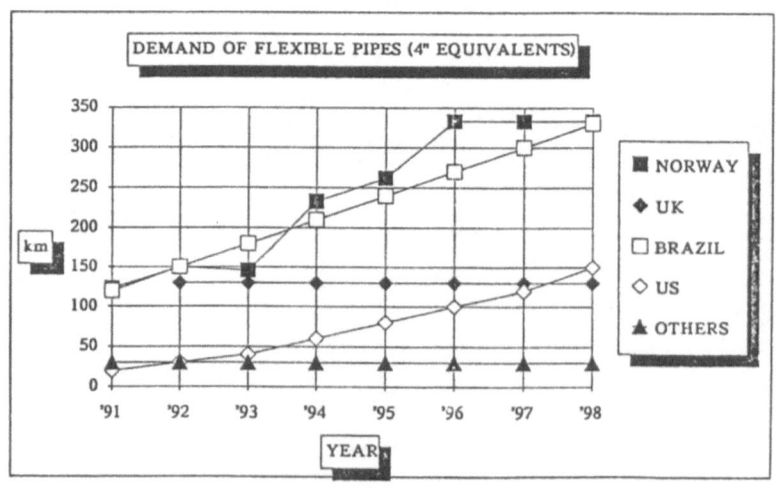

ADVANCES IN PIPE TECHNOLOGY

Technological advances in product design and construction largely reflect market needs and the amount of effort expended on R&D together with the level of investment made. In the early 1980s prior to the oil slump a number of manufacturers were actively engaged in development of flexible pipe products. Testing largely centred on determining material properties and mechanical strength. Investments were made into improvements in quality control, increasing plant capacity; expanding production lengths; improving reinforcement and armouring machinery reliability, and developing improved end termination techniques. As overall production capacities increased, manufacturing experience consolidated, and the corresponding quality of manufacture improved. It should be noted that product rejection rates (in the plant) on average generally remained below 5% of all products produced. This rejection rate is highly favourable when compared to similar rates in other industries, such as cable making (8-10%), flexible hose (10%), or umbilical manufacture (17%, or greater). The achievement of such reliability was largely realised due to the high R&D investment levels, strong quality control, and intense client participation especially in joint industry studies. R&D in the 1980s principally addressed the following areas of interest:

. assessment of pipe material properties
. mechanical testing of pipe capacities
. fatigue lifetime assessment
. pipe stress level assessment
. development of numerical models
. in-situ testing of riser systems.

In situ-scale testing such as that to determine principal loadings as shown in Figure 3a was largely confined in the 1980s to one open water site as part of the Flexible Arch project (1983). The project was financed by Mobil Exploration Norway Inc, Elf Norge, Total Marine Norsk, the French IFP, and by an EEC subsidy. The purpose of the experiment was to verify the behaviour of a flexible pipe in a seabed-to-surface connection mode in what was then deep water (106m). Figure 3b illustrates the general arrangement of the flexible arch system. According to Goodfellow Associates (1990 Pennwell) the test was carried out over one year (1982-3) on a 135m, single 10-inch i.d., 4500 psi working pressure flexible pipe filled with water. The pipe remained under pressure and was suspended between a surface buoy and a subsea float arranged in a catenary mode. Sensors located at each pipe end and at the lowest point of the arch provided a continuous measurement of key parameters. The sensors measured:

. Movements of the upper end of the flexible pipe
. Movements of the lowest point of the catenary and the subsurface of the flexible pipe.
. Wave and tidal heights
. Tension of the pipe
. Radius of pipe curvature
. Pipe internal pressure
. Relative angular deflection of pipe to surface buoy.

The lack of more than one in-situ test could be attributed to the low volume of flexibles installed outside Brazil in the early 1980s and a shortfall in client awareness.

During the early 1980s approximately 70-75 % of all flexible pipes placed in service went into Brazilian waters, a total of some 900 km. Since the mid 1980s approximately 30-35% of the world total has been installed outside Brazil with this share likely to increase to 40-45% in the coming five years (1990, Confidential Market Study).

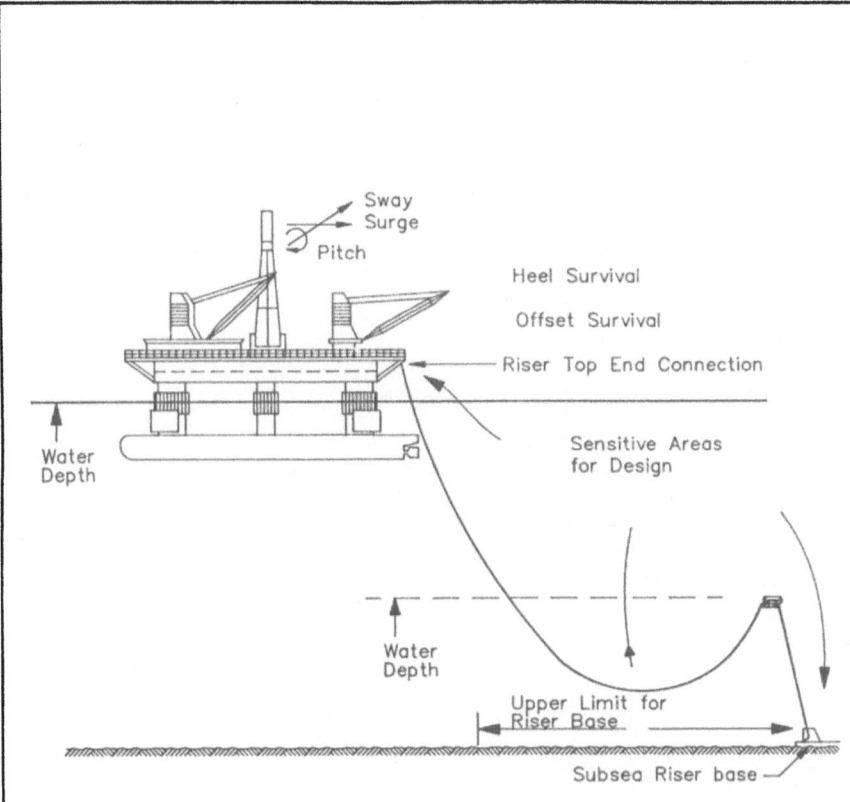

Figure 3a : PRINCIPLE MOTIONS AFFECTING A RISER SYSTEM

MS963/M58

Figure 3b : Flexible Arch North Sea Riser Layout 1983

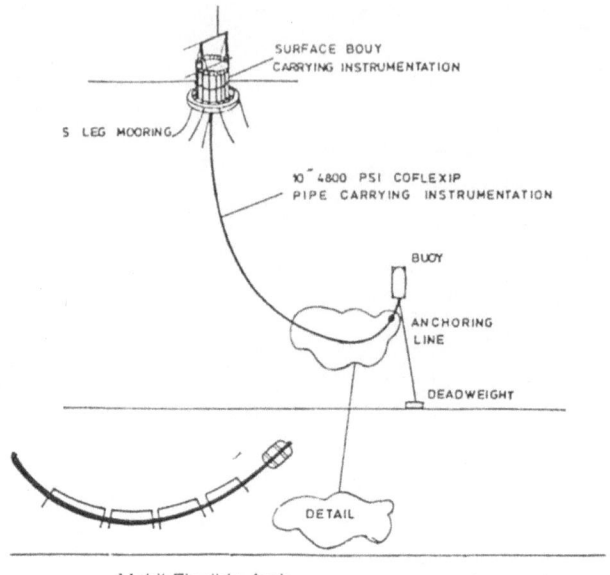

Mobil Flexible Arch

As a result of increasing regional interest particularly by North Sea Operators, several new riser monitoring programs have since been initiated. These are:

. Norsk Hydro on Petrojarl
. Statoil's Veslefrikk catenary riser monitoring system;
. Sun Oil's Balmoral riser top end measurements;
. Petrobras Marimba riser angular and top tension measuring system.

Guidelines and Integrity Monitoring

In addition, the past decade has resulted in an increased awareness by Operator and pipe designer alike of flexible pipe's composite construction but has also culminated in a realisation that there is a requirement to develop dedicated non-destructive testing techniques. In this context the oil industry has identified a need for defect recognition and industry standards and has correspondingly instituted new guidance notes. Several guidelines specifically written for the design and construction of flexible pipes are now available, notably: API's RP17b (1988) and Veritec's Guidelines (1987). Not lagging far behind, the Institute of Petroleum has recently amended provisions in Part 6 of the "Model Code of Safe Practice," while BP and Bureau Veritas are currently producing their own standards and Guidelines respectively for flexible pipe.

The need for such guidelines cannot be overemphasized largely because flexible pipe is made from alternating layers of steel, polymers, fabrics, forming a 'composite' construction. The composite construction by design aims to achieve material durability and reduce internal wear while reinforcing the strengths of individual components. The use of specialised materials such as Kevlar or Aramid reinforced fabrics can offer significant product benefits by combining high fibre strengths with good material resistances to corrosion or chemical degradation.

In order to achieve such benefits, however, this composite construction brings with it a correspondingly complex rather than simple mechanical behaviour. Such behaviour in practice means, that (see Neffgen, Pigging & Inspection 1990):

. bending moments and strains cannot be easily calculated
. some component materials exhibit non-linear behaviour
. differences exist between component elastic moduli which must be analytically explained
. strain distribution around the pipe is axi-symmetrical.

In order, therefore, to determine the integrity of such a complex structure and to detect the presence of significant defects, i.e. those that affect structural capability, specialised inspection techniques must be applied.

Efforts in the 1980s largely focused on understanding by numerical means how pipes behaved and related this information and analytical results to operational experience. During 1990 a joint industry initiative specifically concerned with the integrity determination of flexible pipes commenced under the sponsorship of a number of Norwegian and foreign oil companies. The 3-year program is being co-ordinated by the Norwegian science foundation, SINTEF.

In a recent paper (Neffgen, Pigging & Inspection, 1990), an overview was given of the relationship between off-the-shelf NDT methods and their reported ability to recognise significant defects. Table 3 below illustrates this relationship and identifies several methodologies which are currently state-of-the-art with respect to flexible pipe defect detection. Further progress in this area will undoubtedly be made as a result of the joint industry efforts.

Operational Experience

Someone once stated "experience is the best teacher" and this old adage can hold true for flexible pipe technology. The 1980s saw a break through in the use of flexible pipes in oilfield applications largely as a result of one operator's experience. In the early half of the decade most flexible pipes supplied were installed as flowlines in static applications with less than 20% of pipes produced being installed in dynamic applications. By the end of the decade however, more than 90 risers were installed in Brazil alone in water depths ranging to 400m (1300 feet). In the same period dynamic riser systems were installed on a number of floating production systems in non-Brazilian waters to depths ranging to 540m (1770 feet).

Pioneer end users in this field were not the oil majors, but less conservative oil operators such as Petrobras, Texaco, Sun Oil, and Hamilton Brothers. Their confidence can in part be attributed not only to early Brazilian successes, but also due to the fact that their field development scenaria necessitated formulating more innovative solutions. Such operators could also draw on field experience gained in West Africa and the Far East where flexible flowlines and risers were and are extensively utilised with SPMS (single point moorings) and tanker based production and storage systems (FPSO).

Table 3 : Relationship between pipe defects and recognition by various equipment.

Parameter Monitored	Geometry changes	Material degradation	Cracks & breakage in steel comp.	Cracks in polymer layers	Disbonding
Method					
Thermography	X	X	X		X
X-ray and gamma radiography	X	X	X		X
Acoustic methods		X	X		X
Tracing isotopes			X	X	X
Cable-based leak detection			X	X	X
Magnetic induction			X	X	
Eddy current		X	X		
Photogrammetry	X				
Boroscopes	X	X	X	X	
Ultrasonic inspection		X	X		X
Holography	X			X	X
Impedence		X	X		

Brazil

Early experience in using flexibles in Brazil was largely gained in the development of small producing wells where individual well flow rates ranged in the hundreds of barrels per day.

In order to link up distant wells scattered over many separate fields in the Campos Basin, flexible flowlines generally of less than 6 inch i.d. were laid.   The largest stresses experienced by the pipe were seen during installation rather than operation as well temperatures, pressures, and flowrates were relatively low.   Pigging and TFL (Through Flow Line) operations were not practised.   Instead Petrobras made strenuous efforts to standardise product design and manufacturing techniques.   With over 300 subsea wells installed offshore Brazil, they attempted to develop pipe installation and even flowline recovery techniques which remain the industry standard approach today. During the 1980s Petrobras pioneered the "lay-away technique" for installation of trees and flowlines using a drill rig and DSV (Diving Support Vessel) a method which was successfully used on a number of occasions (see Neffgen, "Advances" 1990).

By contrast, flexible pipes were painfully introduced in the North Sea and US Gulf of Mexico in earnest during the latter part of the 1980s. This can be explained in part due to the conservative business bias of western oil companies, due to stricter legislation, and due to different driving forces in the way North Sea and U.S. Fields are developed.   The prospects exploited during the 1980s were largely giant fields in what is today considered 'shallow water', i.e. less than 200m (660 ft). Production rates per well were in '000s of barrels per day and reservoir temperatures often approach 100°C – 130°C at seabed trees.

North Sea crudes can be particularly aggressive fluids having high percentages of H_2S and CO_2 (to 7 mole percent) making the use of ordinary carbon steels less attractive. Because of the presence of complex infrastructures and the availability of diving support most fields could be linked using rigid pipelines with possibly a combination of flexible spools or tie-in jumpers. The necessity to rely on flexible flowlines could thus be averted.

By contrast with Brazil, the real introduction of flexible risers into production applications began with the Argyll development (See Figure 2a) in the UK during the first quarter of the decade. Multi-riser applications were only brought to prominence by Sun Oil's Balmoral floating production system. In contrast to the vast majority of Brazilian riser systems in the 1980s, the Balmoral system was perhaps the first example of a multi-riser system arranged in an S-configuration. Brazilian risers by contrast are largely arranged in free hanging catenaries as pipe excursions and vessel maximum offsets tend to be less than similar excursions and motions for North Sea floaters.

The Balmoral riser system (Goodfellow, 1990, pp 47-51) consisted of four bundles - two production and one sales line on the aft side and one water injection bundle on the fore side of the floating production vessel. The bundles consisted of the following risers having design pressures ranging from 1650 psi for production/annulus/kill lines to 2900 psi for the sales line:

. Bundles 1 & 2 :   1 x 8 in. + 2 x 4 in. production/
                    annulus/kill/gas lift risers,
. Bundle 3 :   1 x 8 in. sales line
. Bundle 4 :   2 x 8 in. water injection line

The arrangement of the risers was laid out not too dissimilarly to the idealised simple catenary design shown in Figure 5 (Goodfellow, p. 40, 1990).

From this development concept, a new wave of development of floating production systems commenced. In the latter part of the 1980s floaters or tensioned leg platforms which utilised flexible risers were as shown in Table 4.

Table 4 : Shortlist of Floating Production Systems using Flexible Risers

Client	Field	Configuration
Texaco	Tartan/Highlander	S - Shape
Hamilton	Argyll, Duncan	S - Shape
Sun Oil	Balmoral	S - Shape
Amerada Hess	Ivanhoe/Rob Roy	S - Shape
Conoco Inc.	Green Canyon	Free - Hanging
Petrobras	Pirauna	Free - Hanging
Petrobras	Marimba	Free - Hanging
Statoil	Vesslefrikk	Simple Catenary
BHP	Challis/Jabiru	S - Shape
Sovereign/MSR	Emerald	S - Shape

Operational experience must also serve to point out trends in field developments and such trends can be illustrated by two floating production configurations - that for the tethered semi-submersible drilling/production vessel and that for a satellite development. Figures 4 and 6 show examples of the above two configurations currently pioneered on the Vesselfrikk field (Statoil) and by Petrobras on their deepwater Marimba field development.

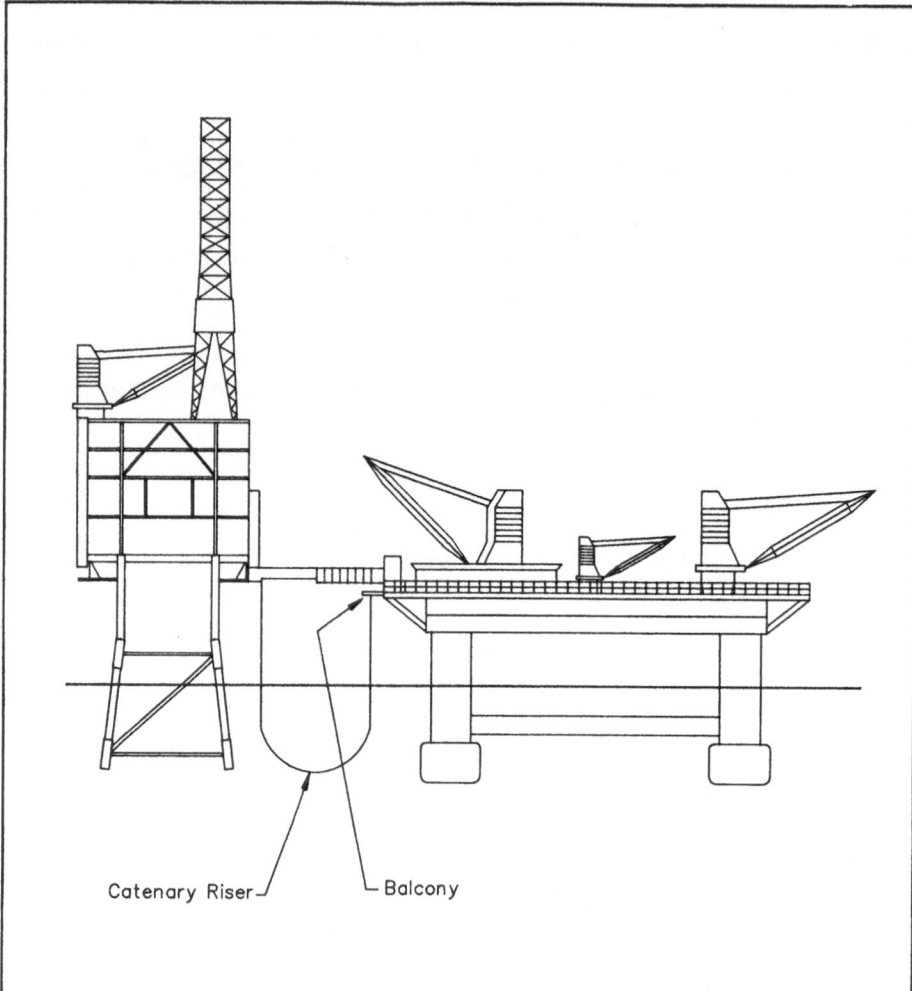

Catenary Riser —   — Balcony

Figure 4. CATENARY RISER SYSTEM FOR TETHERED
DRILLING/PRODUCTION

MS964/M58

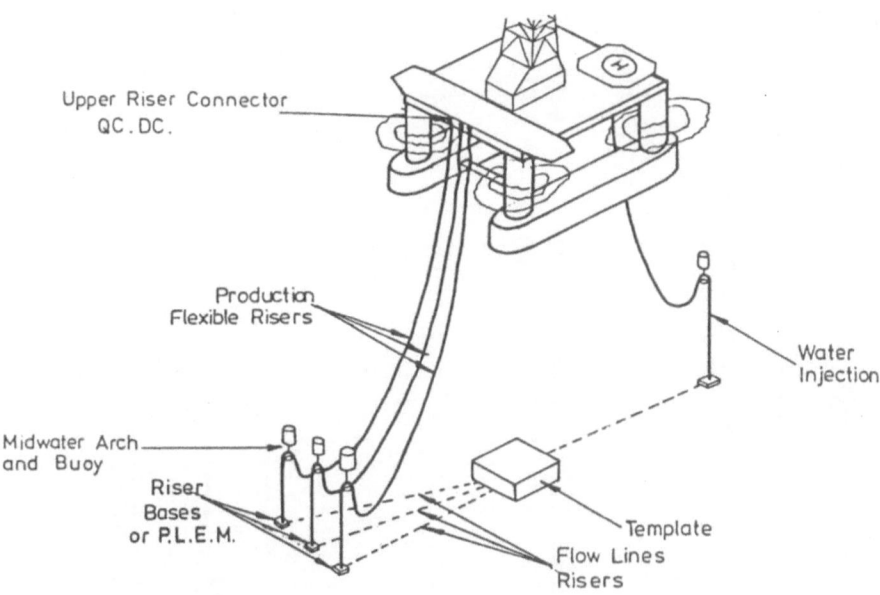

Upper Riser Connector
QC.DC.

Production
Flexible Risers

Water
Injection

Midwater Arch
and Buoy

Riser
Bases
or P.L.E.M.

Template

Flow Lines
Risers

Water depth 460-480 Feet.
Total of 9 dynamic risers ranging
from 4" to 8" diameter and pressures up. to 5000 P.S.I.

Fig. 5   Simple Catenary Design

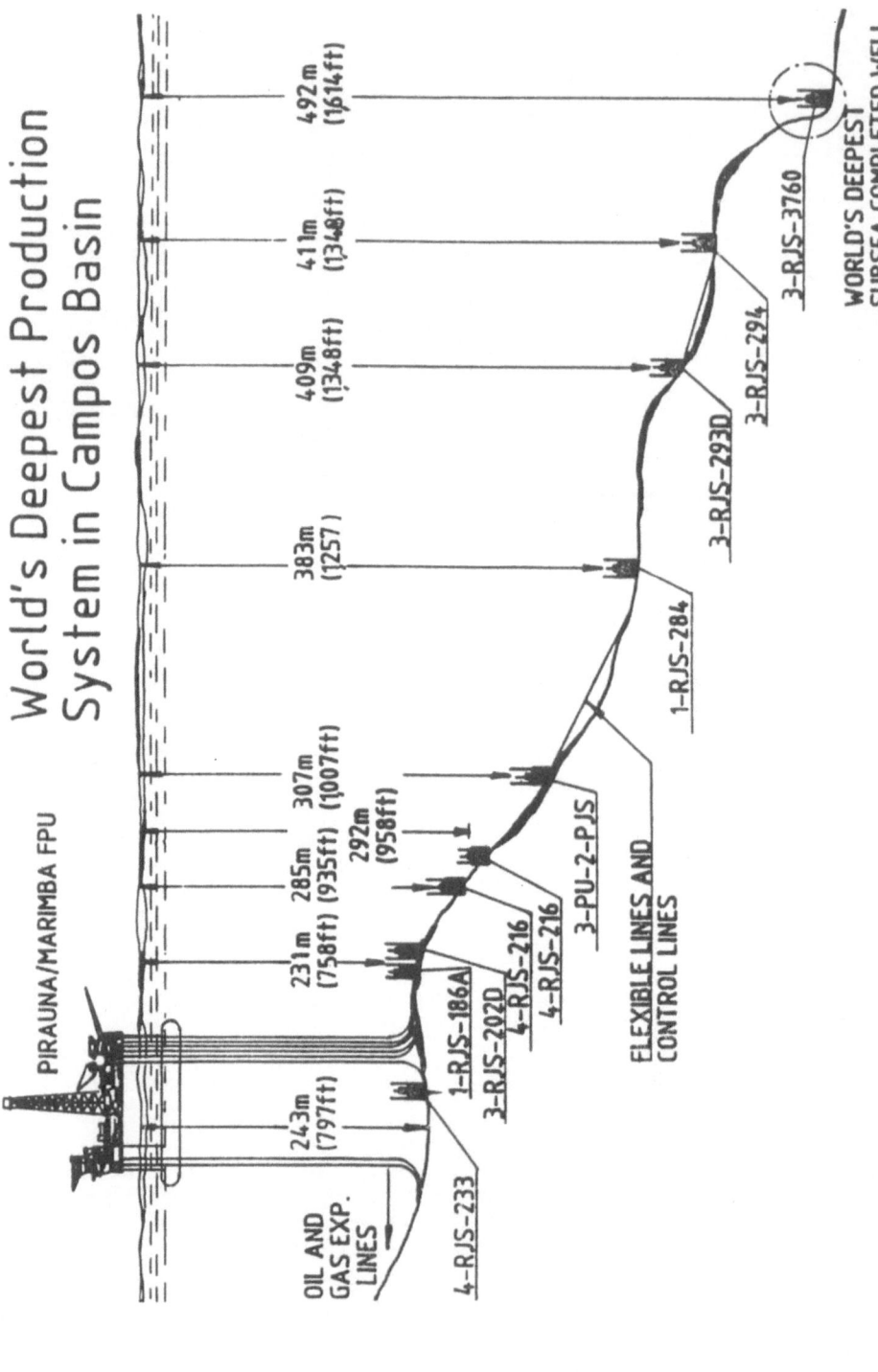

FIGURE 6 : PETROBRAS' DEEP WATER SATELLITE DEVELOPMENT (BY COURTESY OF PETROBRAS)

In the tethered semi-sub application, such as was used on Vesselfrikk, early production can be achieved via the use of a floating production/mud kill facility tethered to a fixed, lightweight wellhead platform. As a variation on this theme, Shell UK for their Gannet development will use a tethered drilling assist (TDA) vessel combined with a steel platform to perform drilling operations over a limited number of years. This concept is also proposed for BP's Bruce Field, UK and can lead to considerable savings in reducing platform weight and deck loading as many facilities can be transferred to the larger semi-submersible.

In the Marimba development which is again echoed in the way Petrobras intend to develop their pilot production system for Marlim in very deep water (600 - 1000m), the use of flexible risers play an integral role. To date subsea wells have successfully been completed and commissioned to 492m (1614 ft.) on Marimba as shown, and recently to 600m (2000 ft.) on the Marlim Field.

Installation Experience

One unique feature of any such field development is the ability of available installation contractors to install major field components in a cost-effective manner. This criterion is largely related to the contractor's resources, both vessel and personnel, and his knowledge of specialist or interface engineering aspects of the work. This latter aspect is especially important where supply and installation contracts are to be separated. As the largest stresses flexible pipe is likely to experience often occur during installation, particular diligence and expertise are required. Contractors can with respect to assessing installation and operational stresses avail themselves of professional software services, developed by a number of institutes and companies over the past decade. Such programs and expertise are now in the public domain and are available to non-pipe manufacturers.

During the 1980s the vast majority of flexible flowlines and risers were installed by one pipe supplier largely because a) of their dominance of the pipe supply market, and b) they were the only company to have made the significant investments in capital equipment and installation technology, and c) reluctance on the part of clients. In particular, because of the requirement to maintain pipe minimum bend radius at all times during transport and installation, large circular reels are required.

Radius of pipe curvature primarily governs reel drum diameter, but pipe unit weight, bending stiffness, and the presence of stiffened appurtenances at end terminations governs capacity. Given that fact limitations are placed on manufactured pipe length, the limiting factor being the maximum continuous pipe length handled for a particular reel, and the ability of an installation vessel crane to handle the fully laden reels. Reel dimension of 7.8 x 7.8 x 8.5m are common for the carriage of flowlines or large diameter (> 8 in. i.d.) risers. Ultimate reel capacities today reach 210t of pipe while the reel plus power and drive units together can weigh an additional 90 tonnes. As most vessel cranes are rated for 100t or less s.w.l. for offshore lifts, the handling of such loads requires considerable planning and expertise.

In the latter part of the 1980s new competition to the market leader with regards to installation arose. Several DSV operators made the initial capital investments required to safely handle, and lay flexible pipes in an equivalent manner to that of the dedicated flexible pipelay spread. Since 1987, in fact, STENA OFFSHORE has successfully completed approximately fifteen major flexible pipelay and/or trenching projects. STENA are currently installing flexible flowlines and risers in the U.K., Australia, and Brazil to water depths of 625m (2000 ft.).

In order for contractors to undertake 'turn-key' projects, it is often a client imposed requirement to complete all of the subsea pipelining and construction work utilising a spread of resources. Flexible pipes can be laid by reel lay vessels, such as STENA APACHE shown in figure 7 which has a capacity to install 1800 tonnes of pipe on one trip. The advantage of this vessel is that it possesses as permanently installed equipment a variety of reels up to 23m (75 ft.) in diameter and can perform simultaneous pipelay of rigid and flexible pipe.

Horizontal turn tables or vertical powered reels can also be made 'portable' with respect to mobilising such equipment onto DSVs (dive support vessels). As such vessels are more numerous than dedicated flexible pipelay vessels and they have both the necessary cranage and the required open after deck space, installation by DSV is a very viable and cost-effective option.

The sequence illustrated in figure 8 shows how a dynamically positioned DSV can successfully transport and instal long length flexible risers as part of the hook up of one major north sea floating production system. The engineering and management of such projects require a strong degree of operational and construction experience together with specific knowledge of relevant codes, break points, and mechanical interfaces with subsea or surface facilities.

FUTURE DEVELOPMENTS AND CONCLUSIONS

In the past decade a number of technological advances have been made with regards to subsea pipelining and floating production. As a result of the successes gained offshore Brazil in the use of flexible flowlines and risers, many operators are considering the benefits of using flexibles or a combination of rigid pipelines and flexible tie-in jumpers to develop smaller and deeper water fields.

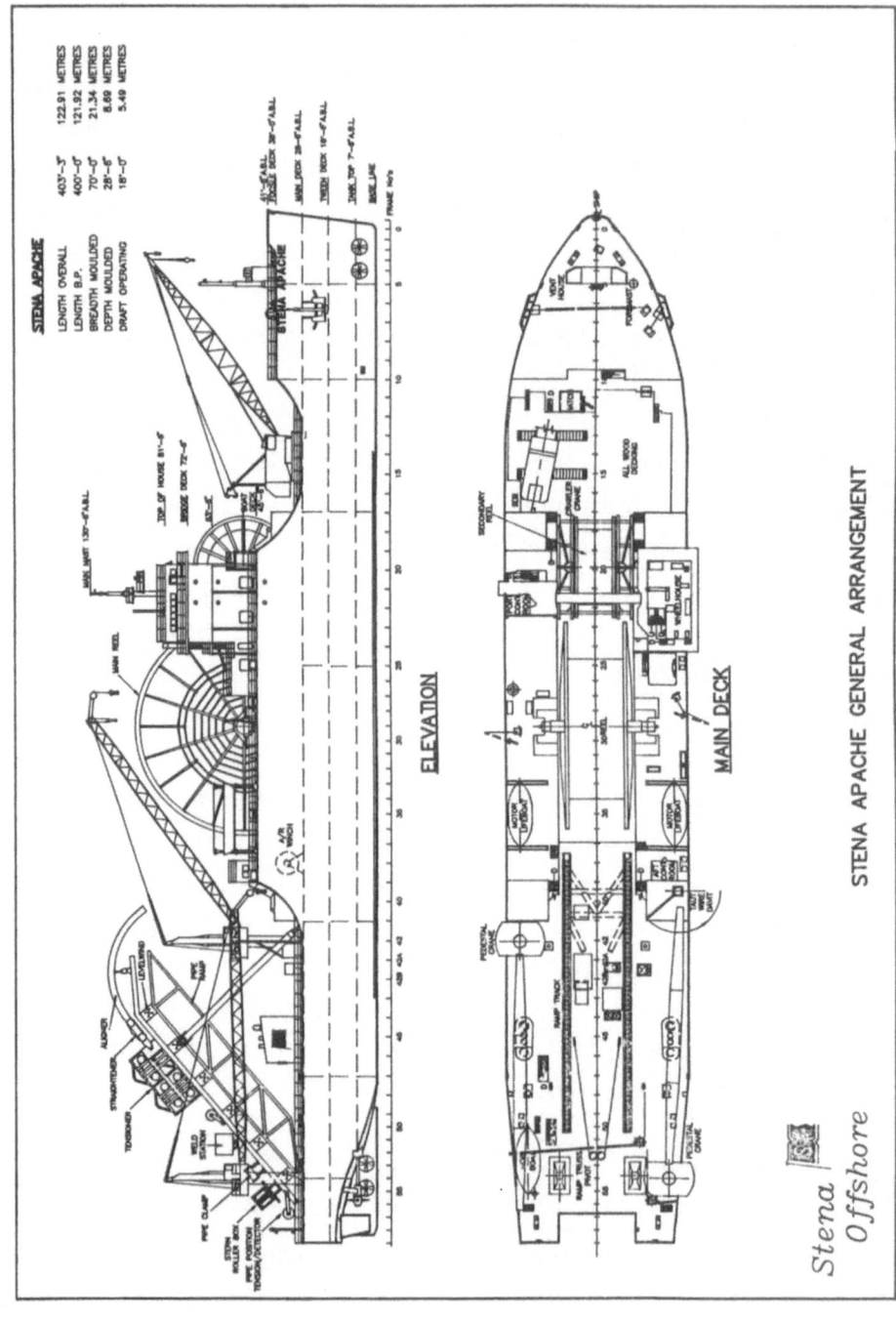

ELEVATION

MAIN DECK

STENA APACHE GENERAL ARRANGEMENT

Stena Offshore

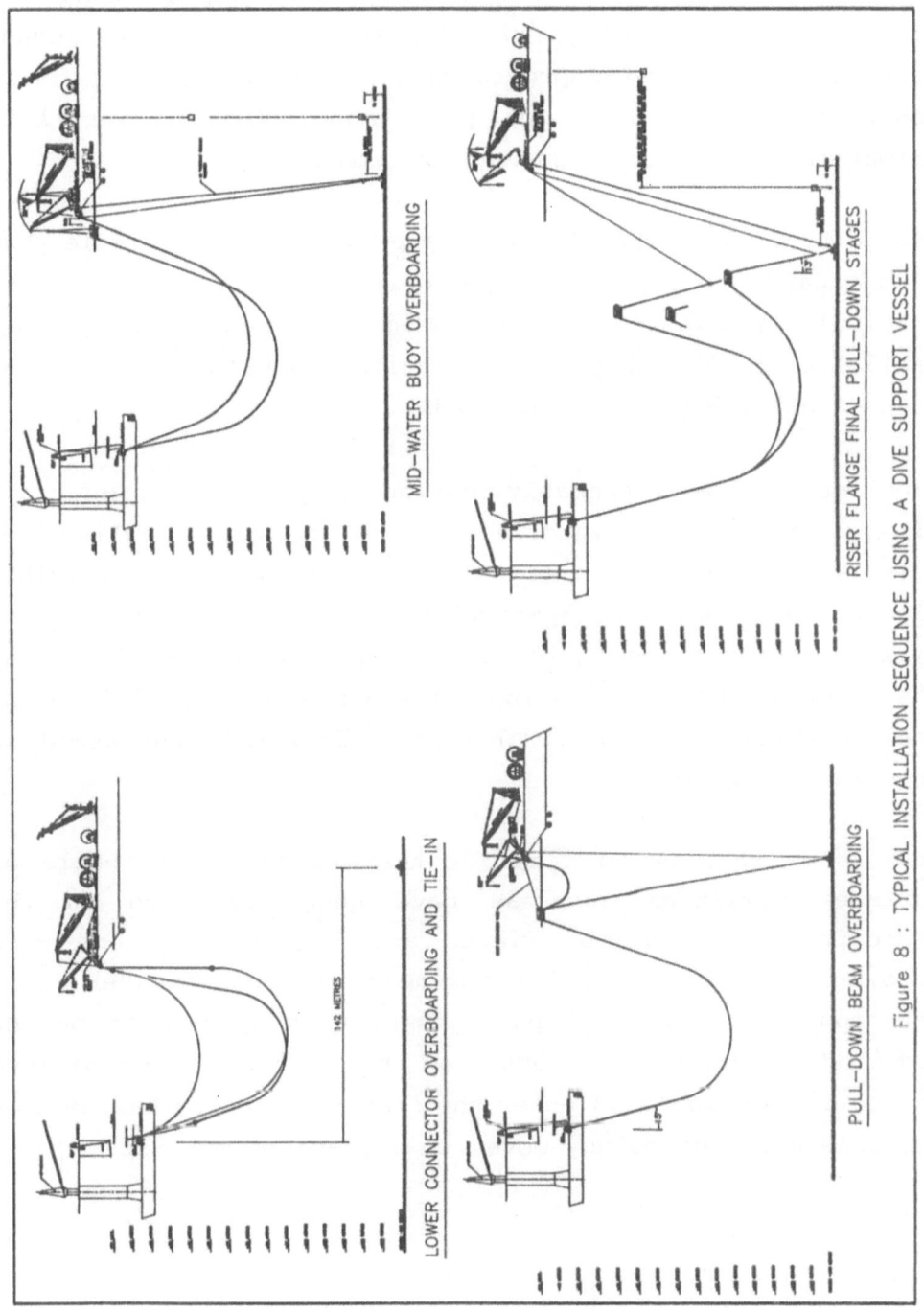

Figure 8 : TYPICAL INSTALLATION SEQUENCE USING A DIVE SUPPORT VESSEL

Advances in material engineering and improvements in manufacturing technology have resulted in reduced product rejection rates and improved operational performance. By contrast, the number of pipe manufacturers has been dramatically reduced as has production capacity.

Hybrid field developments would appear to offer a middle road with regards to pipeline technology whereby flexible pipes, particularly risers will be used to adapt fixed seabed facilities to floating surfaces units and will cater for the consequential high inertial loads.

Pipelines will most probably continue to be a mixture of rigid pipe with flexible interfaces. Intrafield flowlines of relatively short length (i.e. 1 to 6 km) made from flexible pipe will continue to compete with rigid pipe particularly for small diameter pipes (< 10 in. i.d.). For larger diameters, the material costs, armouring constraints, and limitations of reel capacities may constrain further flowline development in this market sector.

New configurations to link floaters to satellite wells or surface facilities such as have been installed in the Vesselfrikk and Marimba fields are likely to become more widely accepted, especially for marginal or deepwater field developments. 'Turn-key' packages of work appears to be the order of the day for the 1990s and it is anticipated that more marine contractors will enter the flexible pipelay business as flexible pipe technology moves to a stage of maturity.

REFERENCES

1. Alvheim, N. (1990). "Installation of Flexible Pipe", in Procs. "Flexible Pipes and Hoses Offshore", Norwegian Civil Eng., 8-10 Jan., Trondheim.

2. American Petroleum Institute (1988). <u>API RP 17b Recommended Practice for Flexible Pipe</u>, API, Washington.

3. Ashcombe, G.T. and Kenison, R.C. "The Problems Associated with NDT of High Pressure Flexible Pipes", <u>in</u> <u>Advances in Subsea Pipeline Engineering and Technology</u>, Kluwer Academic Publishers, Vol. 24, pp 179-204, Dordrecht.

4. Calquhoun, R.S., Hill, R.T., and Nielsen, R. (1990) "Design and Material Consideration for High Pressure Flexible Flowlines" <u>Advances in Subsea Pipeline Engineering and Technology</u>, Kluwer Academic Publishers, Vol. 24, pp 145-178, Dordrecht.

5. Ellis, B.J. (1943) "Improved Method of and Means for Installing Pipelines" Patent Specification, 601103, Sept, London.

6. Fagen, C. (1990) "DNV's Requirements and Recommendations" in Procs. "Flexible Pipes and Hoses Offshore", Norwegian Civil Eng., 8-10 Jan., Oslo.

7. Golan, M. (1988) "Technology, Products and Applications of Flexible Pipes", Procs. "Flexible Pipe Technology", Norsk Pet. Forening, Sept. Oslo.

8. GoodFellow Associates (1990). <u>Applications of Subsea Systems</u>, Pennwell Publishing Co., pp. 45-60, Tulsa.

9.   Huse, E. (1990). "Model Testing of Flexible Riser
        System" in Procs. "Flexible Pipes and Hoses Offshore,
        "Norwegian Civil Eng., 8-10 Jan., Trondheim.

10.   Lotveit, S.A. and Often, O. (1990) "Increased
        Reliability through a Unified Analysis Tool for Bonded
        and Non-Bonded Pipes", in Advances in Subsea Pipeline
        Engineering and Technology, Kluwer Academic Publishers,
        Vol, 24, pp, 79-110, Dordrecht.

11.   Meland, T. (1990) "Polymer Materials for Flexible Pipes"
        in Procs. "Flexible Pipes and Hoses Offshore", Norwegian
        Civil Eng., 8-10 Jan., Trondheim.

12.   Neffgen, J.M. (1990), Pigging and Inspection of Flexible
        Pipe", Procs. Pipeline Pigging and Inspection
        Technology, Gulf Publ. February, Houston.

13.   Neffgen, J.M. (1990) "Advances in Flexible Pipe Design
        and Construction", in Advances in Subsea Pipeline
        Engineering and Technology, Kluwer Academic Publishers,
        Vol. 24, pp 111-144, Dordrecht.

14.   Perzborn, V., Gropper, H., Sehramm, T. (1987)
        "Experimental Tests on the Behaviour of Steel Reinforced
        Elastomer Pipes for Offshore", VDI Berichte NR. 631, pp.
        375 - 387, Hamburg.

15.   Veritec (1987). Guidelines for the Design and
        Construction of Flexible Pipes, Joint Industry Report
        JIP/GFP Rev. 02, Olso.

A REVIEW OF PAST AND FUTURE DEVELOPMENTS IN BUNDLED
PIPELINE INSTALLATION BY THE MID DEPTH TOW METHOD

COLIN MCKINNON
J P Kenny and Partners
Thames Plaza, 5 Pine Trees
Chertsey Lane
Staines, Middlesex, TW18 3DT

ABSTRACT. 1991 has seen the installation of bundled pipelines by the
mid depth tow method emerge as a realistic alternative to
conventional offshore installation techniques through the entry into
the market place of a second credible installation contractor, namely
the Land and Marine/Field Enterprises Joint Venture, the first being
Rockwater. This paper presents a review of developments in bundled
pipeline installation by the mid depth tow method from the 3 element,
conventionally insulated bundles of the early 1980's to the multi
element, gel insulated bundles complete with towheads with looped
piping for round trip pigging, provision for control modules for
wellhead operations and isolation valves. The large number of small
diameter lines that can be installed within a single bundle permits
the direct connection of subsea wellheads with the platform topsides
without the need for complicated subsea manifolding systems. The
paper identifies the design and economic issues that should be
considered when evaluating the mid depth tow method against
conventional techniques. The paper concludes with an assessment of
the future trends in bundled pipeline design with particular regard
to the use of bundled pipelines as a part of step out developments
from existing facilities.

63

SUBSEA REPAIR OF CONCRETE WEIGHT COATING ON 30" NORTH SEA PIPELINE

ERIK REKER
Dansk Olie & Naturgas A/S
and
HANS CHR DAHLERUP KOCH
Topp. Tech A/S

The 30" North Sea Gas Pipeline in the Danish sector had suffered damage to the concrete weight coating and the corrosion coating in several spots on an exposed part just outside a subsea valve assembly protection cover, located approx. 1.5km from the Tyra gas field. The damage was probably caused by fishing activities such as trawlboards and other heavy fishing gear. To repair this damage several methods of repair were considered to find one which could provide both the necessary mechanical protection, as well as adequate corrosion protection of the bare metal areas on the pipeline and with a minimum of cleaning of the pipe surface. After considering all the alternatives, it was decided to carry out a repair using a new product called DENSIT. DENSIT is a composite material mainly made by mixing Portland cement with micro-silica and superplasticizer. The ultrafine micro-silica particles fill the voids between the cement grains, thus creating an extremely dense paste, giving the material a number of unique properties. The material can be pumped through long lengths of hose and will cure underwater. The paper will describe the actual design, planning and repair work carried out on two sections covering the full surface of 18m and 20m of the above mentioned pipeline, and the results of laboratory examination of the capability of the product to act as a corrosion protection to bare metal areas.

THE FULL PAPER WILL BE ISSUED AS A SEPARATE INSERT AT THE CONFERENCE.

Volume 27: Subtech '91, 65.
© 1991 *Society for Underwater Technology.*

DEVELOPING THE TROLL PROJECT

O.R. MARTINSEN
A/S Norske Shell
P.O.Box 40
4056 Tananger
Norway

ABSTRACT. Since the TROLL FIELD was declared commercial, many different development concepts have been considered based on a fully integrated offshore drilling, production and accommodation platform. Norske Shell has, as development operator and with Partners consent, decided to change this concept to include three main elements. The new concept consists of a less complex offshore platform, pipelines for transport of wet gas to shore, and an onshore processing and compression plant. In the processing plant the gas will first be treated and then compressed for export by pipeline to the European Continent. Landbased processing was chosen to improve safety, increase the ability to meet a growing demand for TROLL gas and improve the economic potential of the project. Figure 1.

1. INTRODUCTION

A/S Norske Shell discovered hydrocarbons in Block 31/2 west of Bergen in 1979. Figure 2. In the early eighties several appraisal wells were drilled, indicating substantial volumes of gas with some underlying oil, particularly in the West. Neighbouring blocks were also explored and together blocks 31/2,/3,/5 and /6 became known as the Troll Field and was unitised in 1986 and is the largest offshore field in Europe. That same year Norske Shell was appointed Operator for Phase I gas development in Troll East gas province. Figure 3.

The 'troll' is a creature which often appears in Norwegian folk-lore. The modern troll, however, is located on the Norwegian continental shelf in the North Sea. When gas begins to flow from the Troll field in 1996 it will provide for more than ten percent of Europe's total energy needs far into the 21st century. In fact Troll gas will be exported through several transport systems to a number of countries on the European Continent.

The Troll field can be divided into three areas: Troll East gas, Troll West gas and Troll West oil. Oil and gas production from Troll has already started. In 1986 it was decided to export gas from Troll East to the Oseberg field from a subsea production module to enhance oil production from that field. The TOGI (Troll Oseberg Gas Injection) project came on stream in January 1991 and will deliver a total of 25 billion cubic metres of Troll gas for the next 10-12 years. Norsk Hydro is operator for TOGI. In 1989 the Troll Unit Partners decided to test oil production in Troll West. Two horizontal wells have been drilled - the first well was 500 metres and the second was 800 metres long. Both tests indicate that some of the large oil accumulations in the West can be exploited. The oil produced during these tests has generated income to the Partnership and partly covered the cost for testing. Norsk Hydro is also Operator for the oil test project.

67

Volume 27: Subtech '91, 67–82.
© 1991 *Society for Underwater Technology*.

2. TROLL UNIT PARTNERS

The production license for block 31/2 was originally awarded to a group of companies comprising Statoil(50%), Norske Shell(35%), Norske Conoco(5%) and Mobil Exploration Norway(5%). The three adjacent blocks 31/3,/5 and /6 were awarded to Statoil(85%), Norsk Hydro(9%) and Saga Petroleum(5%). When it became clear that the licenses covered a single large field, the parties made an agreement to coordinate the development of the Troll field. This resulted in a redistribution of owner interests in the field. The present licensee structure for Troll is as follows:

Den norske stats oljeselskap a.s. - Statoil   (74.576%)
A/S Norske Shell   (8.288%)
Norsk Hydro Produksjon a.s.   (7.688%)
Saga Petroleum a.s.   (4.080%)
Elf Aquitaine Norge A/S   (2.353%)
Norske Conoco A/S   (2.015%)
Total Marine Norsk A/S   (1.000%)

The Norwegian Government's direct economic involvement represents 62.696%. Figure 4.

3. THE ORGANISATION

Until September 1989, the project organisation of Troll Phase I was located in Stavanger. The project moved into its own offices at Sandsli near Bergen in the autumn of 1989. The Norwegian authorities' approval of the Troll Phase I Gas development, is based on Norske Shell being the operator in the development phase, whereas Statoil shall take over as operator during the operational phase. This is manifested in a cooperation agreement between Norske Shell and Statoil.

4. THE TROLL PHASE I CONCEPT

The present development concept for Troll Phase I include three main elements. An offshore production platform, pipelines to shore to transport wet gas, and an onshore processing and compression plant. With the gas treatment facilities located onshore will make the whole production platform less complex. The platform substructure will be made concrete and will have four main shafts rising up from a large cellular base. The deck will be an integrated steel deck. The function of the platform is to produce gas from the reservoir and send it by pipelines to shore. Figure 5.

Most of the processing equipment used to treat the gas and the export compression will be located onshore. This simplifies the platform equipment, reduces the manning levels and therefore increases the safety considerably. The deck will now weigh approximately 14.000 tonnes, as compared to 40.000 tonnes for a full processing deck.

5. PRODUCTION AND OPERATION

From the Troll platform, 39 production wells and one observation well are due to be drilled within a 500-metre diameter immediately below the platform. Figure 6. Each well can produce in excess of 3 million cubic

metres of gas per day. During normal production, the daily rate will be
limited to 2.8 million cubic metres. The offshore platform and the
processing plant on land will be designed for a daily production rate of
approximately 84 million cubic metres. This corresponds to an annual
production of more than 20 billion cubic metres of gas.

The well stream will contain hydrocarbons in gas and liquid form
(condensate), as well as water. Before the gas can be piped ashore, some of
the water is removed from the gas during inlet separation. The condensate is
returned to the well stream in the export pipeline(s), while the water is
thoroughly cleansed and released into the sea. To prevent the gas from
freezing in the pipeline, glycol is added. In addition, corrosion of the
line is inhibited by the glycol, which will be recovered at the processing
plant and returned to the platform via a smaller diameter pipeline.

Reservoir pressure is sufficiently high during the first years of
production to deliver the gas through the 60 km long pipeline to the onshore
plant. By 2010, however, the production pressure will have dropped to a
level where it becomes necessary to install offshore compression facilities
to maintain the required inlet pressure to the onshore plant.

Around 20 staff members will be required to run the platform during
normal operations, but more personnel are needed during the initial
drilling of the production wells, and for maintenance work. As a result, the
platform is designed to provide accommodation for 75 people. The living
quarters will meet day-to-day crew requirements for privacy, recreation,
social activities and catering. The work environment will generally be of a
high standard.

6. TRANSPORTATION OF WET GAS

One of the major technological challenges in connection with the Troll
development has been to develop new technology which allows the
transportation of wet gas through pipelines over long distances. The current
plan is to transport the wet gas through two or more 36-inch pipelines from
the offshore production platform to the land based treatment facilities.

The seabed outside the plant location is extremely uneven and considered
too difficult to lay pipe across. Therefore a 3 km long landfall section of
pipeline will be laid inside a subsea tunnel. Under transport from offshore
to onshore, the well stream will cool down to seabed temperature. Water
vapour and heavier hydrocarbon fractions will condense out and form a
separate liquid during this cooling. At the onshore plant, these liquids are
first separated from the gas. At normal production rates, the velocity of
the gas in the pipeline is sufficiently high to keep these liquids dispersed
and they are carried along evenly distributed in the gas stream.

During interruptions in production, however, these liquids will gather
up and form distinct pools in low spots in the lines. On resuming production
these 'slugs' will be pushed ahead by the gas until they are removed in a
slugcatcher before they reach the processing plant. Prediction of the size,
number and location of these slugs has been a problem which has now been
solved. By solving this problem, a better basis for sizing of both pipelines
and slugcatcher has been achieved.

The gas stream contains both water and carbon dioxide which can corrode
the pipeline. It is therefore very necessary to be able to control and
measure the corrosion rate, which has not been previously possible. Recent
research has shown that glycol inhibits corrosion and can therefore be used

as a corrosion controller. In addition, a new generation of inspection
devices,'intelligent pigs', has been developed which allow better estimation
of pipeline condition. These tools are sent through the line at regular
intervals and inspect the pipe wall by means of ultrasonics. Shell has been
at the forefront of the research and development of both the use of glycol
and online ultrasonic pipeline condition monitoring.

The availability of the new technology in flow modelling, corrosion
control and inspection has allowed the concept of long-term transport of
multiphase flow over long distances to be realised.

7. PROCESSING FACILITIES

The land based processing facilities will cover circa 2 million square
metres and will be dimensioned to handle a daily flow of around 84 million
cubic metres of gas. In the gas processing plant, the wet gas from Troll
will be dehydrated to leave minimal quantities of water vapour. Through
pressure reduction and cooling, the heavier components of the gas are
separated in liquid form. Thereafter the gas is compressed for
transportation through pipelines over a distance of about 1.100 km to
Zeebrugge in Belgium. A schematic layout of the plant is shown in Figure 7.

Approximately 2.000 tonnes of condensate annually will be transported
with the wet gas from Troll. This is a relatively small quantity, and will
hardly make a basis for any separate industrial activity in the vicinity.
The condensate will be separated from the gas and transported by ship to the
customer via a terminal close to the plant.

The land area and plant lay out takes into consideration the need for
future expansion and handling of additional gas from the Troll field.

8. CONTROL SYSTEM

The Troll Phase I development will consist of three operational units:
Production platform, pipelines and processing plant. The three units will be
coordinated by remote control from a Control Centre on shore.

The system shall control and monitor the operation of these units
through two data loops – one on land and one offshore. These will be
integrated with the aid of a fibre-optic cable, which is incorporated in the
power supply cable between shore and platform. Immediate warning of abnormal
events will be given to the control centre on land. The operators will
decide which counter measures best fit the occasion, but provision is also
made for automatic shut-down in particular critical circumstances.

If the system breaks down and communication is lost between the offshore
and land based sections, the two systems can continue to operate
independently. The aim is to create as simple and reliable systems as
possible, compatible with operational security and safety in general.
Measuring instruments will be selected from among the most advanced
technology available. Since the operational units will be remotely
controlled from an onshore control room, the offshore personnel can be
reduced to a minimum.

9. EXPORT SYSTEMS

A gas pipeline system (Statpipe/Norpipe) already exists from the Norwegian
continental shelf to the European markets, but this capacity needs to be
increased considerably to handle the gas deliveries committed under the
Troll Gas Sales Agreements. Gas will be delivered already under this
agreement from Sleipner in 1993; three years ahead of gas deliveries due
from Troll in 1996.

As a result, a new system of altogether 1.300 km of pipeline will be
laid. One 40-inch pipeline will go from the Sleipner platform to Zeebrugge
in Belgium, and one 30-inch line linking Sleipner to the Statpipe/Norpipe
system at the riser platform 16/11-S (Zeepipe Phase I). Two pipelines will
be laid from Oygaarden to the Sleipner and Heimdal platforms respectively;
the gas from Troll can therefore be transported to the continent by two
separate pipeline systems. Figure 8.

Further optional volumes have been exercised by the European buyers
which require additional transportation capacity. In this respect a third
pipeline (Europipe) is being planned installed to handle these additional
volumes. It is strategically important to deliver the Troll gas by separate
transportation systems. This is due to greater flexibility, and the
increased reliability in the gas deliveries.

10. SCHEDULES AND INVESTMENTS

The schedule for the three main elements in the development of Troll Phase I
is shown in Figure 9.

The Troll development involves major investments and operating costs.
The long lifetime of the field means that the cumulative operating costs
will, after some time , draw near investments costs. As a result, efforts to
reduce investments have been paralleled by heavy emphasis on cutting
operating costs.

One of the economic benefits of the land based alternative is that part
of the investment spending can be postponed. This is because a large
integrated production and processing platform offshore must be built to
handle peak production from day one. A plant on land can be expanded
according to the production rate of gas to fit with offtake requirements.
Since the build up of production is gradual, full processing capacity will
not be required in the first few years.

Investment for Troll Phase I up to 1997 is estimated at less than NOK 22
billion. Total investment in the Gas Sales contract period up to year 2022,
is estimated to be between NOK 24 and 28 billion, depending on production
level and reservoir characteristics. Figure 10.

Troll is by all standards a huge project and will have for that reason a
significant impact on the Norwegian economy. The State and Statoil own 74.5%
of the Troll licence. When the field comes on stream, approximately 85% of
the income from the Troll field will go to the State (when including taxes,
duties paid by the other licence owners).

Shell's aim is to award a large share of development contracts to
Norwegian industry, provided that Norwegian suppliers are competitive. This
will have economic effects in the form of increases employment and
industrial activity.

11. EMPLOYMENT

The need for labour in connection with the development of Troll will be largest during the building period. A work force of around 1.300 will be required for the construction of the processing plant and work will give business opportunities to local companies. A land based project will also positively affect the surrounding municipalities.

The operational period of the processing plant will consist of two stages. When the first development stage ic completed in 1996 the plant will offer some 100 permanent jobs. When the plant is extended to full capacity this number can increase to approximately 120 positions. Throughout, the plant will require additional support from contractors and this will vary from time to time.

Gas will flow through the plant for at least 50 years and it is assumed that this activity will give benefits to the local communities far into the 21st century.

12. QUALITY, SAFETY AND ENVIRONMENT

Efforts are being made to secure high standards for the physical and social work environment in all phases of the project. Other important objectives are to make the Troll installations as reliable and environmentally safe as possible. As a result of close cooperation between Norske Shell as operator for the development phase and Statoil as operator for the operational phase the Troll project has been in a position to draw on Statoil's experience from the North Sea as well as Shell's international experience.

Safety and quality are continuously being improved throughout the project. Norske Shell's policies and requirement for high standards in these areas will also be reflected in all engineering and building work that takes place before the Troll installations are completed in 1996.

Developing the Troll field presents an important challenge in the fight to preserve the environment. According to Norske Shell's philosophy, the use of modern technology will reduce the discharge of environmentally damaging components, from offshore platforms as well as onshore plants. A processing plant on land involves discharges of hydrocarbons to both air and water. Emissions to air will largely take the form of exhaust fumes from gas turbines; modern technology makes it possible to reduce these discharges.

The supply of electrical power by cable from land is being planned. This eliminates the need for large offshore power plants based on gas turbines that emit carbon dioxide, carbon monoxide and nitrous oxide with their exhaust gases. Consideration has also been given to the installation of electrically-powered fire pumps offshore, with a set of diesel-driven pumps as reserve; this will again reduce emission of sulphurous compounds normally associated with diesel-powered equipment.

Water-based drilling mud will be used as opposed to oil-based drilling mud, which has been more common up to now. Because no oily cuttings will be discharged into the sea pollution will be reduced, and the seabed will be restored more quickly than in the case when cuttings containing oil is used.

Acknowledgement.

The author wish to thank Kaare Holm, and the Public Affairs Department, Troll Project for their help making this paper.

Figure 1.   The Troll Concept. Showing the three main elements.
              The Offshore platform, the Pipelines to shore and
              the Land based process facilities.

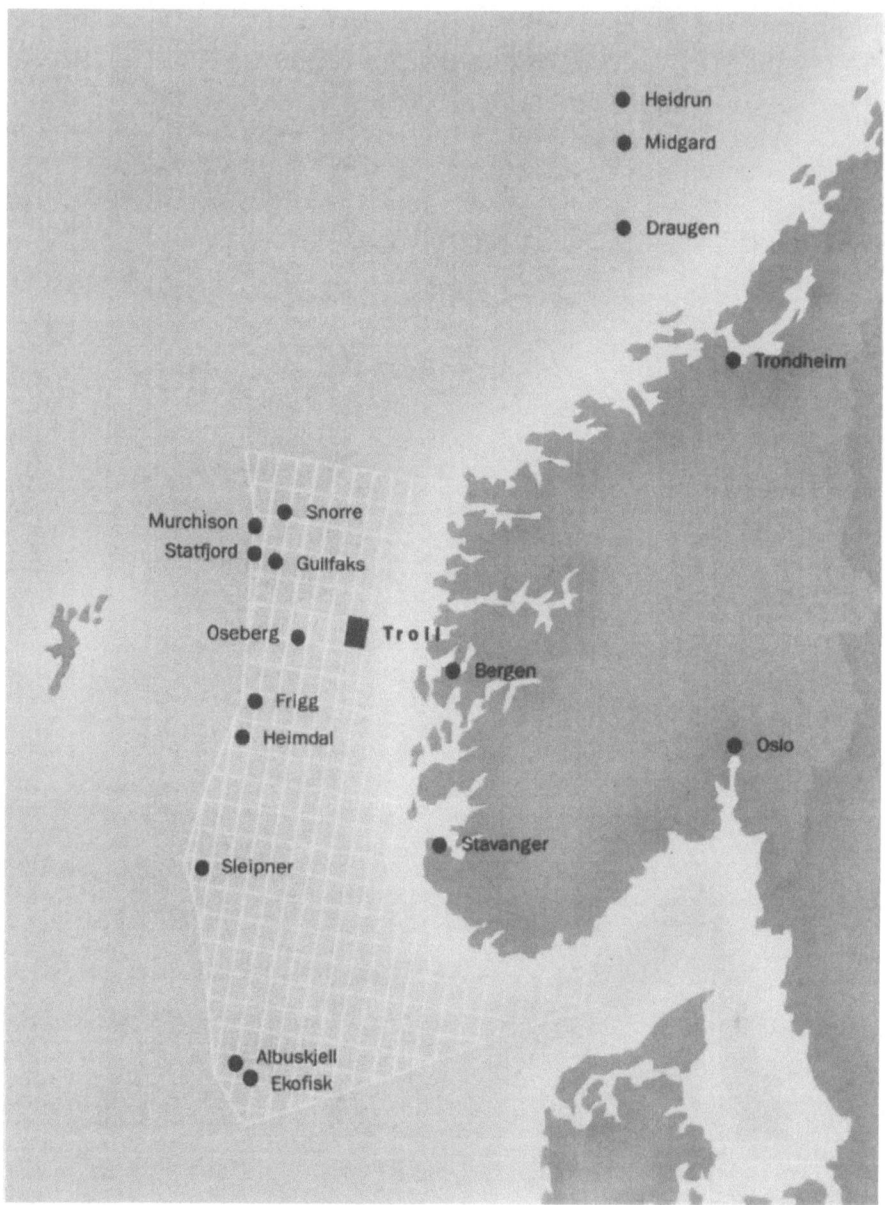

Figure 2. Discovering Troll. The Troll field is located in the northern part of the North Sea, approximately 80 km northwest of Bergen. The first gas was discovered by Norske Shell in block 31/2 in 1979.

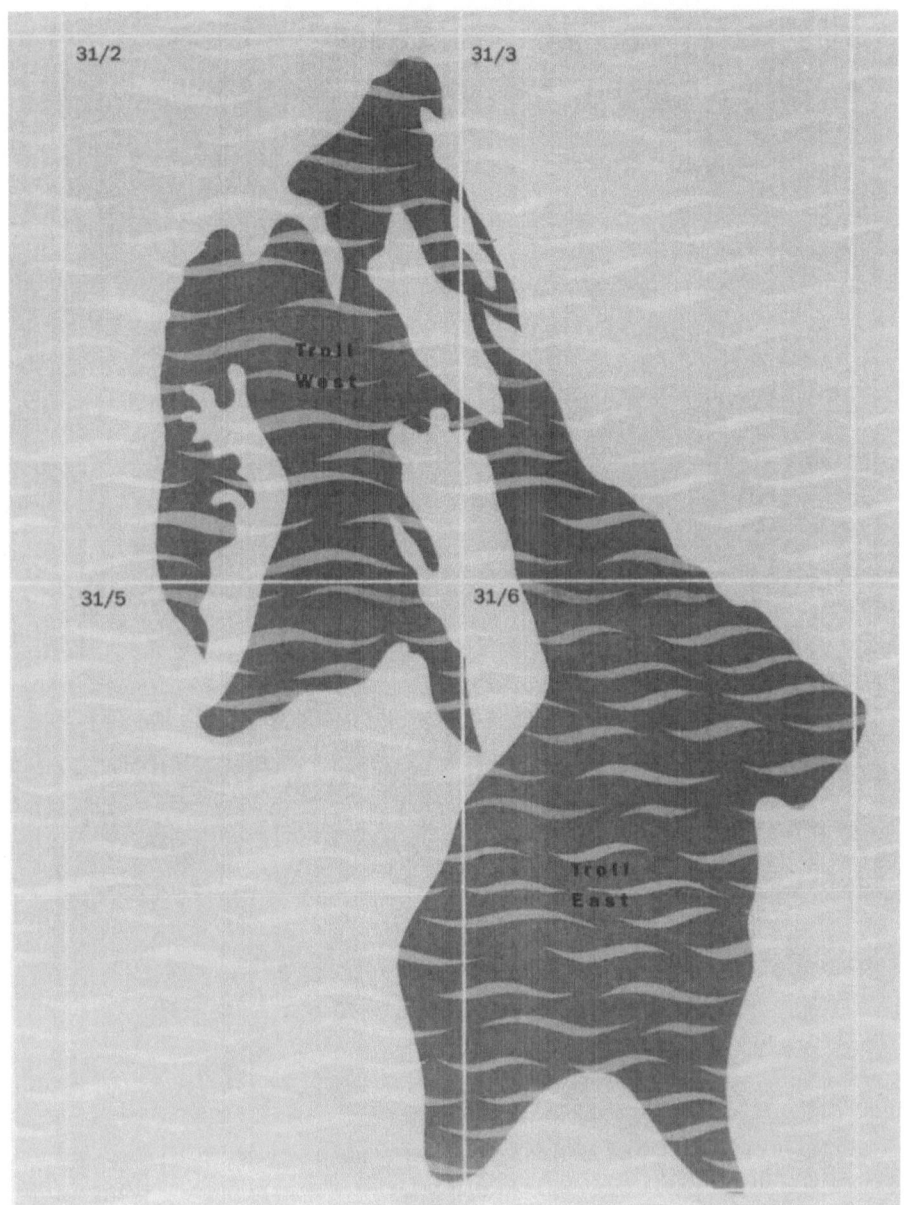

Figure 3. The extent of the Field. Recoverable gas reserves in the field is approximately 1200 billion cubic metres. This makes Troll the largest offshore gas field in Europe.

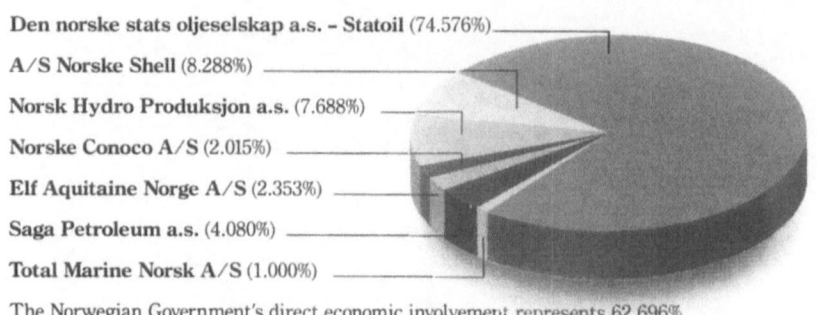

Den norske stats oljeselskap a.s. – Statoil (74.576%)

A/S Norske Shell (8.288%)

Norsk Hydro Produksjon a.s. (7.688%)

Norske Conoco A/S (2.015%)

Elf Aquitaine Norge A/S (2.353%)

Saga Petroleum a.s. (4.080%)

Total Marine Norsk A/S (1.000%)

The Norwegian Government's direct economic involvement represents 62.696%.

Figure 4.   The Troll Unit Partners. After unitization the figure
shows the existing ownership relationship.

Figure 5.   The offshore platform. The structure is located in
             305 metres of water. It will be a concrete gravity
             base structure. The wells will be located in the two
             central shafts and the facilities supported by an
             integrated steel deck.

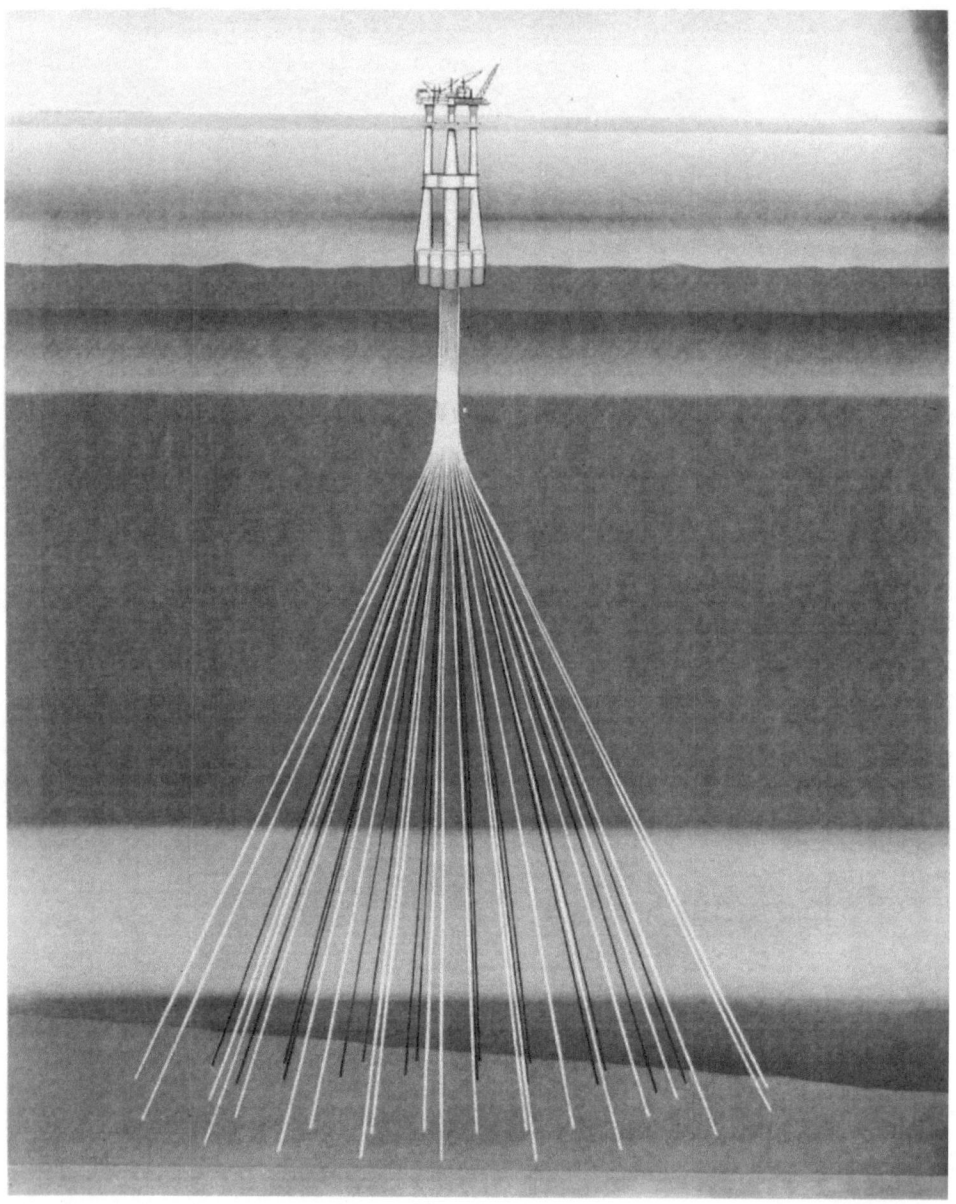

Figure 6.   Production wells. 39 wells will produce the gas from the reservoir 1400 metres below the sea bed.

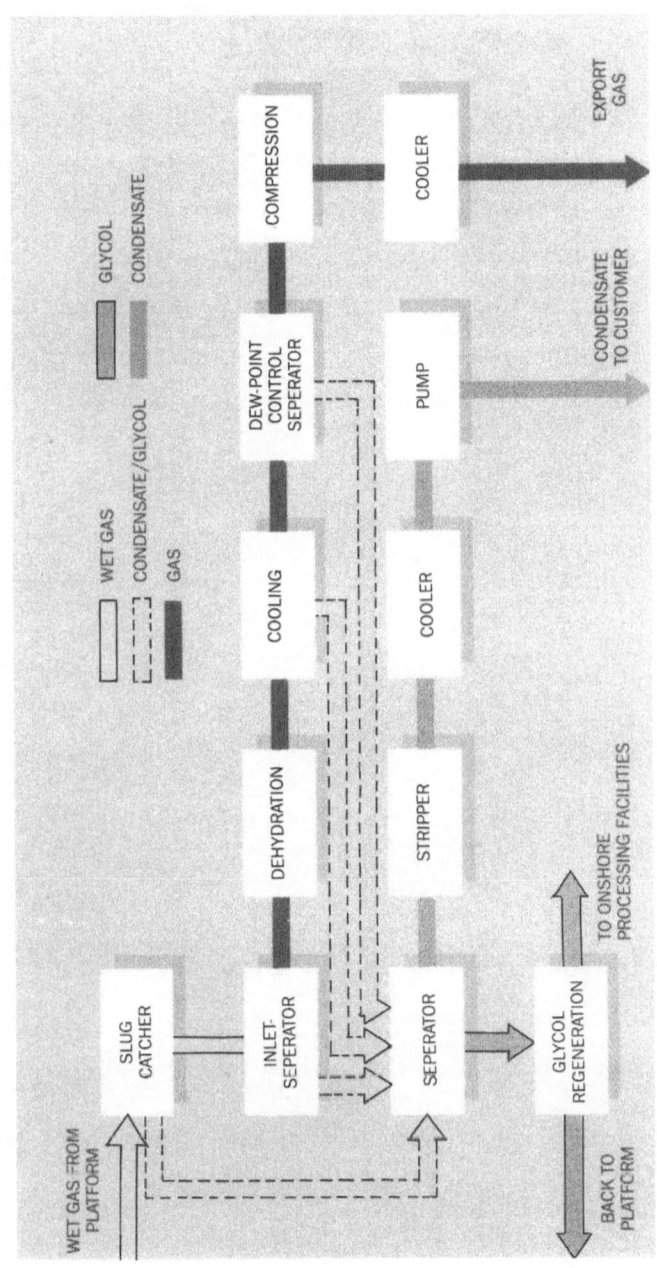

Figure 7. Process facilities. Schematic layout of the land based facilities and main plant components.

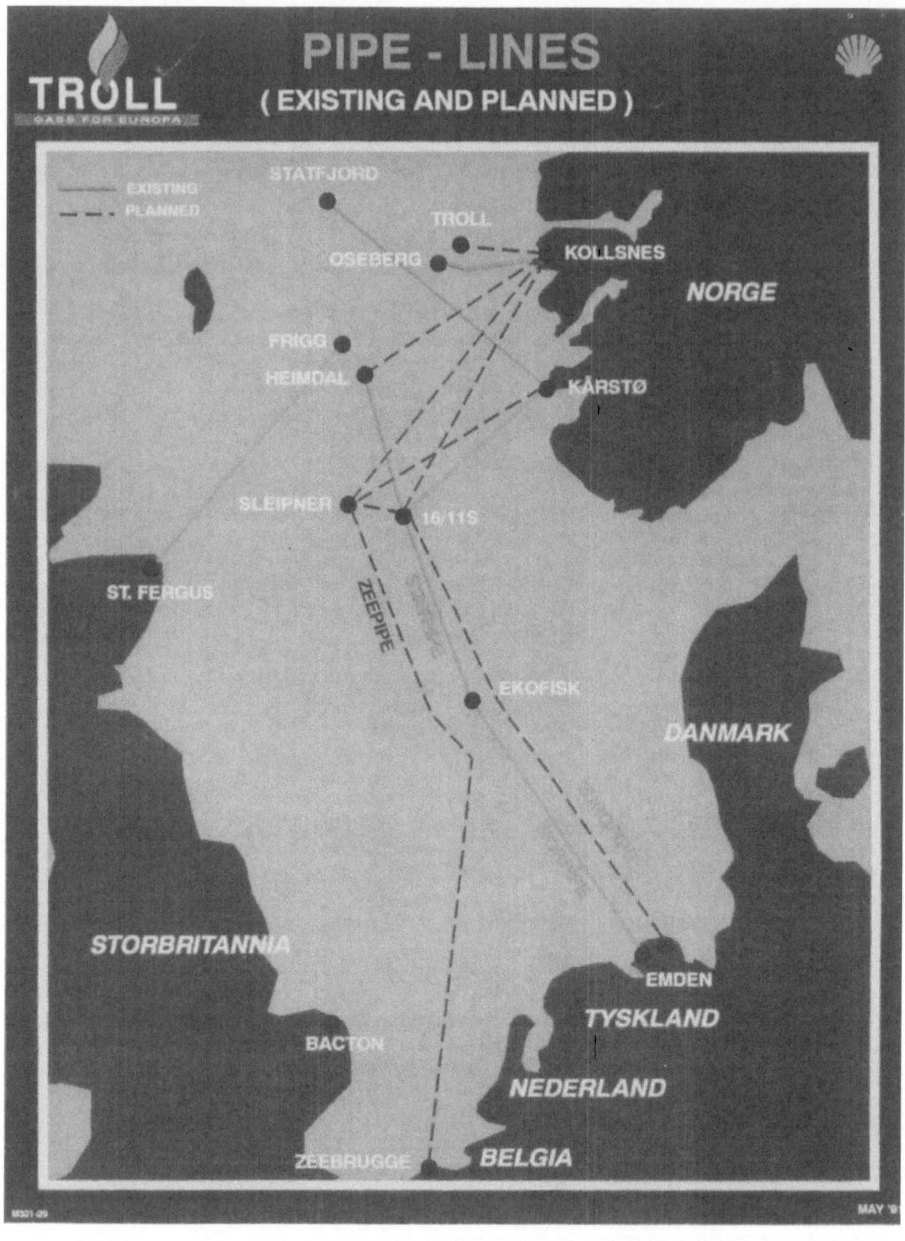

Figure   8.   Transportation systems. An overview of existing and
               planned pipe-line systems for transporting gas to
               Europe.

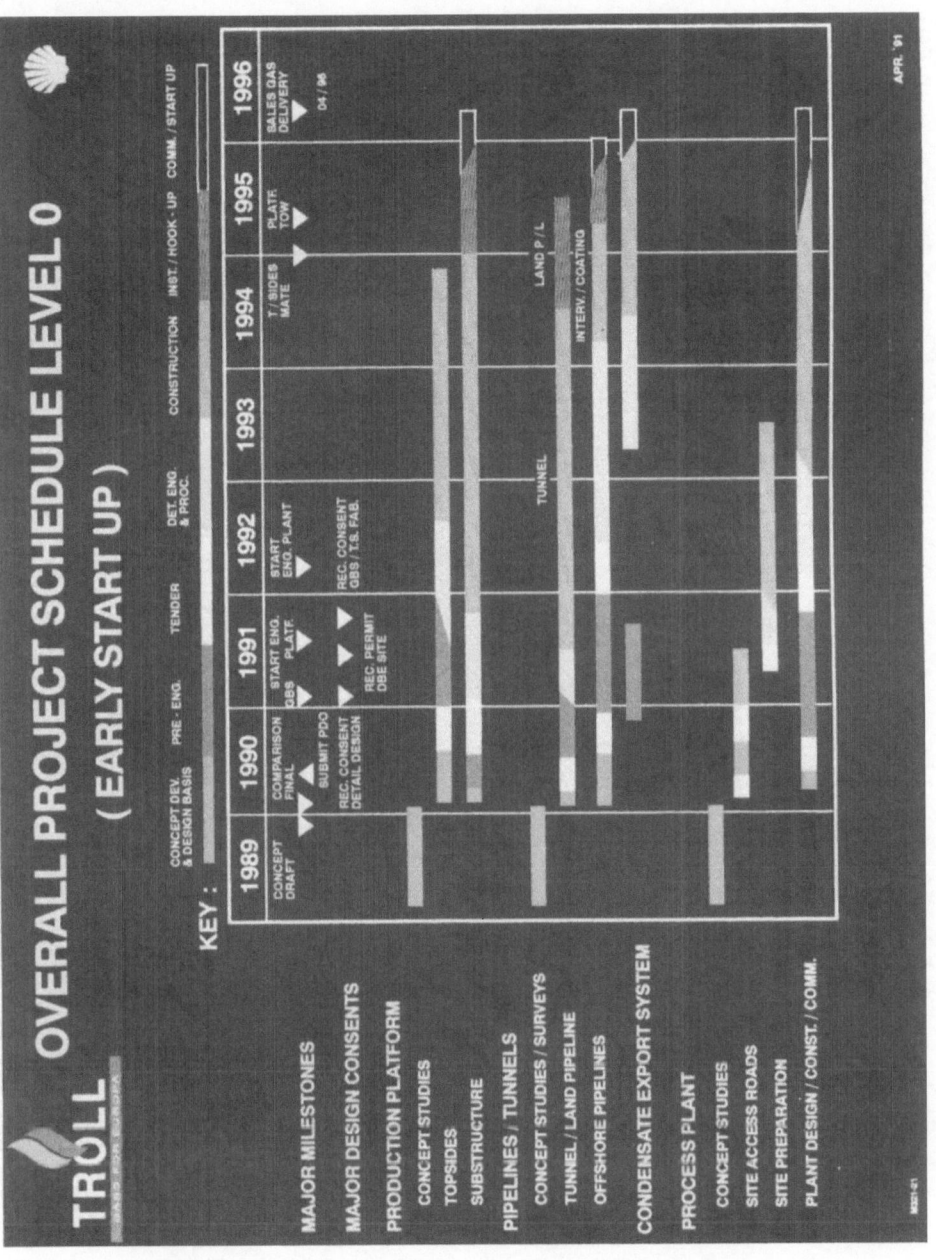

Figure 9.   Project schedule. The current schedules for the three main project elements.

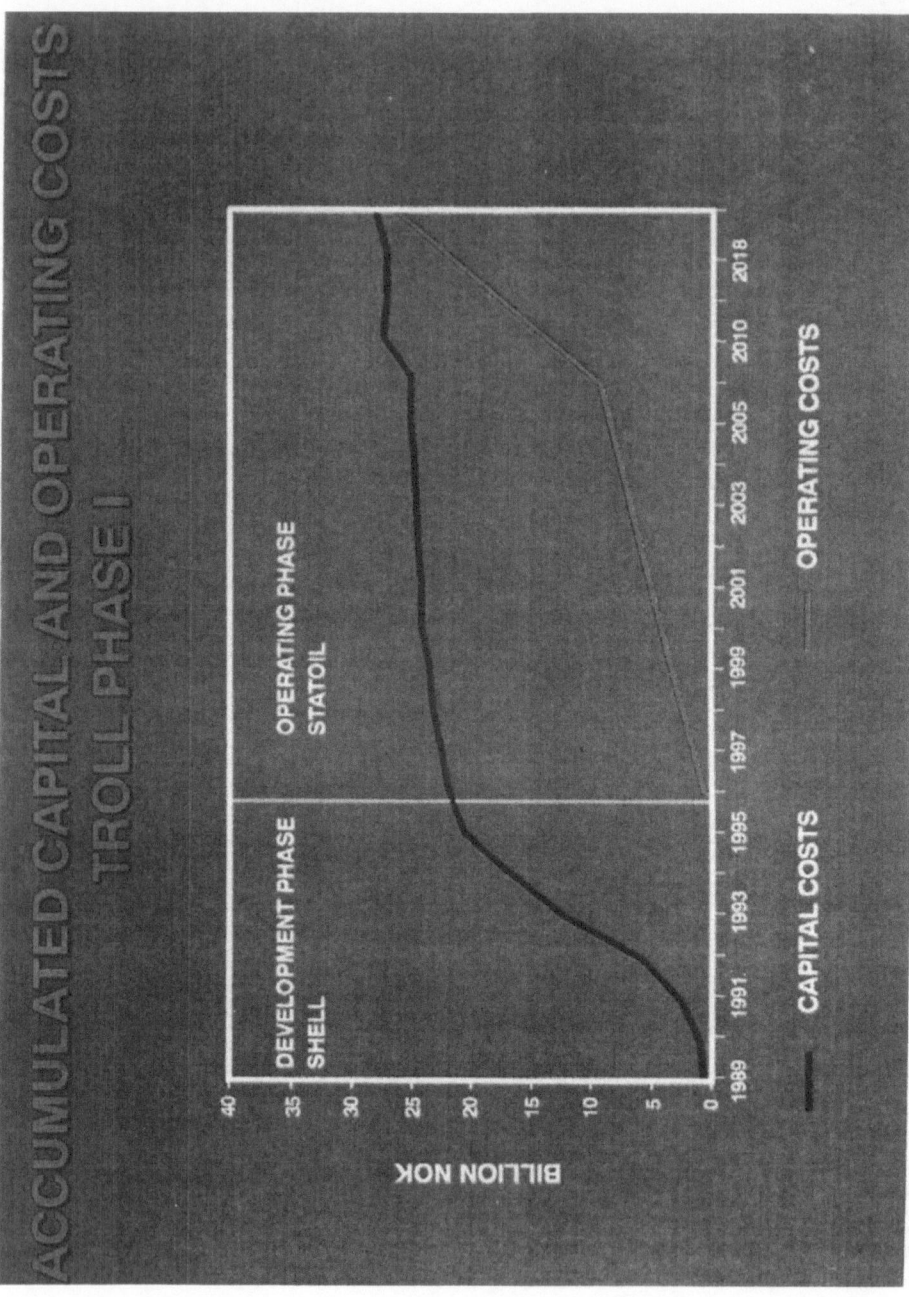

Figure 10. Capital and operating costs. The cash flow in the project in the development phase and operating phase.

WELLHEAD PROTECTION AT GANNET B

D. HEAL
Anchortech Ltd.
Stena House
Westhill Industrial Est.
Aberdeen
AB3 6TQ

L. NAUGHTON
Shall U.K. Exploration and Production
1 Athens Farm Road
Nigg
Aberdeen
AB9 2HY

Anchortech Ltd. is a wholly owned subsidiary of Stena Offshore Limited.
Shell U.K. Exploration and Production operates on behalf of Shell, Esso and other co-venturers in the UKCS.

ABSTRACT. This paper describes the Gannet development and its extensive use of subsea production facilities. It describes the consultations carried out with other seabed users, and the agreement on a protection philosophy for subsea wells using a combination of fishing exclusion zones and non-exclusion zones with over-trawlable structures. The functional requirements and design loads used for detailed design are outlined. The paper also describes the structures and piles and how they meet the functional requirements. It covers the model testing carried out to assess over-trawlability, and what changes were made as a result. The installation of the structures is also briefly described.

INSTALLATION AND MAINTENANCE OF SUBSEA ISOLATION VALVES

Dr. R. K. Jain
Brown & Root Ltd.
5 St. Georges Road
Wimbledon
London SW19 4DG

ABSTRACT. Lord Cullen's report on the Piper Alpha disaster does not recommend a mandatory requirement for all hydrocarbon pipelines to have subsea isolation valves(SSIVs). However, the requirement of SSIVs will be determined through a Formal Safety Audit. This paper briefly reviews the impact of the Cullen report on the pipeline system design and the requirement of SSIVs including preliminary proposals to dispose off the trapped hydrocarbon inventory. An overview of the SSIV system design identifies the cost effective approach without affecting the SSIV reliability. After identifying testing, inspection and maintenance requirements, the paper discusses their impact on the design. Finally, a conceptual design of a maintenance frame is presented which should help in reducing diving time during SSIV maintenance.

1. INTRODUCTION

Unfortunate events in the North Sea over the past few years lead the U.K. Government to review safety policies and practices of North Sea operators and regulatory bodies. As a result of this review, Statutory Instrument No. 1029 was introduced which makes it mandatory for all offshore installations which import or export hydrocarbons to incorporate Emergency Shutdown Valves into the risers at locations above the splash zone. These valves are required to have a positive fail-safe control system to close the valves which detects and responds to an emergency. There are, however, no mandatory requirements for pipeline operators to incorporate subsea emergency/isolation valves.

The report by Lord Cullen following the Public Inquiry into the Piper Alpha disaster makes comprehensive recommendations for improving overall safety of North Sea installations. The whole safety process should start at the conceptual design stage by way of a Formal Safety Audit (FSA) and finally lead to a Safety Case which also includes safety management systems. The inquiry looked into the hydrocarbon pipeline operations and makes certain recommendations. The Cullen report endorses Statutory Instrument no 1029 and does not recommend a blanket requirement for subsea isolation valves(SSIVs). The need of such valves would depend upon the Safety Case developed for each offshore installation.

Although neither the Department of Energy nor the Cullen report spell out mandatory installation of SSIVs, it is expected that most pipelines transporting gas and many transporting oil will be fitted with some form of barrier system. Many new and operational pipelines have already been fitted or planned to be fitted with isolation valves.

The purpose of this paper is to briefly review the impact of the Cullen report on the pipeline system design and to identify situations where SSIVs can lead to an improvement in the safety of the pipeline systems and hence that of the installation and the personnel working on it. Some modifications to riser design have been identified in this paper which would lead to rapid evacuation of the product, particularly the gas, trapped between the SSIV and the riser topsides ESDV during an emergency shutdown.

The paper presents a brief overview of the SSIV design and associated control and protection systems. An important aspect of the design should be its simplicity without sacrificing the integrity and operability. This would in general lead to reduced down-time and also to easily maintainable and testable system.

The paper discusses the inspection, maintenance and repair of SSIVs and finally presents a conceptual design of a maintenance frame which is intended to minimise major underwater repair time of SSIVs

Volume 27: Subtech '91, 85–95.

2. CULLEN REPORT REVIEW AND ITS IMPLICATIONS ON PIPELINE SYSTEM DESIGN

2.1 CULLEN REPORT RECOMMENDATIONS

The report by Lord Cullen of the Public Inquiry into the Piper Alpha disaster in the North Sea in 1988 documents events leading to the unfortunate accident and makes recommendations for improving the safety of offshore installations. It is presented in two volumes and its 106 recommendations are contained in Chapter 23 of volume two. The United Kingdom Government have accepted all the recommendations.

The recommendations have major implications on the design and operations of offshore oil and gas installations. This section briefly reviews the recommendations in general and discusses the impact on pipeline and riser system design.

The most important recommendation contained in the Cullen report is that the operator should be required to submit to the regulatory body a Safety Case for each of its installations. The Safety Case should demonstrate that the safety management systems of the company and that of the installation are adequate to ensure that the design and operation of the installation and its equipment are safe. It should further ensure that the potential hazards to the installation and the risks to personnel have been identified and appropriate provisions are made to deal with any emergency and the safe evacuation of the personnel who may be affected by it. The safety management system should set out the objectives and means of achieving these goals. The performance of these systems are to be monitored by audits. It would appear that Lord Cullen considers the Safety Case as a document which is developed through the conceptual, detailed design, commissioning and operation phases and is kept up-to-date. A very important part of the Safety Case is the Formal Safety Assessment(FSA) which is to be carried out at the outset of design development and would form the basis of detail design, equipment selection, installation, testing, commissioning and operations of an installation. The FSA will include any connecting pipeline systems in the assessment.

As far as pipeline systems are concerned, the Cullen report recommends that:
a. operators should maintain, regularly review and keep up to date the pipeline emergency procedures and manuals. These procedures must interface and be consistent with the emergency procedures for all the connecting installations.
b. similar procedures, as above, should be developed and regularly updated to shutdown production if an emergency occurs on another installation which is connected by a pipeline to the producing installation and if continuing the production would exacerbate the emergency.
c. the Emergency Pipeline Valve Regulation, SI 1029, should continue to be in force until incorporated into new regulations.
d. there should be no blanket requirement for subsea isolation valves. The need of such valves will depend on the Safety Case for the particular installation.
e. the number of pipeline connections to a platform should be minimised.
f. studies should be carried out to improve reliability and reduce cost of SSIVs.
g. methods should be developed to achieve passive fire protection of risers without aggravating corrosion.
h. studies should be carried out into the dumping of oil in an emergency in an environmentally acceptable manner to reduce the fuel available to a fire. Note that the report does not recommend any action for the gas inventory.

2.2 IMPLICATIONS ON PIPELINE SYSTEM

The Public Inquiry by Lord Cullen into the Piper Alpha disaster did not make any new recommendations regarding the safety in design of subsea pipelines which may be taken to imply that basic design practices normally followed for subsea pipelines are satisfactory and there is no major action or change of direction called for. This does not mean that review of these design practices should not be carried out as and when required.

It is well known in the industry that operators have large amounts of design and operational data for their installations and pipelines but only a few have readily available structured 'As-Built' documentation.

Lord Cullen's recommendations regarding procedures and manuals can not be achieved without access to easily available and up-to-date as-built design information. In a number of cases, operators have already taken steps to compile as-built data manuals for some pipeline systems.

One impact on new pipelines, particularly trunk lines, will be to consider, during the design phase, possible future fields which could be tied into the pipeline through subsea Tee or Wye connections to reduce the number of pipelines going onto a platform. This needs optimisation of pipeline utilisation through mutual cooperation of operators and long term planning.

Riser design criteria normally include its protection as a basic requirement which is achieved by careful routing, location, design and analysis. The recommendation of passive fire protection of the risers by coatings, etc., should include by their design and operations the means and procedures for regular monitoring of internal and external corrosion by different methods including wall thickness measurements. Other recommendations to evacuate the riser of its hydrocarbon inventory and of SSIV cost optimisation are discussed in some detail in the following sections.

2.3 INVENTORY DISPOSAL

A major problem during an emergency situation on an installation is how to evacuate the hydrocarbons contained in the pipeline system which may be either transporting the product to other facilities or importing it from other installations. By introducing ESDVs on the riser topsides and shutting the production on the affected platform or the satellite installation, supplies of hydrocarbons to the affected platform can be minimised. However, there is a possibility that the product left in the pipelines and elsewhere could still be sufficient to pose serious problem for the safety of the platform and the personnel on it.

The amount of inventory available to fuel a fire should be limited by the design of the riser/pipeline system. The riser ESDV provides the facility to cut off supplies but it will not be able to control the product release if the riser were to be damaged below the ESDV unless a second cutoff is incorporated into the pipeline/riser system away from the high risk area. This point is discussed in the following section.

Crude oil is not as major a fire hazard as gas but it can still lead to serious consequences. In case of an oil pipeline/riser, a limited quantity of oil can, in an emergency, be released into the sea with minimal impact on the environment. The amount required to be released should generally be small as it will be necessary only to equalise the riser/pipeline internal pressure with the external pressure. It is proposed that, for an oil pipeline/riser, a small valve may be located at a suitable position, on the riser between the ESDV and the SSIV, which opens automatically in an emergency when the riser ESDV and SSIV close and thus dispose of the oil. After the emergency, the sea borne oil can be treated with chemicals or collected as required. The location and size, and controls for such a vent valve should be considered during the detail design to achieve optimum product evacuation time.

It may be pointed out that the Cullen report has not addressed the disposal of gas in a pipeline/riser system. In the case of gas risers, one suggestion would be that a bypass small diameter pipe from the riser should carry the gas direct to a flare in case of an emergency. This bypass should be at a suitable point below the riser ESDV and is always shut with a valve which opens only automatically when the riser ESDV shuts in an emergency. It is important that the bypass pipe to the flare should be well protected and routed through the least vulnerable areas on the platform.

The author would like to record that the foregoing proposals need careful consideration from the long term safety and the design must ensure that weak links are not introduced in doing so. The design must ensure that in case of an accident, the vent valves and associated piping do not become the cause of accelerating the damage. The oil release and gas bypass valves must be integrated into the overall control systems for the ESDVs and SSIVs.

2.4 REQUIREMENT OF SSIVS

As stated earlier, the Cullen report does not make it mandatory to install SSIVs into the pipeline system but their need should be investigated during the Formal Safety Audit and preparation of the Safety Case. The author does not question the findings of the Cullen Inquiry and other reports on the subject but would submit that short term economics should not dictate the decisions regarding the SSIVs. All offshore development are costly, running into several hundred million pounds, and SSIVs at a fraction of the total development cost provide a second but significant tier of safety and further insurance. A pipeline is nothing but a very long pressure vessel containing very large quantities of highly inflammable hydrocarbons, and any feature which will effectively limit the supply of fuel to a fire should be welcome.

It has been reported in a number of studies that the time required to dispose of the inventory, between riser ESDV and the SSIV, in a large diameter pipeline could be as long as one hour depending on the size of outlet and the length between the riser ESDV and the SSIV. By careful design considerations and providing relatively safe means of disposal of entrapped oil or gas between the riser ESDV and the SSIV, it should be possible to minimise this inventory evacuation time. This should enhance safety of the installation and that of the personnel working on it. It is essential that consideration is given to the incorporation of SSIVs into all long gas pipelines. Simplified SSIV designs can both reduce the cost and probably improve long term reliability. The author would recommend all gas pipelines should have SSIVs unless other means are available to put a stop to the supply of gas to a fire in an emergency. The benefits of SSIVs in oil lines should also be considered in FSA.

The next Section presents a brief overview of SSIV design and proposes some optimisation of costs.

3. AN OVERVIEW OF SSIV DESIGN

The purpose of this section is to present an over view of the SSIV design; readers are referred to other literature quoted in the reference section for further information.

3.1 LOCATION OF SSIVS

The criteria which dictate the location of SSIVs are as follows:
a. these should be preferably outside the effective radiation area of expected fire;
b. minimum time for the entrapped inventory evacuation through rupture and other outlets such as the flare for gas in case of an emergency;
c. simplicity, reliability and effectiveness of control systems;
d. SSIV protection
e. installation, inspection, testing, maintenance and repair.

Considering the above criteria, SSIVs should be installed as close to the platform as possible but outside the pool radiation area or where falling debris from a platform can pose a threat to the SSIVs or the pipeline outside of the SSIVs. It is therefore considered that SSIVs should be within 500 metres of the Statutory platform safety zone, with the preferred distance being between 150 and 350 metres.

3.2 VALVES AND VALVE CONFIGURATION

Several different types of valves are available but ball, gate and check valves are considered to be the most suitable for subsea duty.

Ball and gate valves require actuators and require a control system but check valves in general do not require either actuators or controls. With reference to ball valves, internals for the top entry and split body can be maintained and repaired whereas an all-welded body does not allow this facility. The gate valves suffer from their large size for large diameter pipelines, hence these are useful for small diameter pipelines. Check valves are designed to close automatically if pipe flow stops or reverses. Swing check valves are most commonly used. Check valves are not suitable for import pipelines or where positive control of the clapper is required. Subsea maintenance of top entry ball and check valves is possible. Top entry ball valves have dominated the SSIV designs to-date. It is pointed out that several variations of the top entry valve are available and have been addressed in references.

The SSIV system may comprise one or more valves of the same or different types. If more than one valve is used, only one valve is normally designed to perform emergency isolation function, the other valves are primarily for maintenance purposes. Two-valve SSIV systems are generally recommended. Depending on the pipeline operating conditions, expansion offsets may be required to absorb pipeline expansion. The SSIV configuration should be designed to be convenient for maintenance and should allow free passage of pigs.

3.3 ACTUATION AND CONTROL

The actuators considered suitable for subsea application for both ball and gate valves are of hydraulically powered piston type. These can be Single Acting or Double Acting. For pipelines above 10" or 12" diameter, single acting actuators may become impractical due to their size. Double acting actuators will require subsea hydraulic accumulator banks kept in fully charged state from the hydraulic power unit (HPU) located at the surface facility. For a single acting actuator, the hydraulic power is provided by the HPU on the surface facility.

The type of hydraulic fluid has significant effect on the speed of valve closure. The choice is between mineral oils or water based fluids. Both have been used in subsea hydraulic control systems. Selection will depend on a number of factors discussed in the literature quoted in the references.

A number of different control systems have been designed and installed in the North Sea. Some are simple but effective while others are quite sophisticated to perform a number of functions which ,in this author's opinion, are not necessary. The control system must be able to detect an emergency and should react to close the valves. It should be integrated into the overall control systems for the platform.

The umbilical is the lifeline of the control system and as such it must be carefully designed to provide the intended service.

3.4 PROTECTION OF SSIVS

The SSIV system including the umbilical must be protected from damage. That is why its location is so important. The umbilical should be trenched and/or covered with rock or similar to protect it from dropped objects or other subsea activities. The valves and control systems must be safeguarded by suitable protection housings which allow access for inspection and maintenance of the SSIV system. The valve spool should preferably be installed onto a support frame.

3.5 COST EFFECTIVE DESIGN

The SSIV design, installation, operation and maintenance should be simple and cost effective but reliable. To this effect following considerations may help in reducing the cost:

a. control system should be simple with only limited data transmission necessary to indicate full open/close positions of the valve and pressure in the accumulators; fewer components in general means better reliability,

b. valves should be selected to suit the system function requirements,

c. the protection structure should be designed to allow easy access for inspection, maintenance and repairs.

d. the system weight should be kept to a minimum so that conventional diving support vessels (DSVs) can be used for installations,

e. the system should be tested on land prior to installation subsea,

f. minimum amount of subsea work should be required to complete SSIV system.

A typical two-valve SSIV system is shown in Figure 1.

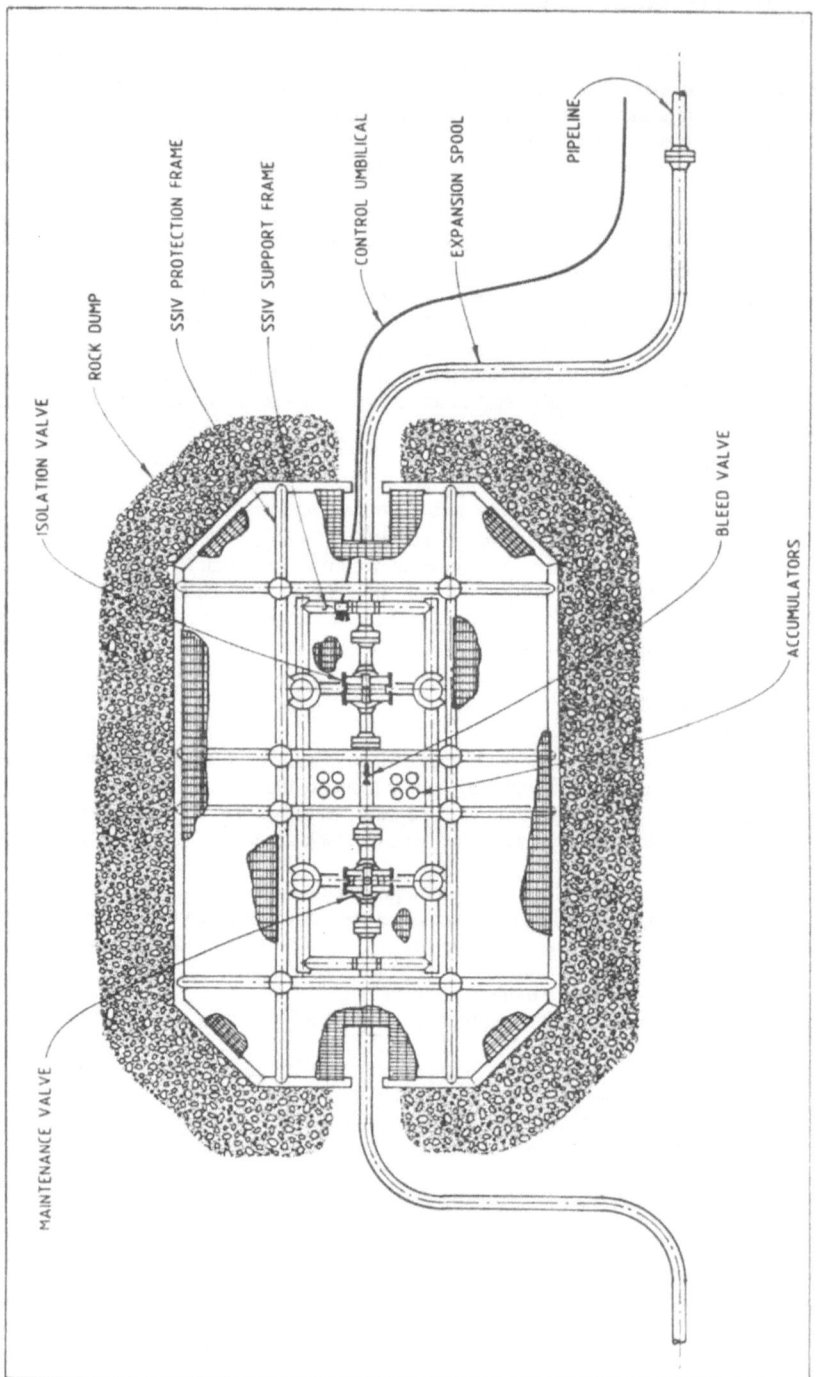

Figure No. 1 Typical Two-Valve Subsea Isolation Valve System

4. INSPECTION AND MAINTENANCE

Over the past two years, a number of SSIVs have been designed and installed in the North Sea. It would appear from the designs that more attention could have been given towards inspection, in-situ maintenance, recovery of defective components and reinstatement, particularly removal and replacement of the complete valve. This section presents the basic requirements and design considerations to achieve these.

4.1 INSPECTION AND TESTING REQUIREMENTS

There are no statutory regulations for routine inspection and testing of SSIVs on the lines of SI 1029. However, it is proposed that testing of valves, where possible, should be carried out at the same time as the riser ESDVs. In case of check valves, controlled closure and opening of the valves is not possible thus SI 1029 requirement can not be applied to such a system. It is not possible to state comprehensive testing and inspection requirements since these depend on the design of an individual system. These testing and inspection procedures should be developed during the design phase to meet manufacturer, statutory and operator requirements. It is proposed that the following testing and inspection may be performed:

a. partial closing and re-opening of the main isolation valve every six months,

b. full closure and re-opening of the main valve within two to four months of carrying out testing in (a),

c. accumulators should undergo periodic examination and testing every five years by removing these from their subsea position,

d. pressure in the accumulators should be checked and recorded daily to ensure that these are not leaking. This can be carried out by monitoring pressure indicators located in the control room. If leak is suspected, inspection of the accumulators should be carried out and remedial steps undertaken,

e. other control components should be tested and inspected in accordance with the manufacturer's procedures or operator requirements.

f. control umbilical should be inspected by ROV/divers to ensure that it is not exposed in sections where it is designed to be in the trench or buried. The untrenched and unburied sections should be inspected for any visible damage.

h. maintenance valves, if present in the SSIV system should also be tested, possibly once in two years to ensure that when needed, these will work.

4.2 MAINTENANCE REQUIREMENTS

Routine maintenance is not normally required for the valves as these are designed for long life but these may still need occasional intervention to rectify any problems. Valve manufacturer should identify periodic maintenance, if required. Note that all welded body ball valves can not be maintained, these will need to be replaced if problems occur.

The accumulators should be maintained once in five years as a minimum. This is a Statutory requirement for land based accumulators in accordance with the Pressure Systems and Transportable Gas Containers Regulations 1989 and this should be a good basis for the subsea accumulators.

Other components, such as the actuators, limit switches, pressure transducers, hydraulic trip valves, etc, should be maintained in accordance with manufacturer's instructions.

Maintenance of subsea installed components is an expensive exercise, hence it is imperative that the design of SSIV system is as simple as possible and components should be designed to require minimum maintenance.

4.3 DESIGNING FOR MAINTENANCE

Maintenance aspects of the SSIV systems must be considered at all stages of design and specification of the system components. Some criteria to be considered are discussed hereunder:

a. The layout of control system components should ensure that all components are accessible to divers/ROV for inspection with minimum underwater manipulations.

b. The actuators are fitted with limit switches to indicate the valve status and the signals displayed on the surface. In addition, valve status can be physically ascertained by divers if actuators are fitted with position indicators. This in only required occasionally to confirm the working of limit switches.

c. The accumulators, if installed subsea should be laid out in such a way that these can be inspected with minimum hazard to diver/ROV.

d. The subsea accumulator bottles should be retrievable individually or as a bank with minimum subsea work and with no disruption to effectiveness of the SSIVs. It would mean that the system should be designed with sufficient redundancy or replacement accumulators to be in place before any bottle or accumulator bank is removed.

e. The protection cover should provide access points for divers/ROV and should be designed to allow retrieval of all SSIV components including the complete valve and any associated pipe, Figure 1.

f. To facilitate valve retrieval, the valves should preferably be flanged.

g. If welded end valves are used, the protection housing should be designed to accommodate hyperbaric welding unit or whole protection cover should be removable to allow hyperbaric welding of replacement valve.

i. Top entry valves can, in principle, be maintained in-situ. If such valves are used, the valves and protection housing should be able to accommodate a specially designed maintenance frame to remove the valve internals. Neles and TK Valves offer such frames. The TK frame uses guide sleeves on the valve body to locate the frame and internals can be removed or replaced after servicing.

j. Breaking of valve flanges, particularly for large pipelines, is quite involved. The valve spool should be designed to provide enough flexibility through expansion offsets to facilitate flange breaking to remove the whole valve. Figures 2 and 3 show how valve flanges can be broken for gasket replacement or valve retrieval for valve spools with and without expansion offsets. Note that the latter method involves cutting and destruction and should be avoided. It may be possible to spring flanges sufficiently the replace gaskets without pup piece destruction.

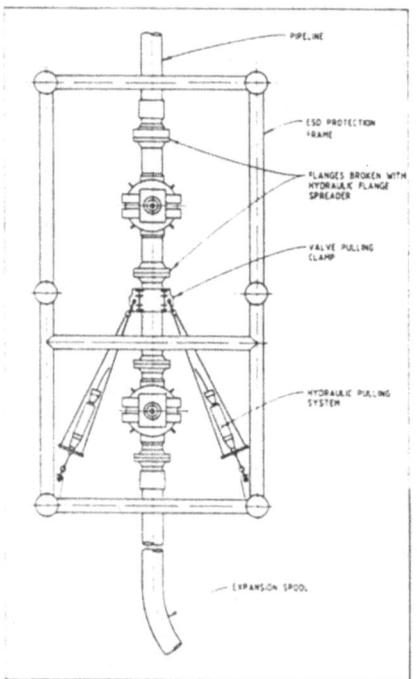

Figure No. 2 Conceptual Arrangement for
Flange Spreading of SSIV with expansion offset

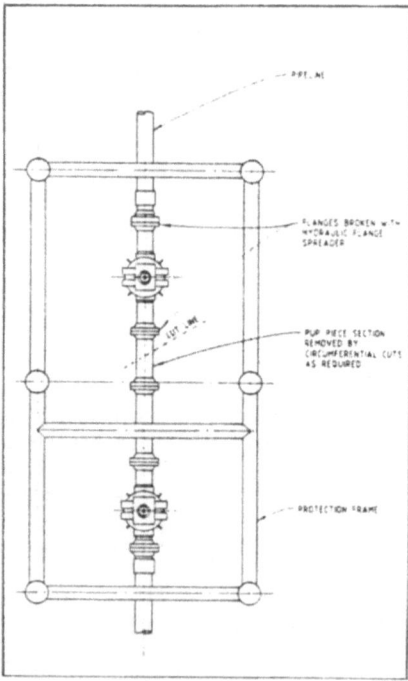

Figure No. 3 PUP Piece to facilitate
Flange Spreading of SSIV without expansion offset

k. If removable roof cover panels are used, the design and maintenance procedures must ensure that the roof panels or other equipment do not drop on to the SSIV or other components. A simple lifting frame for this purpose is shown in Figure 4.

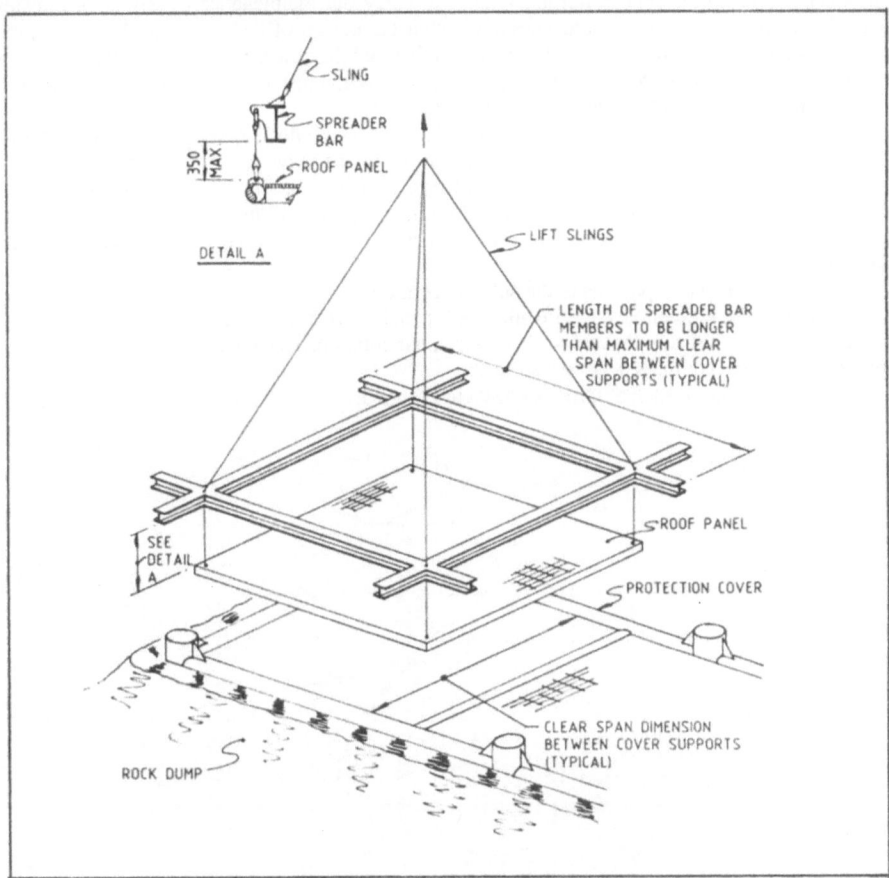

Figure No. 4 Lifting Frame for Roof Cover

l. To minimise cleaning and recommissioning work following valve internal servicing or valve replacement, the SSIV system should incorporate maintenance valve(s).

m. Whenever working on the signal cables, the procedures must ensure that the electric supply is disconnected to that particular cable during that period.

n. All underwater hose and electric cable connectors should be water tight for the hydrostatic head as a minimum. It should be possible to connect and disconnect these under water without polluting the hydraulic fluid or causing short circuit of electric cables

5. CONCEPTUAL DESIGN OF A MAINTENANCE FRAME

A major problem with maintenance of SSIVs is the cost involved with the offshore spread and possible loss of production due to the time it normally takes to carry out maintenance of any subsea installed equipment. Therefore it is of paramount importance that the design of SSIVs should be simple, easily maintainable, components should be well tested for their reliability, materials to be suitable for subsea applications and the complete SSIV system is subjected to comprehensive testing on land including, if possible, a dry run on procedures for the maintenance of SSIVs.

In case a valve mal-function is suspected, it may be necessary either to service the valve in-situ or recover the internals of a top-entry valve or even recover the whole valve for servicing/replacement. The time it would take to recover the defective parts and then to take replacement parts down to the seabed could have significant impact on production. Therefore it is important to minimise this time by reducing the number of diving runs. To achieve this, a maintenance frame is proposed which will perform the following functions:

a. transport the replacement components to the subsea location
b. dock and lock onto the protection structure after roof panels have been removed
c. remove and "park" the old components while replacement components are installed
d. transport the recovered components to the surface.

The conceptual design of such a frame is shown in Figure 5.

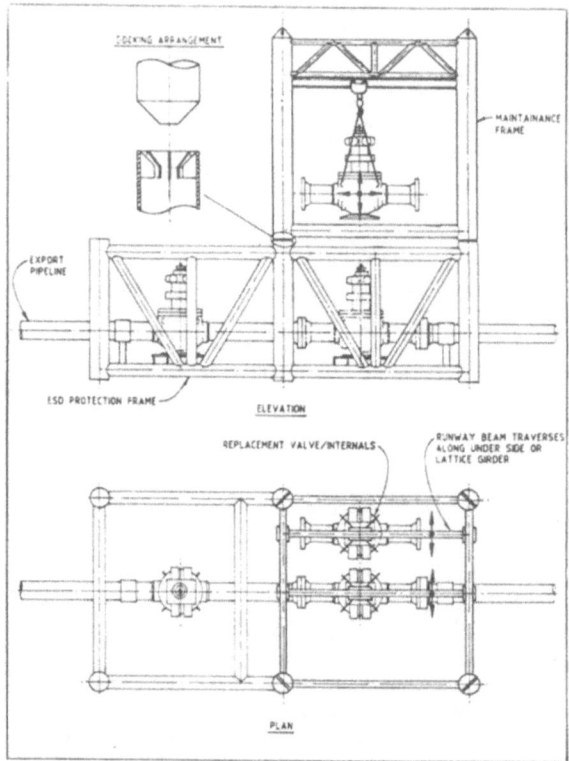

Figure No. 5 SSIV Maintenance Frame

The maintenance frame consists of a space frame structure with two gantries each carrying a mobile lifting facility which could be mechanically or hydraulically operated under water. The frame will be installed on top of the SSIV protection structure directly over the valve to be serviced. The frame could be designed to replace components other than the valve components.

It is anticipated that such a frame designed to withstand wave and current loads will weigh in order of 30 tonnes and thus for most common size pipelines, the cumulative weight including replacement components should not exceed 60 tonnes which is within the lifting capacity of normal maintenance DSV.

It is pointed out that the maintenance frame and the protection structure have to be designed to be compatible. The design of protection structure and its foundations must take into account the loads and additional forces imposed during maintenance with such a frame.

The maintenance frame can be installed with the help of guide wires. The detailed procedures will depend on the design of the protection structure and the maintenance frame.

Acknowledgments

The author thanks the management of Brown and Root Ltd for its permission to prepare and issue this paper. Any opinion expressed herein does not necessarily represent the policy of Brown and Root Ltd.

References

Subsea Emergency Shutdown Systems, Proceedings of one day symposium, June 1990, London. Organised by Subsea Engineering News.

Jain, R.K. (1990) 'Emergency shutdown facilities for new and operating pipelines', presented at meetings of the Pipeline Industries Guild, London. Also published in the International Pipe Line Industry, July and August 1990.

Subsea Isolation Systems (1990) Proceedings of one day seminar, London. Organised by Subsea Engineering News

Jain, R.K. (1990) 'Impact of subsea ESD valves on pipeline integrity monitoring' Proceedings of Pipeline Pigging and Integrity Monitoring Conference, Aberdeen

REVIEW OF SUBSEA ISOLATION VALVES INSTALLATIONS IN THE UK NORTH SEA

HUGH MCINTYRE, PROJECT ENGINEERING MANAGER, ROCKWATER LIMITED

INTRODUCTION

When planning the configuration of an offshore production facility certain
client requirements and field layouts resulted in specifying a subsea pipeline
isolation valve in close proximity to the platform. In the UK sector of the
North Sea only 7 valves of this type were installed at platform construction
phase prior to July 1988.

The catastrophic result of the Piper Alpha disaster in 1988 highlighted
the benefits, with regard to platform safety, that could be realised by
addition of this type of isolation system whereby the inventory of pipeline
product that may be released during an emergency situation, could be greatly
reduced. Shortly after this incident the UK Department of Energy issued
directives to operators requiring the installation of Emergency Shutdown Valves
(ESV) on platforms as close to riser top as possible however no guidelines were
issued specifically relating to Subsea Isolation Valves (SSIV).

At this point in time most facilities planned for installation were
revised to incorporate SSIV's in their construction and several operators
decided to commence retrofit of SSIV's to their existing facilities. These
actions were taken in anticipation of prescriptive regulations being
introduced, insisting on SSIV installation. On publication of the Cullen
Report into the Piper Alpha disaster it became apparent that prescriptive
regulations regarding installation of SSIVs would not be enforced and that all
operators would be required to submit a Safety Case to the regulatory body
which would demonstrate, among other aspects, that "hazards arising from the
inventory of hydrocarbons in risers and pipelines connected to the installation
have been minimised." Being more specific the report states that "the operator
should demonstrate in the Safety Case that adequate provision has been made,
including if necessary the use of SSIV's, against hazards from risers and
pipelines."

It therefore follows that SSIV installation is not necessarily a
requirement but the cost/benefit of this type of system will be assessed along
with other methods of protection such as riser relocation, enhanced
accommodation protection, resiting of accommodation etc. It may be a natural
course of action for operators to look at alternative methods of complying with
the Cullen requirements in the perception that retrofit SSIV's are likely to be
the most expensive option for existing facilities. However we believe that the
industry has been offered a rather unique opportunity to ensure that economic
proposals for essentially repetitive installation of retrofit SSIV assemblies
to be developed by capitalising on previous experience. Only by this approach
will the undoubted safety advantages of SSIV's be fairly assessed.

This paper will review recent retrofit SSIV installations in the UK
North Sea and offer proposals to allow the experience gained to date to be
utilised to ensure practical and economical proposals are evaluated for
consideration for enhancement of platform safety.

Volume 27: Subtech '91, 97–104.

COMPONENTS OF SSIV ASSEMBLY

Figure 1 illustrates the component parts of a typical retrofit SSIV assembly and can be summarised as below:-

Subsea Isolation Valve - The specification and selection of the correct valve is obviously of great importance to ensure long term suitability for intended purpose and this selection process must review the pipeline product, maintenance requirements, reaction time, compatibility with actuator and control systems. Protection of these components during installation must also be considered.

Valve Actuator - This item is critical to guaranteed operation in an emergency and of those installed to date the medium used to actuate closure has been spring or hydraulic accumulator powered. Selection of this item again must review reaction time and compatibility with other components.

Control System - it is accepted that the functions to be monitored and controlled for SSIV assembly are not particularly complex in relation to other existing subsea control systems however it is again critical to the operation of the facility that compatibility of all components is rigorously checked at design stage and other considerations relating to platform operations must be reviewed. In particular the integration of the valve operation control system to the platform shutdown system varies greatly dependant on operator requirements and platform age.

Umbilical - In all cases an umbilical requires to be installed for operation of the system which may also dictate the requirement for a retrofit J-tube. The design and installation of composite electro-hydraulic umbilicals has not been trouble free in recent North Sea experience and consideration of installation and protection of umbilical at design stage is critical.

Protection - As the location of the valve assembly by its nature requires to be in close proximity to the platform, a protection structure to primarily withstand dropped objects will be required. Consideration to design criterion for impact, ground conditions, tidal scour etc. all affect final design and as installation may be carried out during pipeline shutdown, the final configuration should take account of installation time and ease of transportation as a key parameter.

VALVE ASSEMBLY CONFIGURATION & POSITION

Figure 2 illustrates the basic valve combinations utilised for North Sea installations.

The single valve option while meeting the immediate requirements of isolation does not take account of any maintenance work that may be required which in this case would entail full flooding of the pipeline. This configuration is only evident in older installations incorporated during field development.

The most common solution adopted to date is to install a valve combination consisting of an actuated valve nearest the platform with a manually (diver operated) valve immediately on the pipeline side of this to allow maintenance and possible replacement of the isolation valve by only flooding the short section of pipeline up to the manual valve.

More complex configurations have been proposed (and used on small diameter lines) whereby full redundancy of the system is achieved by installing two isolation systems on a pipeline bypass which could allow production to continue while carrying out maintenance on one assembly.

The decision as to what configuration is adopted will mainly be driven by operator requirements and pipeline use with particular analysis of maintenance effects on production.

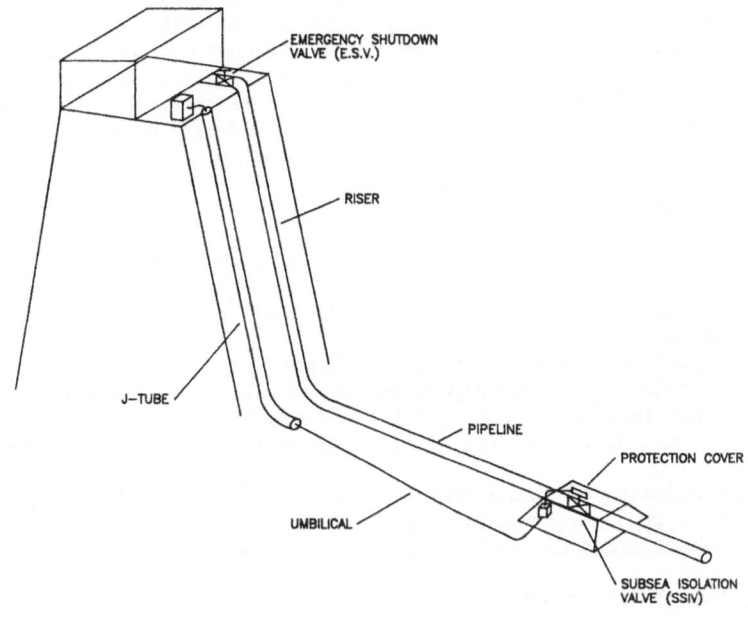

FIG 1 – COMPONENTS OF SSIV ASSEMBLY

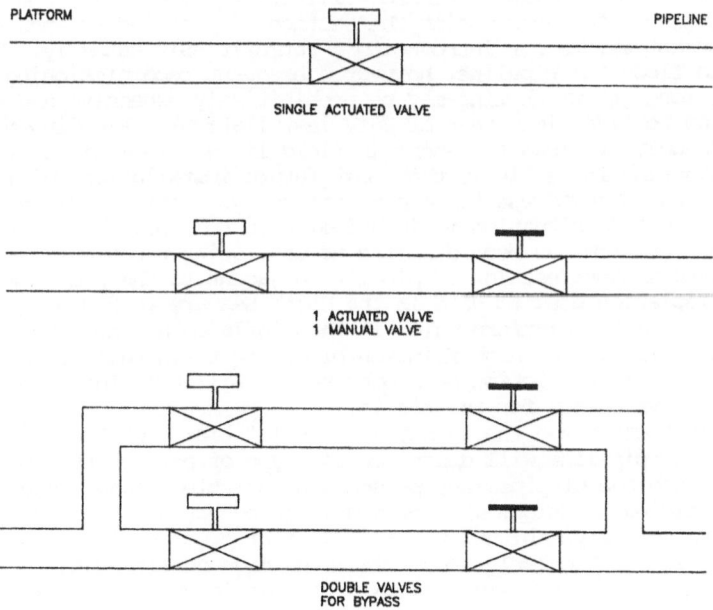

FIG 2 – VALVE CONFIGURATION

The position of the valve assembly will normally be dictated by results
of the Safety Case which would predict the maximum time that critical elements
of the platform could survive an emergency fuelled by the pipeline/riser
contents. This would result in a maximum survivable inventory being identified
which would dictate the furthest position of the valve from the platform. The
minimum distance of the valve from the platform would be calculated taking
account of possible damage from dropped objects, the effects of a gas cloud on
the platform from a pipeline rupture downstream of the valve, pipeline routing
in the vicinity of the platform etc.

The final position of the assembly within these limits should then be
considered with a view to ease of installation taking account of existing
pipeline connections (i.e. closing spools etc), pipeline protection (i.e. rock
dumped, trenching transitions) and any other features that may effect on
installation schedule.

Figure 3 outlines the normal configurations utilised for valve assembly
location in relation to the pipeline. While the in-line and off-line
assemblies may be similar the latter may be achieved with shorter pipeline
shutdown as certain activities may be carried out prior to pipeline
decommissioning.

The existing pipeline configuration may indicate optimum position of the
assembly e.g. replacing existing spoolpiece with a valve skid to utilise
flanged connections installed during the initial construction and tie-in phase.

SSIV ASSEMBLY INSTALLATION

As noted in the previous sections successful and timely installation of any
retrofit assembly (thus reducing pipeline shut-down) can be greatly assisted if
the installation progress is carefully considered at time of design. Many
factors relating directly to the installation of an SSIV assembly have a
considerable affect on the economics and scheduling of such a project.

Isolation of Pipeline:- Figure 4 illustrates 3 of the techniques that
have been employed to decommission the section of pipeline to ensure that a
safe environment exists for introducing a retrofit SSIV assembly. The simplest
method is to flood the pipeline, however subsequent recommissioning and drying
of a large, long gas trunk line may be prohibitively expensive and time
consuming (up to five times that of SSIV installation). Flooding of pipelines
is therefore only utilised for short in-field lines. To minimise the effects
of lost production it is likely that most future installations will be carried
out during planned shutdowns for other reasons eg. topsides maintenance.
However even in this situation it is undesirable to flood the entire line and
considerable research has been directed at developing techniques that can
safely isolate a short section of pipeline adjacent to the platform. Two
methods of isolation used to date in the North Sea are to deploy an inflatable,
tethered pig from the platform which is then inflated and monitored through the
umbilical or to introduce high friction bi-di pig train that can be guaranteed
to develop sufficient friction once past the installation location to ensure a
safe working environment for the tie-in.

Tie-In technique - the selected method of tie-in of the valve assembly
to the existing pipeline will depend on the type of product transported, size,
construction and age of pipeline, presence of existing flanges etc. The
hyperbaric welding technique whereby a welded connection is performed by
welder-divers in a dry habitat at seabed ambient pressure is commonly used for
gas line tie-ins particularly those requiring high quality metallurgical
properties. A tie-in utilising standard flanges is obviously relatively quick
and inexpensive however this assumes that flanges exist at the chosen tie-in
location. The cold forging technique can be employed to introduce flanges
subsea to the bare ends of the pipeline at tie-in location however several
pipeline parameters such as seam weld, age of pipe, type of service may render
this technique inappropriate for the operators needs. Figure 5 illustrates the
techniques described above.

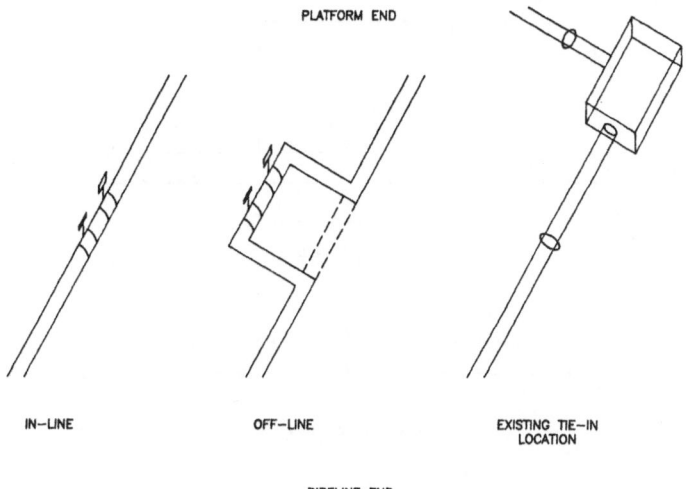

FIG 3 — VALVE ASSEMBLY LOCATION

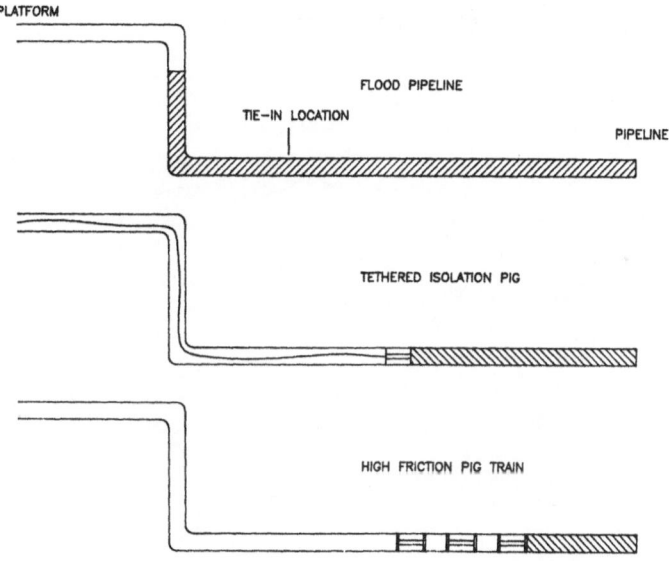

FIG 4 — ISOLATION TECHNIQUES

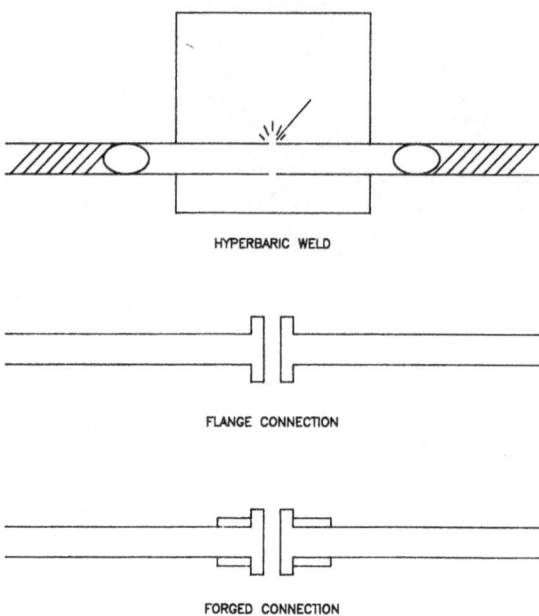

HYPERBARIC WELD

FLANGE CONNECTION

FORGED CONNECTION

FIG 5 — TIE—IN TECHNIQUES

Umbilical Installation & Protection - Planning the successful installation and protection of the control umbilical is critical to achieving a dependable isolation system. Most current control systems require a composite electro-hydraulic umbilical and as noted previously the design, manufacture and installation of such an item has given rise to several problems. A recent survey has concluded that 55% of all umbilicals installed in the North Sea over the past 5 years have experienced problems during commissioning. This item along with top sides control system, and most likely J-tube installation; requires considerably more detailed interfacing with platform operations than any other components of an SSIV assembly. This point should not be overlooked at the planning stage to ensure successful commissioning on schedule. It is therefore critical to consider the installation process and protection options (i.e. trenched, rockdumped etc.) at an early stage.

Installation Resources - The detailed design of the SSIV assembly, tie-in method and umbilical installation will determine the resources required for offshore works. The critical factors in this will depend on offshore lift capacity, expected weather conditions, extent of diver works (i.e. if welder divers required), umbilical protection specified etc. As the duration of offshore activities will determine the minimum pipeline shut down time, it has been normal practice to specify a semi-submersible DSV for support of these operations. Other activities that can be performed prior to shutdown (e.g. J-tube installation) may be undertaken by more weather sensitive monohulls. It can therefore be seen that design parameters may result in limiting the number of vessels capable of successful installations and their availability may be a key factor in project scheduling.

PROBLEMS ENCOUNTERED DURING SSIV INSTALLATION

The foregoing text has highlighted certain problem areas that require detailed analysis during the design and installation phase of any valve installation.

We outline below by way of example the type of problems that have been encountered to date on actual installations that have contributed to exceeding planned schedules.

Pipeline Isolation - Planned isolation unable to be achieved as inflatable plug did not maintain integrity during pigging operations.

SSIV Installation - Excavation for SSIV support structure extended due to unforeseen bed conditions.
- Designed weight of installation restricting choice of installation vessel.
- Interference from other marine works requiring rescheduling of planned programme.

Umbilical/Control System - Umbilical failed to perform to specification after installation.
- Delay to commission of control system due to problems interfacing with platform.

CONCLUSIONS

The problems noted previously are not unique to SSIV installation, indeed disruption/alteration to offshore schedules caused by similar problems has been encountered and overcome in offshore construction over the past 25 years of field development.

To date all retrofit SSIV installation projects have been administered by a contract philosophy as used on most construction works in relation to field development i.e.

- client commissions independent (or in house) detailed design
- hardware, procurement, fabrication progressed.
- when all technical details finalised installation contractors invited to tender.

While there is certainly an impetus for movement away from this "classical" contracting philosophy even in large field development works, retrofit SSIV installations do have certain unique aspects that should encourage a more integrated approach from all parties:-

- SSIV installation by its nature is repetitive.
- Installation costs are proportionally higher percentage of total costs in comparison to other subsea construction activities.
- The effects of programme delays on pipeline shutdown periods can have significant effect on overall costings when lost or delayed production accounted for.

We therefore conclude for SSIV retrofit installation to be fairly considered as a viable option under an operators Safety Case then an integrated contracting approach with an early involvement from installation contractors is required to ensure maximum input at the earliest stage which will result in an economical solution with minimum disruption to production.

RECENT DEVELOPMENTS IN THE DESIGN OF SUBSEA ISOLATION VALVE INSTALLATIONS

J.RITCHIE, P.G.BROWN, J.H.A.BAKER, S.A.WARD
J P Kenny Caledonia Limited
485 Union Street
ABERDEEN
AB1 2DB

ABSTRACT: This paper examines the criteria which must be assessed when developing the philosophy for designing a Subsea Isolation Valve (SSIV) Installation. By reference to actual designs, it is shown that satisfying these criteria for different applications can produce different solutions to what is essentially a similar problem. Consequently it is deduced that the design of such installations should be undertaken on a project by project basis.

1. Introduction

In the wake of the Piper Alpha incident, and in the light of Lord Cullen's enquiry report [1], it is recognised that an emergency shutdown valve located anywhere above sea level may not always provide sufficient protection against un-isolateable hydrocarbon leaks in a pipeline system. As a result, operators are now considering the installation of seabed emergency shutdown/isolation valves, in close proximity to the platform, thereby isolating the topsides from the main pipeline inventory.

2. Requirement for Subsea ESDVs

Before addressing the design of SSIV installations some SSIV requirement criteria will be briefly discussed. The decision as to what system to adopt and where, will be based primarily upon the nature of hydrocarbon product in the pipeline, the requirement being more critical for gas lines.

The occurrences on Piper Alpha and subsequent studies have demonstrated that ESDVs located topsides cannot be fully relied upon to contain a major explosion since they themselves may be disabled. Given this scenario, an effective way of limiting the supply of hydrocarbons through a major pipeline to a disabled platform is by a "subsea" isolation valve.

Volume 27: Subtech '91, 105–119.

Equally, should a dropped object, or an anchor, rupture a pipeline near a platform, a fundamental method of limiting the remaining pipeline inventory of hydrocarbons from leaking to the oil slick or gas cloud is by a "subsea" isolation valve. Since the dropped object and anchor hazards decrease with distance from the platform, the position of such an SSIV should ideally be at an appropriate distance from platform, but within the 500m exclusion zone (see Figure 1).

At this time the actual requirements for mandatory SSIVs are determined as an integral part of the Full Safety Assessment in the UK sector of the North Sea, but the discussions above clearly demonstrate their necessity where large pipelines serve a platform. Consequently operators are now installing them in both new and existing facilities.

3.   Design Criteria

In developing a shutdown system design, a number of aspects must be assessed in order to arrive at an optimum solution for each application. This paper discusses the principal criteria, presenting some associated advantages and disadvantages of each aspect.

Depending upon the application, a number of varying solutions may be developed for what is, essentially, a similar problem.

The following criteria should be considered in the design of any SSIV facility:-

- o   Proximity to the platform;
- o   Valve type;
- o   Configuration;
- o   Testing/commissioning;
- o   Protection philosophy;
- o   Installation.

All these are discussed below, under relevant sub-headings.

3.1   PROXIMITY TO THE PLATFORM

In the event that an incident were to expose the platform topsides to an "un-isolateable hydrocarbon leakage", the Subsea Isolation Valves minimise the exposure of the topsides facilities to the pipeline inventory.

Accordingly, the optimum location of the SSIV system is determined as follows:-

- o   System requires to be sufficiently distant from the platform to negate risk due to un-isolateable hydrocarbon leakage occurring on the pipeline side of the shutdown valve;

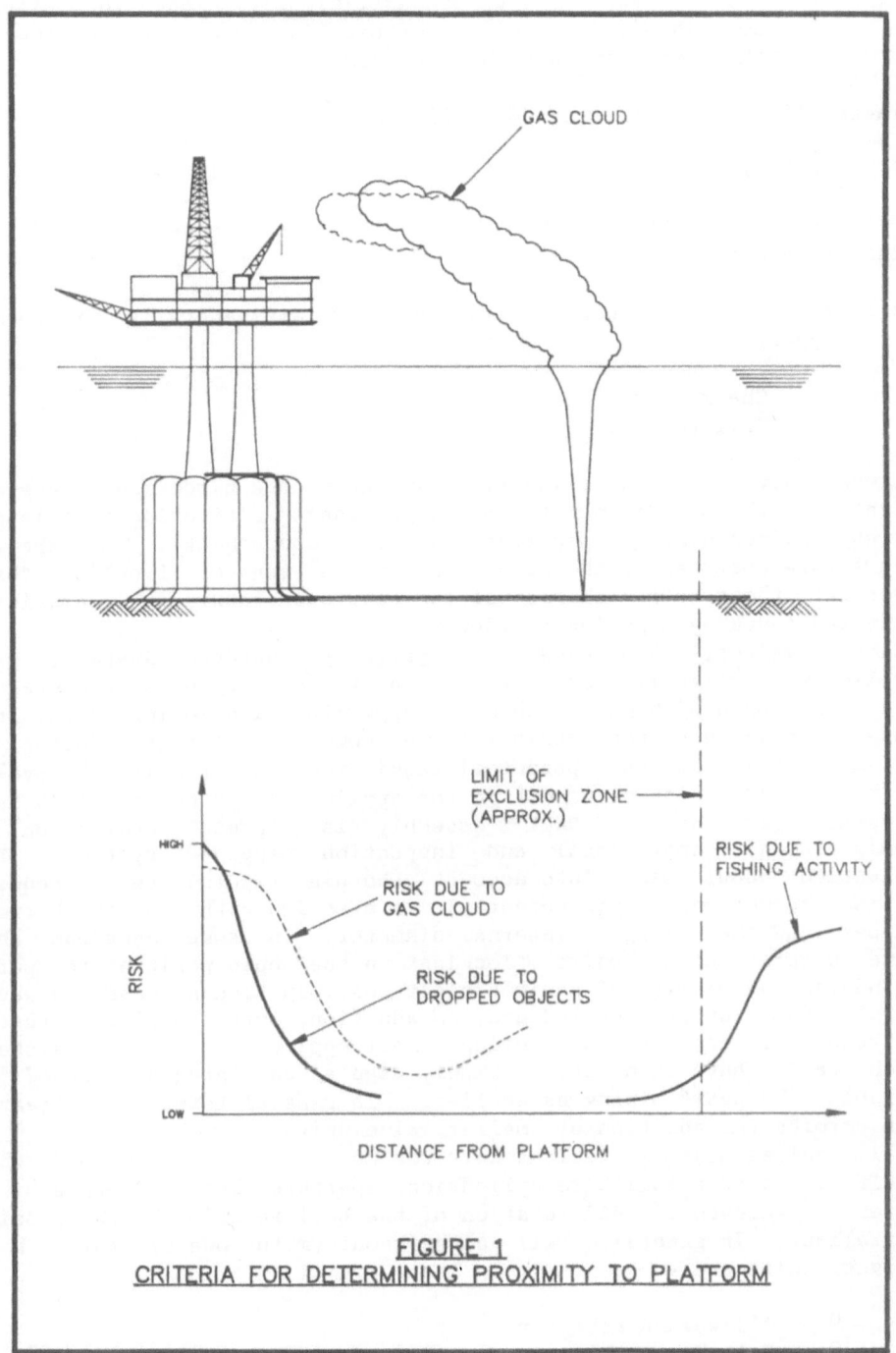

GAS CLOUD

LIMIT OF
EXCLUSION ZONE
(APPROX.)

RISK DUE TO
GAS CLOUD

RISK DUE TO
FISHING ACTIVITY

RISK DUE TO
DROPPED OBJECTS

HIGH

RISK

LOW

DISTANCE FROM PLATFORM

FIGURE 1
CRITERIA FOR DETERMINING PROXIMITY TO PLATFORM

o   System requires to be sufficiently close to platform
     to minimise the volume of pipeline inventory between the
     shutdown valve and the platform.

These effects are illustrated in Figure 1.

3.2   VALVE TYPE

Under this heading both the valves themselves, and their
actuators are considered:

3.2.1 *Valves*   which can be used for SSIVs are generally of two main
       types:

o   Check valves;
o   Ball valves.

Check valves, which allow flow in only one direction, generally
operate on the principle of flowline pressure activating a "clapper"
thereby maintaining the pipeline aperture. Consequently, upon the loss
of pressure upstream of the clapper or in the event of flowline "back-
pressure" the return mechanism of the aperture closure is automatically
activated invoking pipeline shutdown.
Check valves, when used for emergency shutdown systems, offer
simplicity, reliability and fast acting response upon system pressure
loss. In common with most mechanical apparatus with moving parts, it is
prudent to address the possibility of component failure during the
system lifetime and the potential requirement to "repair" the valve.
Another consideration pertinent to the application of check valves as an
integral part of an SSIV assembly is careful evaluation of
commissioning, operational and inspection pigging systems. This
evaluation should take into account the pig type(s), their required
driving speeds and their potential interaction with the check valve
clapper and the enlarged internal diameter. In some cases the check
valve clapper will require "locking" in the open position to permit
unhindered the passage of various pig types. In such a case, however,
the SSIV function is disabled and, in addition, more complex actuation
is required. To date, considering their application to SSIV systems,
check valves have been most commonly used as an integral part of non
piggable "by pass" piping assemblies. In general terms, this feature
also permits the addition of smaller valve units.
Ball valves house a steel sphere sealed within the valve body which
itself contains a full bore cylindrical aperture whereby closure of the
system is achieved via 90° rotation of the ball relative to the pipeline
centreline. In general, ball valves conform to one of three basic
designs, which are:-

o   All-welded assembly;
o   Bolted assembly;

o   Top entry.

The "all-welded assembly" is a sealed welded unit with no access to
the internals via the valve body. The bolted assembly is a unit
assembled by bolt arrays. The top entry valve (which may be flanged or
welded in line) permits in-situ access to the ball valve internal
workings, including the stem, seats, seals and the ball itself. Top
entry is achieved by "pulling" the actuator and the bolted valve bonnet
vertically clear of the assembly. In consequence, therefore, the top-
entry ball valve can theoretically undergo in-situ repair as compared
with bolted and welded valves which require to be removed from the line
entirely and repaired or replaced. In addition, all the above types of
ball valve may be readily designed to facilitate the unhindered passage
of all types of pigging systems.

At the design stage, valve type and vendor selection will be subject
to an appraisal under an array of criteria which would typically include
the following:-

o   Technical merit;
o   Lead time;
o   System weight;
o   Product;
o   Configuration;
o   Reliability;
o   Cost;
o   Replaceability.

3.2.2 *Actuators* are not required for Check valves, unless a means is
required for locking them open for pigging runs.

Ball valve actuators are of two main types:

o   Single acting;
o   Double acting.

The former utilises hydraulic pressure to hold the valve open against
a mechanical spring. Active steps to interrupt the pressure (on the
platform), or loss of umbilical integrity, will result in the spring
driving the valve closed. This mechanism is widely used in subsea
production systems and presents a "fail safe" facility. It should be
noted, however, that, for large bore valves, the actuators become very
large and heavy.

Considering the double acting actuator, the mechanical spring is
replaced by hydraulic accumulators which exert pressure on the "close"
side of the actuator. The control system diverts the pressure to be
exerted on the "open" side of the actuator. Loss of primary pressure
which is normally delivered via the umbilical, causes the accumulator
pressure to drive the valve shut. This is a more complex option, and
the hydraulic accumulators require regular monitoring and systematic
change-out. However, for large valves, in many cases, this is
realistically the only practical option.

3.3   CONFIGURATION

A number of factors influence the selection of the overall SSIV configuration. These include:

- o   Local Isolation;
- o   By-passes;
- o   Redundancy;
- o   Accessibility.

3.3.1 *Local   Isolation* valves are required if the SSIVs are to be serviced without decommissioning the whole pipeline. Whether to fully decommission/recommission is essentially a function of pipeline length, product composition and anticipated costs for a full decommissioning and recommissioning cycle.

3.3.2 *By-passes* have been referred to above as a means of incorporating a non-piggable check valve as the principal SSIV. A by-pass line can offer the operator the potential advantage over in-line systems of allowing continued production via the by-pass during mainline valve downtime.

3.3.3 *Redundancy* may be required to allow for valve failure. A typical method of providing redundancy is to incorporate a by-pass with a second SSIV (ball valve or check valve). This route will not be piggable.

One other alternative is to incorporate identical valves and actuators for local isolation so that the SSIV function can be transferred to one of these other isolation valves. The original SSIV would then be locked in the open position.

3.3.4 *Accessibility* is a requirement regardless of the number and types of valve incorporated. As a general principle, fitting the pipework around three sides of a skid base optimises the size of the skid, which is an important consideration with regard to installation. This also facilitates access to the valves for local override by diver or ROV, and enables the SSIV structure to be incorporated into an expansion loop. Vertical access can be provided via removable lids. It is clearly important to decide at the design stage whether intervention is to be by diver or ROV (or both).

A range of configuration options is depicted on Figure 2.

If valve replacement is considered to be a requirement, habitats and/or other customised work frames may be required and the protection structure should be designed to facilitate this. The top-entry ball valves can thus be serviced in situ.

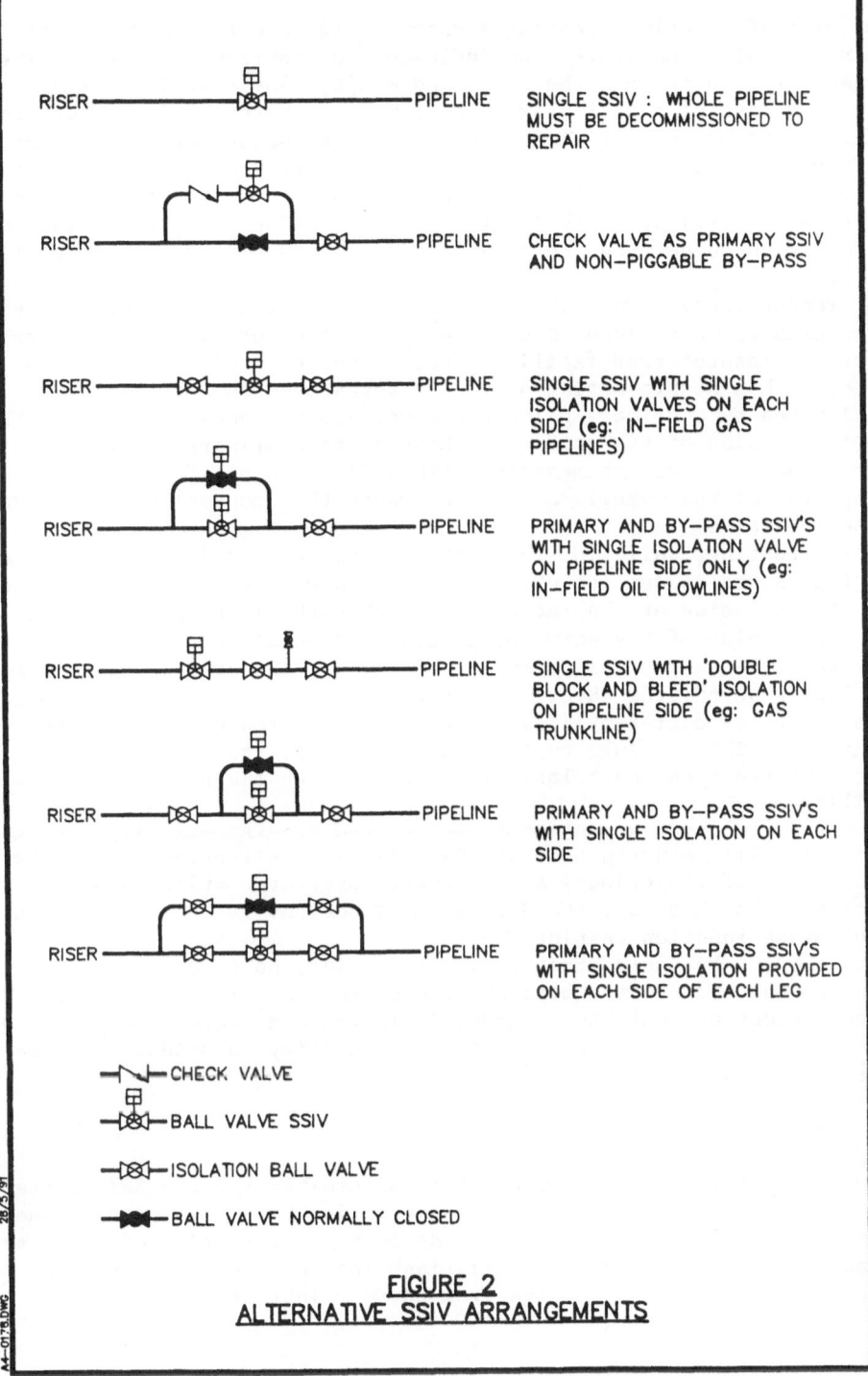

FIGURE 2
ALTERNATIVE SSIV ARRANGEMENTS

3.4   TESTING

Two ways of function testing a check valve are to shut off the flow whilst a diver monitors the indicator to confirm that the valve has closed, or to blow down the riser and monitor any rise in pressure.

In the case of a ball valve, the simplest approach without monitoring with divers or ROV is to close the valve, blow down the riser and monitor the pressure to determine that the valve has closed. This may, however, have major implications with respect to loss of production and pressure variations. In this light, testing of this type would typically be undertaken during planned shutdown or maintaining of the system.

A recent innovative means of SSIV testing which does not interfere with export, nor cause excessive pressure drops is to incorporate a "partial closure" test facility. Such a system would comprise an SSIV fitted with a double-acting hydraulic actuator. An accumulator pressure is diverted by a primary pilot pressure, via the umbilical, to work on the "open" side of the actuator. Loss of this primary pilot pressure, due either to active measures taken on the platform, or loss of integrity of the umbilical, would cause the accumulator pressure to close the valve.

When testing, while in the normal operating condition, a secondary pilot pressure is applied which in turn causes pressure to be applied to the "close" side of the actuator. With both pilot pressures applied, the "open" side of the actuator pressure is equal to the "close" side, and thus the valve remains fully "open". If the primary pilot pressure is then removed, the hydraulic control circuit controls the pressure such that a differential in pressure is created across the valve actuator. The pressure on the "open" side is reduced causing the now higher pressure on the "close" side to initiate the partial closure test facility.

A partial closure arrangement is depicted diagrammatically in Figure 3, and it will be noted that the SSIV is also fail-safe during testing since loss of the primary and secondary pressures still cause the valve to close. In this way, the SSIV still functions to protect the platform even during function testing.

It should be noted that there is, as yet, no firm requirement to be able to function test subsea SSIVs once they are fitted. However, such a requirement may well be introduced, in the near future, and it would appear that the availability of this facility provides a number of advantages.

3.5   PROTECTION PHILOSOPHY

Once an acceptable location has been determined for the SSIV system and the valve components and piping configuration have been established in principle, it is necessary to provide in-situ protection of the system components from damage. Facilities located on the seabed are, potentially, subject to damage from three main sources, namely:

        o   Anchors;
        o   Dropped Objects;
        o   Fishing Gear.

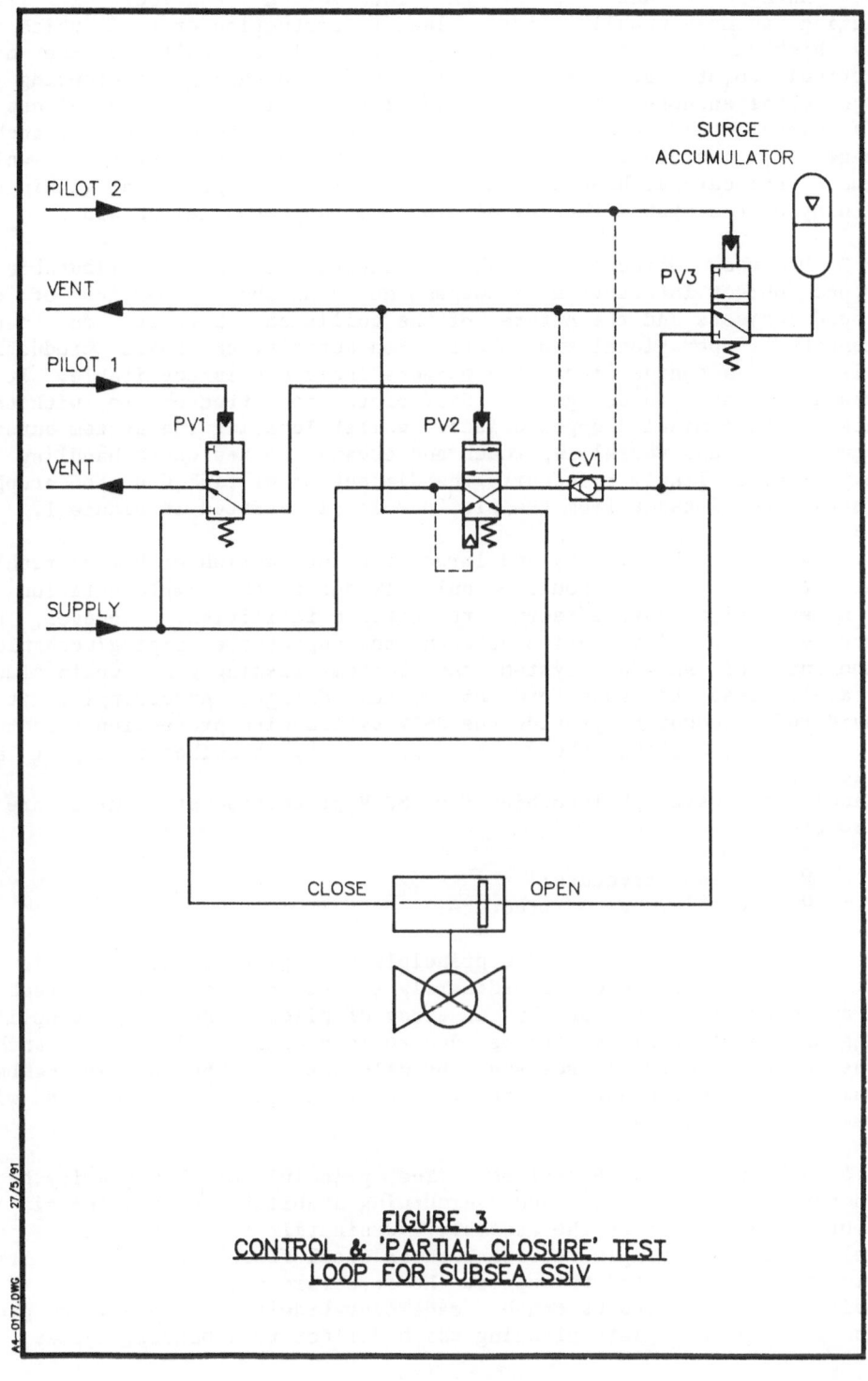

FIGURE 3
CONTROL & 'PARTIAL CLOSURE' TEST
LOOP FOR SUBSEA SSIV

3.5.1 *Anchors*.   Analysis of the interaction behaviour of large
dragging anchors usually tends to lead to protection criteria which are
very high which in turn does not generally lend itself to the more
practical solutions. The forces imparted by dragging or dropping of
large vessel anchors are such that localised failure of the pipelines or
SSIV system would be a predictable consequence. Protection from anchor
damage is, therefore, usually achieved by "passive" measures which
promote the careful handling and control of anchor placement within the
vicinity of the system thereby eliminating the potential hazard.

3.5.2 *Dropped Objects*.   Risks to subsea equipment attributable to
dropped object interaction is dependent upon the properties of the
dropped objects and the nature of the collision. However, due to the
intensity of operational and construction activity on board production
facilities, a degree of dropped objects incidence is inevitable. It is
necessary, then, to design the SSIV protection structure to withstand
contact with nominal dropped objects, whilst locating the system outwith
crane radii and vessel approach and escape routes where handling of
heavy loads is likely to occur. The diminution of risk due to dropped
objects with distance from the platform is illustrated in Figure 1.

3.5.3 *Fishing Gear*.   By and large, the interaction of bottom fishing
gear with SSIV systems should be unlikely due to the implementation of
500 m exclusion zones adjacent to platform facilities. However, the
consequences of interaction between the unprotected piping/mechanical
components of an SSIV system and bottom fishing gear would almost
certainly lead to some form of system damage. Accordingly, it is
considered prudent to provide the SSIV system with protection structure
capable of resisting the forces imposed by trawlboard impact and
pullover.
    Candidate design philosophies for SSIV protection structures are as
follows:-

        o   piled structure;
        o   gravity based structure.

3.5.4 *Piled Structure*.   The principle of a piled structure design is
that the structure, which is ordinarily an open tubular framed assembly,
is anchored and stabilised by an array of piles. Accordingly uplift,
snagging and lateral excursions due to interaction with, say, anchor
wires or trawlboards is resisted by pile action. The primary members
within the structure are selected to resist the forces associated with
these design loadings.

3.5.5 *Gravity Based Structure*.   The principle of the gravity based
structure is that sliding and overturning stability is attained via the
on-bottom deadweight of the assembly. Ordinarily these structures are
developed using a simple trussed tubular steel framework with an array
of mudmats incorporated to support the structure weight on the seabed.
Gravity based structures can be made over-trawlable by profiled grout
bags, gratings and plate cladding which deflect trawlboards, cables, etc

up the side walls and away from the structure. Thus, only impact and dropped object loads need be accommodated and snagging loads are avoided. Alternatively, a "lift-light - ballast-subsea" design can be adopted whereby the additional necessary on-bottom weight for stability of a relatively light, open-sided structure is developed by the post-installation addition of cheap, readily available ballast.

Some protection structure options are shown on Figure 4.

3.6   INSTALLATION

It is important to consider installation at an early stage in the design. If the weight in air can be kept low, this allows much greater flexibility and contractual competitiveness when it comes to selecting an installation vessel

Once an SSIV system is designed and the appropriate protection philosophy adopted, the designer can, and should, examine a number of options in relation to installation of the system since such offshore operations are very capital intensive and some innovative "installation" engineering at an early stage can result in considerable cost saving. Some examples follow:

o Minimise the weight and size of the system so that a greater number of candidate vessels can undertake the installation which enhances vessel availability as well as being contractually competitive.

o Integrate skid and piping with the protection structure which, if practicable, minimises the schedule and number of lifts to install the system and hence reduces both its cost and weather sensitivity.

o Adopt a "lift-light - ballast-subsea" philosophy to minimise in-air weight for vessel competitiveness and reduce fabrication costs, and increase the on-bottom weight of the structure as necessary by appropriate subsea ballasting measures.

o The number of removable lids and/or retrofit panels should be kept to a minimum. Statistically, they are a significant potential source of damage during their removal/replacement. Any components of this nature which are incorporated should have the appropriate stabbing/guideline facilities.

o If SSIV systems are to be adopted on adjacent flowlines their respective components might be combined into a single skid and structure thereby obviating the need to fabricate and install two separate systems, as shown in Figure 5.

o Install assembly complete with associated tie-in spools (Figure 6) which, whilst requiring a larger vessel (Tandem lift) offers the advantage of minimising schedule and number of lifts as well as halving the required number of subsea hyperbaric or flanged tie-ins. This feature may represent a considerable saving on larger systems.

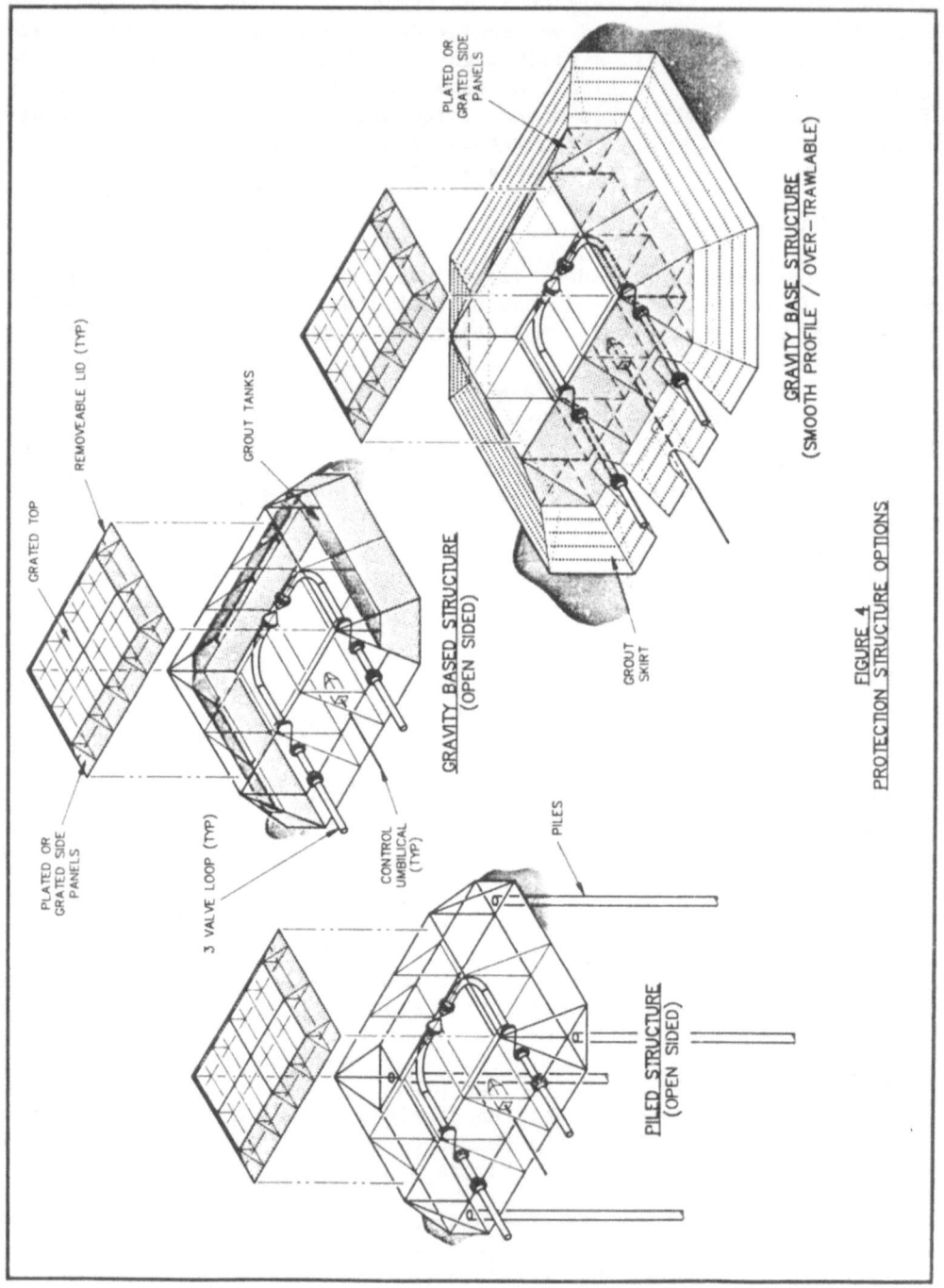

FIGURE 4
PROTECTION STRUCTURE OPTIONS

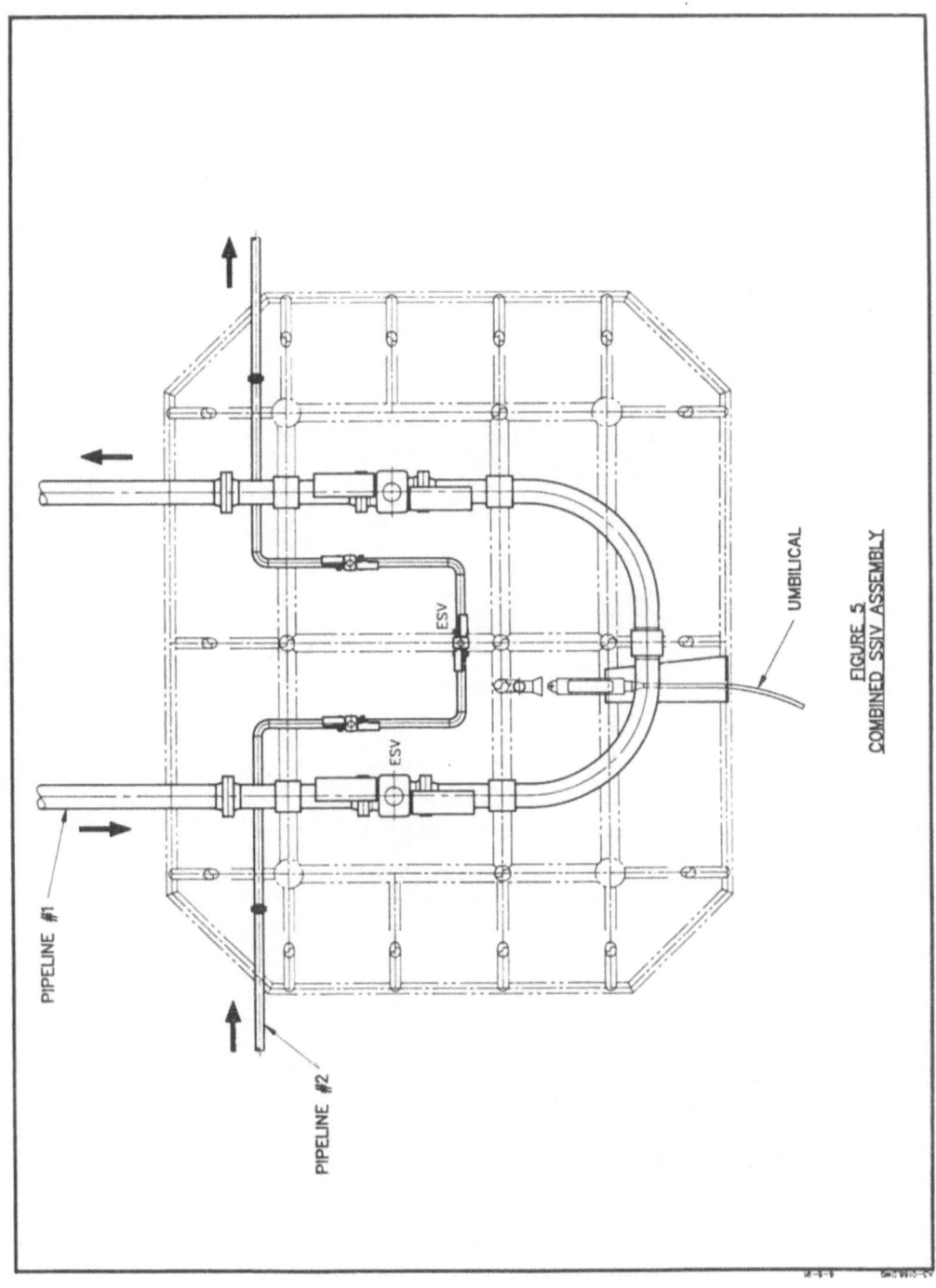

FIGURE 5
COMBINED SSIV ASSEMBLY

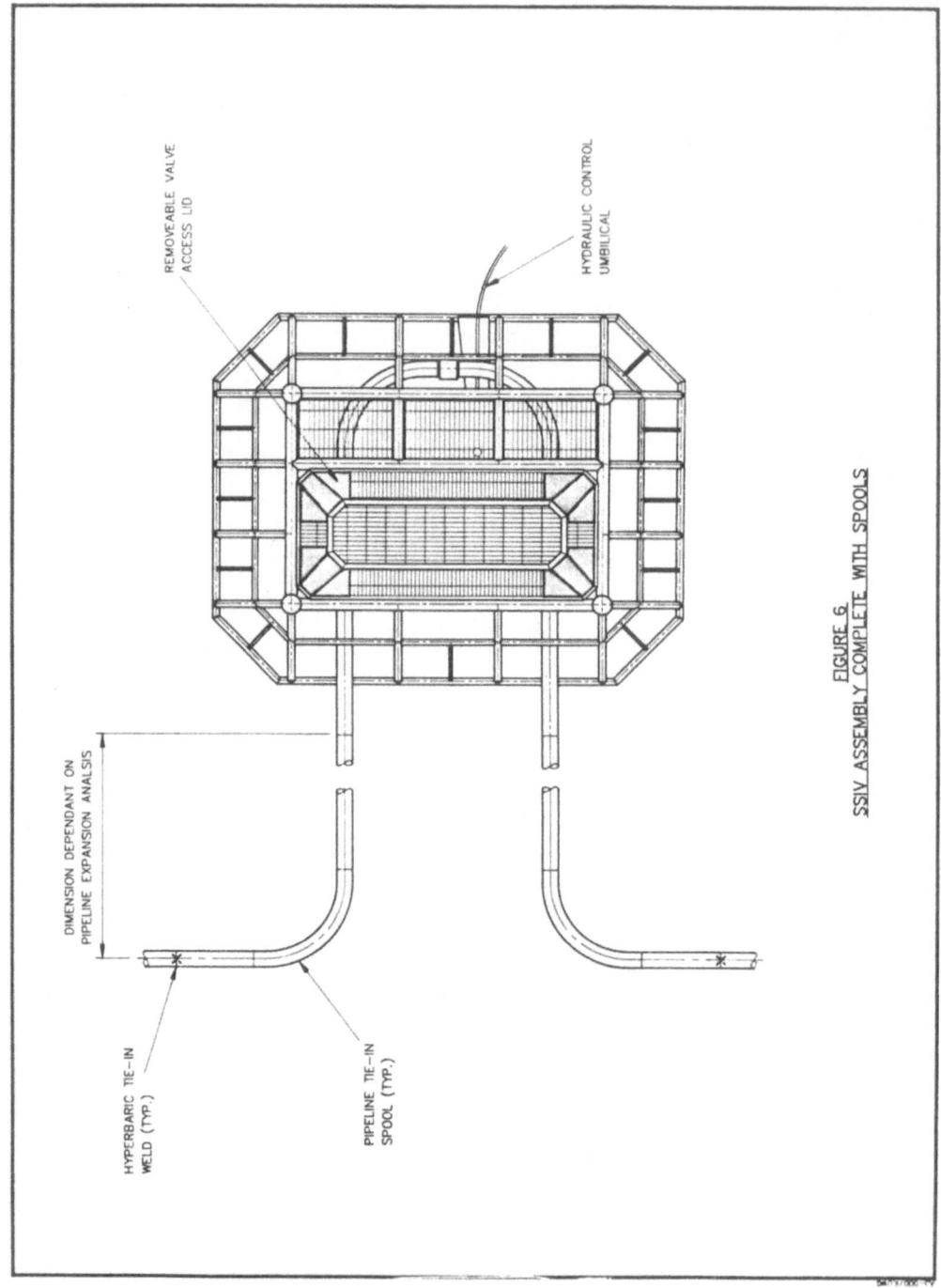

REMOVEABLE VALVE
ACCESS LID

HYDRAULIC CONTROL
UMBILICAL

DIMENSION DEPENDANT ON
PIPELINE EXPANSION ANALSIS

HYPERBARIC TIE—IN
WELD (TYP.)

PIPELINE TIE—IN
SPOOL (TYP.)

FIGURE 6.
SSIV ASSEMBLY COMPLETE WITH SPOOLS

4.   Conclusion

The initial requirement for SSIV systems is determined by safety/riser
assessment considerations which dictate "if" and "where" a system is
required.   Once it has been decided that a SSIV is required, the various
criteria which must be assessed in the design of Subsea Isolation Valves
will be evaluated in turn.   These include the selection of valves and
actuators, the provision of isolation and redundancy, the configuration
of the installation and its proximity to the platform, the provision of
function   test   facilities, protection/structural requirements and
installation.

   It has been shown that application of these criteria on a case-
specific basis will lead to varying solutions to what is, essentially
the same basic requirement. It is finally concluded that, by the
addition of a number of innovative engineering techniques, a system can
be designed and installed with a minimal schedule impact and maximised
cost effectiveness.

REFERENCES

[1]   The Hon Lord Cullen (1990), "The Public Enquiry into the Piper
        Alpha Disaster", Department of Energy, HMSO (ISBN 0 10 113102).

Part 2
Subsea Field Developments

Chapter 5
Urban Field Development

DYNAMICALLY POSITIONED VESSELS - THEIR SUITABILITY AND SAFE OPERATIONS

CAPT. PETER A. COOK, MNI, MRIN,
Principal Marine Surveyor,
Noble Denton Marine Services Ltd,
131 Aldersgate Street,
London.

ABSTRACT.This paper looks back to the early development of dynamically positioned vessels, showing how a system of suitability and safety assessment has been developed. The applications of DP have spread to a wide range of operations, and as the vessels and the systems have become more sophisticated the methods of suitability assessment have changed little, and up to now have been based upon a combination of charterers requirements and specific guidelines. The established means of assessment takes the form of a DP audit and trial, this is not a standard assessment throughout the industry, and in many cases does not take fully into account the suitability of the vessel for the work proposed, only its suitability as a dynamically positioned vessel. The current situation gives a great deal of scope for improvement and proposals outlined in this paper will give details of how a standardised system of suitability and DP safety assessment can be developed and introduced.

1. Introduction - Past to Present

Dynamic positioning systems were originally developed in the 1960's as a means of position keeping for drillships, operating in deep waters and where it was not desirable to have mooring systems deployed. These early systems were normally developed as single computer, non-redundant facilities which in the event of a single failure would normally result in the loss of position. A large number of DP vessels at the time were not purpose built and were often converted from cargo vessels no longer fit for their purpose. As a result the DP systems which were installed had to be adapted to the design of the vessel, inheriting some of the problems which the original hull shape would produce. A major difference

123

Volume 27: Subtech '91, 123–132.

between a cargo ship and DP vessel is the fact that the hulls are designed for different purposes, one is designed for high speed and least resistance, whereas the other is designed to remain stationary with the minimum amount of roll, pitch and heave.

So it can be seen that from the early development of these vessels the best potential of DP system was not being harnessed. Hybrid vessels were being produced to carry out completely new work, under conditions which were highly demanding and potentially dangerous, yet the vessels were not intended nor fully tested for the purpose proposed.

During this early period, DP audits and trials were not fully developed if at all used. As incidents occurred and more was learnt about potential hazards and failures of the systems, the need for regular audits and trials of vessels was identified. The real need for these audits and trials was only really recognised when dynamic positioning was to be used for the station keeping of dive support vessels. Prior to this development, DP vessels had only really been involved in open water drilling operations where the effects of DP failure were, at that time, felt to be minimal and insignificant in comparison with potential drilling hazards.

As DP dive support vessels were utilised more and more the increase in DP incidents became significant, and involved loss of life on a number of occasions. Not only were divers lives at risk, but also the fixed installations alongside which these vessels were working, were being exposed to a high risk of collision. In some instances vessels have lost position, struck the fixed installation and then dragged divers through the jacket structure. It became obvious to both the operators and the regulatory bodies that this declining state of affairs needed to be carefully checked and the situation rectified. Some of the platform operators banned the use of DP vessels alongside their installations, insisting that vessels must deploy mooring systems and in some cases make fast to the platforms themselves. So, as a result of this decline in confidence in the operational capabilities of DP vessels, a set of guidelines for dive support vessels were developed jointly by the UK Department of Energy and the Norwegian Petroleum Directorate, entitled "Guidelines for the Specification and Operation of Dynamically Positioned Diving Support Vessels" dated 1980 and subsequently updated in 1983.

These guidelines were drawn up in order to provide a basis from which those who were building, operating and chartering dynamically positioned dive support vessels could develop the most suitable equipment and operating procedures for each vessel. The guidelines were also intended to provide a yardstick against which the suitability of DP vessels for diving operations could be assessed. In practice they are also used as a guide for the suitability of vessels for a wide range of other operations where dynamic positioning is utilised.

In complying with these guidelines and national regulations, owners, operators and Charterers would before commencing DP operations, satisfy themselves with regard to the vessels suitability for the work planned. Such an assessment would take the form of a DP audit and trial and would cover the following aspects of the vessels DP Operation.

1. Correct siting and mounting of all equipment and associated cabling.
2. Correct wiring of all power supplies, data cabling and equipment.
3. Effective shielding of all potential sources of electrical interference.
4. Software checks.
5. Software tuning for optimum position keeping performance.
6. Correct functioning of all condition monitoring systems and alarms.
7. Correct functioning of all data input systems.
8. Correct functioning of computers and interfacing.
9. Correct functioning of power management systems.
10. Correct functioning of thrust units including response times.
11. Measurement of positioning keeping accuracy using independent means.
12. Correct functioning of all automatic and manual change-over arrangements and procedures, from primary to back-up systems.
13. Correct functioning of offset and heading change control.
14. Position keeping performance in adverse environmental conditions.
15. Satisfactory operation of the DP system, during diving operations, or operations specified by the charter.

In addition the DP audit would include a review of relevant documentation such as:-

1. Operations Manual.
2. FMEA Report.
3. Capability plots.
4. DP log books.
5. Maintenance log books
6. DP operators log books and qualifications.

The DP proving trial would also assess the actual capability of the vessel and the crew, to support DP Operations in both normal and emergency situations.

So it can be seen, that with the introduction of operational guidelines in 1980 the basis for checking vessel suitability and operational safety was established, but the guidelines were open to a broad scope of interpretation and their application varied widely.

It could be assumed that by applying the guidelines to all DP vessel assessments, that vessel suitability and operational safety would be assured and that from 1980 onwards a decline in the number of DP related incidents would be the case.

It was not the case, and a steady rise in DP related incidents was reported, reaching a significant peak in the mid 1980's.

It is also notable, that at this time during the period 1980 to 1988, DP vessels were being developed for a wider range of operations, and utilising more sophisticated DP systems and operational equipment. A list off such applications are given below:-

Heavy Life Crane Vessels   - Monohull and Semi-Submersible.
Emergency Support Vessels   - Monohull and Semi-Submersible.
Accommodation Vessels   - Semi-submersible.
Multi-Purpose Support Vessels   - Semi-Submersible.
Well Servicing Vessels
Pipelaying vessels

This is not an exhaustive list, but it can be seen that the operations being carried out were becoming much more diverse than just diving and drilling. It was often the case that, unless diving operations were contemplated, some of these vessels would commence operations without undertaking a DP audit or trial, a simple pre-operation DP check being acceptable to the parties involved.

So it can be seen that the scope for further DP incidents, and a decline in vessel suitability and operational safety was being greatly expanded, particularly where DP operators were being asked to apply their skills to a wider range of operations and vessels.

2. The Current Situation

From the late 1980's to present, developments have occured which are very much to the benefit of DP vessel operations. Two major steps forward have been made by the Norwegian Maritime Directorate and the Nautical Institute, one with regard to vessel suitability and safety, and the other regarding operator training. Arguably the two most important areas of DP vessel operations.

The Norwegian Maritime Directorate have established their own set of guidelines for dynamically positioned vessels based upon a philosophy which bases reliability upon the effects of consequences, stating that the bigger the consequence the greater the reliability of the DP system must be. In doing so, four consequence classes have been drawn up and are assigned to DP vessels relating to their ability to comply. These are:-

Class 0 :   Operations where loss of position keeping capability is
             not considered to endanger human lives or cause damage.

Class 1 :   Operations where loss of position keeping capability may
             cause damage or pollution of small consequence.

Class 2 :   Operations where loss of position keeping capability may cause personnel injury, pollution or damage with large economic consequences.

Class 3 :   Operations where loss of position keeping capability may cause fatal accident, or severe pollution or damage with major economic consequences.

In addition to the consequence classes quoted, the DP systems must be able to meet requirements regarding single failure situations.

It is notable that only a small number of DP vessels carrying out consequence class 3 operations have yet been allocated this classification, although it must be pointed out that it is only in the Norwegian Sector that the consequence class notation is mandatory.

The Nautical Institute have in recent years introduced a training programme for DP operators based upon course attendances and experience. The early stages of this training programme even allowed for existing qualified operators to attend a short course, eventually leading to the issuance of the "Nautical Institute DP Operators Certificate", something which appears to be highly acclaimed by charterers and operators alike. The scheme unfortunately has been the subject of some criticism due to the fact that experienced DP operators are unhappy that the instructors lack operational DP experience (this is a common complaint throughout marine industry training schemes). As a result, experienced students, and subsequently new entrants to the industry feel that they are simply going through the "mill" in order to be issued with the correct piece of paper.

So we have two significant attempts to assess and secure the suitability and safe operations of DP vessels. One which looks at the consequence of DP failures and the other which introduces a system of certification for DP operators. Both schemes have been taken up by charterers and would in principle remove the current lack of confidence in DP vessel performance and safety.

The NMD consequence system is a very considerable step ahead, but still needs to be fully adopted by IMO and those operators who operate in the areas outside the Norwegian sector of the North Sea. Some owners have taken the initiative and have had their vessels successfully assessed for NMD class 3, and as a matter of principle operate on this basis even when it may not be required, a commendable situation which hopefully will be extended to all consequence classes. Charterers are responding well to the consequence class system, but the authorities outside Norway appear to be reluctant to make it a mandatory requirement.

The Nautical Institute Training Scheme is also very commendable but lacks the confidence of experienced operators who have been required to

attend purely theoretical courses, which are tolerated in order that
attendees will be issued with the "all important certificate".

So we have two advances in standards, both highly commendable, one which
lacks full international recognition and the other which is recognised,
but lacks the confidence of the "hands-on" personnel and is seen merely
as a requirement of the "job". A view which is not supported by the
author of this paper but one which is nevertheless prevalent.

Apart from these two significant steps ahead (which hopefully will be
further developed through consultation and consent) we are left with a
situation where the 1983 guidelines are still open to interpretation.
With regard to the suitability assessment as outlined in section 4.5.2.,
many prospective charterers interpret it as meaning that all DP vessels
must undergo full audits and trials before commencing operations close
to installations or within the designated 500m safety zones. In many
cases this means that all types of DP vessels being utilised for
operations (with some notable exceptions) will undergo a strict DP audit
and proving trial before the commencement of each operation. For vessels
engaged on the spot market or short term charters this could in theory
mean having to undergo audits and trials on almost a weekly basis. In
this light situations have arisen recently where vessels have been
audited by at least three seperate charterers, all at the same time, and
all with differing requirements. Common sense dictates that this is an
unacceptable state of affairs. In comparison, a vessel engaged on long
term charters may not even be assessed once within a 12 month period, or
even longer in some cases. Such an arrangement would indicate that a
vessel on the spot market has to undergo a more rigorous assessment
system than the vessel engaged on long term charters. This need to carry
out audits and trials as and when charterers require it, appears to be
not only one of operational DP safety, but also one of ensuring that the
vessel complies with the charterers interpretation of the guidelines,
and their own operational and safety procedures. In many cases these are
far in excess of those laid down by guidelines and legislation, whereas
others barely maintain the minimum required. An example of the latter
case refers to a situation where an auditor without operational DP
experience was sent to a highly sophisticated vessel with the purpose of
carrying out a DP audit and trials. The auditor in question was not
fully conversant with the requirements of a DP audit and his
inexperience was evident when both he and the charterer were satisfied
by observing "Pre-Dive DP Checks" as being a substitute for a recognised
audit and trial. Once again this is evident of the wide difference
between standards which are being applied to the suitability assessments
of DP vessels and their personnel, and so from this current situation it
can be seen that a level of inconsistancy has been introduced into the
assessment of vessel suitability and operational DP safety.

As stated previously, a situation now exists where the majority of DP
vessels undergo frequent DP audits and trials whereas others are rarely
assessed, maybe due to their area of operation or because they are

engaged on long term charters of say 2 years or longer. There is an argument to say that over frequent audits and trials of DP vessels is detrimental to operational DP safety, one which on the face of it does not hold any credence. However if vessels undergo DP trials as frequently as those on short term charters it is reasonable to expect a higher frequency of equipment failures due to the regular testing at the extreme of the operational envelope. Over recent months there have been examples of equipment failure which can be directly attributed to unnecessary and over frequent trials being carried out. Such failures normally occur to the electrical switching equipment and can result in long and expensive repairs being carried out, even though both the equipment and the vessel responded correctly to the trials, and the vessel was accepted as being suitable for the proposed work. In contrast it is argued that the vessel on long term charter is not subject to any frequent safety monitoring and as such safety standards cannot be assured. Both these arguments are to a certain extent unsubstantiated, but they do highlight the need to standardise the assessment of vessel suitability and operational safety.

Another area of concern which has become evident over the past 2 years or so, is the utilisation of early generation D.S.V's, no longer suitable for diving operations in the North Sea, for other types of DP work such as R.O.V. support, pipelay support, cable laying and a number of other non-diving operations. This, on the face of it, would appear to be a logical sequence of events, but there is concern that the capabilities of these vessels can no longer be assured and as a result their suitability for any form of DP work must be severely restricted. It is well understood that these vessels would not normally be required to carry out operations within close proximity of offshore installations, but as demand for ROV support and other peripheral work increases, then these vessels are being required to carry out operations which under the NMD consequence classes would put them in categories 1 and 2.

So in summing up the current situation regarding DP operations in the 1990's, we can see that considerable steps have been taken by both NMD and the Nautical Institute in trying to assure vessel suitability and operational safety. But we still have an unacceptably high level of DP incidents attributable to both the suitability of the vessels themselves and operator error.

3. Future Developments - Suitability Certification

What needs to be done in the future to assure vessel suitability and operational safety?

It would appear that one of the greatest failings of the existing audit and trial system is that it is inconsistent in both standards and application. This view is not uncommon within the industry and attempts are being made to resolve the situation.

So, it is proposed that a standardised system of suitability assessment and trials be introduced for dynamically positioned vessels, dependant upon the vessel type and the proposed operation. Such a system would introduce suitability certification for all DP vessels and would be subject to terms of validity and published rules. The need for over frequent DP audits and trials would be removed, and a means of monitoring the standards of vessels engaged on long term charters would be introduced.

In establishing such a scheme for DP vessel suitability it would be necessary to determine the frequency of inspection, and certification validity for those vessels entered into it. Consideration would have to be given to the various types of DP vessel and the work they are engaged in. However, in the first instance it is anticipated that all vessels would undergo an extensive DP Audit and proving Trial at least every 12 months for the issuance of a suitability certificate. It would then follow that the maximum period of validity for such certificates would be for 12 months.

To ensure that vessels are assessed as being suitable for particular operations, different categories of suitability would be introduced. These categories would complement the NMD consequence classes but would be directly related to vessel capabilities rather than the effects of system failures.

Vessel types could be categorised:

Diving Support Vessels   :   Category 1
Drilling Vessels   :   Category 2
Heavylift Crane Vessels   :   Category 3
Pipelaying vessels   :   Category 4
Accommodation vessels   :   Category 5
Emergency Support Vessels   :   Category 6
ROV/Submersible Support Vessels   :   Category 7
Survey Vessels   :   Category 8
Rock Dumping Vessels   :   Category 9
Platform Supply Vessels   :   Category 10

These categories are not exhaustive and do not necessarily reflect the levels of complexity involved for the various operations. It can be seen that the types of vessel vary widely and some of these vessels could combine more than one category, in which case this would be reflected in the appropriate suitability certification.

For vessels in all categories the frequency of DP suitability audits and trials would be once every 12 months, but the frequency of intermediate trials (carried out by vessel operators) would depend very much upon the type of vessel and the proposed operations.

For vessels engaged in diving operations a schedule for compliance with the suitability certification could be as follows:-

1. Vessels will be required to undergo a suitability audit and proving trials every 12 months.

2. Vessel owners/operators will be required to draw up an intermediate trials procedure which will be utilised by the vessel prior to commencement of work at new locations or on new contracts.

   Where vessels are engaged on long term charters of 12 months or more then they will be required to carry out Intermediate Trials at periods not exceeding 4 months.

3. Upon completion of an intermediate trial a report will be completed, and be submitted by the vessel to the certifying organisation, in order to ensure the continued monitoring of suitability and safety standards.

4. The owner/operator of the vessel will be required to state and adhere to minimum experience levels and training requirements for DP Operators.

5. Vessels will be subject to 6 monthly visits from a suitably experienced DP Auditor, who will observe operational procedures over a 24 hour period. This will not constitute part of an intermediate trial, but if circumstances permit it would be beneficial for the auditor to observe such a trial.

6. In the event of a DP related incident the suitability certificate will be suspended, and the vessel will be required to undergo a further DP trial before operations can be resumed and the certificate reinstated.

7. Suitability certification will state which categories of operation a vessel is certified for. If a further category of suitability is required then a suitability assessment may be required.

8. Validity of suitability certification will be subject to compliance with all statutory regulations and certification which are pertinent to DP vessel operations.

In establishing a system of DP suitability assessment, it will not only be necessary to have specific requirements for the various categories, but also the trials will need to be specific to the vessel concerned. It is quite clear that when carrying out DP trials on a stone dumping vessel, with limited redundancy and a basic DP system, it would not be expected to comply with the same requirements as a dive support vessel. So when these other categories of vessel are to be inspected and assessed it should be the practice to incorporate specific areas of their operating capability into the DP trial. If for example it is normal for a certain vessel to have a tug made fast on a bridle during close quarters operations then the trials should include exercises to ensure the viability of such an arrangement. Every effort should be made as part of the DP trials to determine the effects of special procedures or equipment on the overall performance and safe operation of the DP system.

DP trials should also be extended to operator training, paying particular attention to emergency response during DP emergency situations. As part of this, owners and operators may consider it necessary to carry out DP emergency drills on say a monthly basis and in conjunction with the existing statutory drills. Such drills would be expected to deal with all envisaged failures, and also to determine the effectiveness of DP contingency procedures specified in the Operations Manual.

It is envisaged that such a system of suitability assessment and monitoring would eventually become an integrated part of both charterers and owners quality assurance procedures, and it is notable that at least 2 vessel owners have already developed their own trial procedures which are specific to the vessels concerned. Such procedures (which in one case are part of the owners quality management system) would fulfil the requirements for intermediate trials as previously mentioned.

The utilisation of Suitability Certification for DP vessels will, hopefully satisfy the requirements within the existing 1983 guidelines, ensuring high and consistent standards of operational safety without duplication of effort, and without the need to test to destruction in order to prove a vessel's capabilities.

References

Cook, Peter A. (1990) "Safety Standards on Dynamically Positioned Vessels".

UK Department of Energy and Norwegian Petroleum Directorate (1983) "Guidelines for the Specification and Operation of Dynamically Positioned Diving Support Vessels".

GETTING THE DP ACT TOGETHER

BY C.A.JENMAN & P.J.NAPIER,
GLOBAL MARITIME.

SYNOPSIS

The paper discusses current methods of testing DP vessels and looks at more effective alternatives.

INTRODUCTION

The proposals in this paper do not supersede the letter of any legislation or can be consideration in any way to go against the spirit of any existing guidelines. In "getting the DP act together" existing legislation and guidelines are used to meet the requirements of field operators, vessel owners and charterers to produce a better system of ensuring a vessel is "fit for purpose".

THE 1983 NPD/DOE GUIDELINES

1) The offshore industry on the UK Continental Shelf is still bound by the "Guidelines for the specification and operation of dynamically positioned diving support vessels". Even non-diving vessels are commonly assessed using these guidelines, for example for hazardous operations close to platforms. DP drilling has its own NPD/DOE Guidelines published in 1983 when the diving guidelines were updated. These have the same fundamental principles as for diving and are similar in the areas we are considering.

2) The Guidelines suggest that the objective of all DP operations is "the vessel should operate effectively and safely". Three main principle checks are recommended for use as a basis in achieving such an aim.

   i.   That systems are checked   -   on installation
                                            -   after modifications relevant to the DP system
                                            -   before starting new charters
                                            -   periodically during use

   ii. That the operational status of the vessel is based on its known capability

   iii. That the procedures adopted should take account of the limitations of the system

       These can be summarised as:-   TESTS AND TRIALS
                                                    CAPABILITY PLOTS AND FOOTPRINTS
                                                  OPERATIONAL PROCEDURES

133

Volume 27: Subtech '91, 133–151.
© 1991 *Society for Underwater Technology.*

TESTS AND TRIALS

3)   Let us first consider what the Guidelines consider as "DP proving
      trials". These are recommended to take place after construction or
      modifications. They are defined as a full series of trials, and stress
      is put on the requirement that the total system must be tested under
      realistic conditions, with the equipment operating in the intact and
      failed states. .

      The list shown in FIG.1 is provided by Guidelines to indicate the tests
      that could be carried out. Clearly, this list is very much about the
      DP computer systems, the control systems on thrusters, conditions
      monitoring and power management. It is worth noting that no mention is
      made of switchboards, bustie breakers or loss of an engine room.

      The statutory rules for diving require that diving contractors and field
      operations should satisfy themselves about a vessel's suitability before
      DP/diving operations are carried out. The Guidelines suggest that such
      an assessment could involve a study of the vessel's documentation, for
      example, the FMEA, Operations Manual, Capability Plots, together with
      a short sea trial. Such a trial should involve assessing the vessel and
      the crew's capability in both normal and breakdown conditions.

      The construction trials are expected to be in order of several days and
      the new charter trials 8 to 10 hours. Obviously the new charter trials
      are not expected to be a re-run of the construction trials. Yet most
      DP vessels have been regularly subjected to tests that were only
      intended to be carried out at construction or after modifications.

      How did this situation arise?

      One answer can be found in the Guidelines, which state that "a
      fundamental principle of all DP diving design and operations is that no
      single failure should cause a catastrophic failure". Furthermore the
      DP system is defined as "all equipment involved in retaining the vessel
      in its required position". This combination of the fundamental
      principle and a widening interpretation of all equipment has led to
      frequent and severe tests of vessel systems. The most controversial
      tests have been on high voltage switchboards. Much of these testing is
      not realistic, sometimes simulations are incorrectly carried out,
      recording of results can be poor and misinterpretation has often led to
      erroneous conclusions and recommendations which have done nothing to
      enhance DP safety. This problem has been compounded by the fact that
      the switchboards are not designed to make testing easy.

      The other answer is that DP auditors have simulated well known failure
      modes which have led to total loss of DP. In such cases it must be
      assumed that construction trials for class surveyors and/or tests after
      modifications were inadequate. We continue to come across cases on new
      vessels where the most basic testing has not been carried out. Such
      cases on old and new vessels demonstrate that the Guidelines'
      requirement that "an FMEA of the main components should always be
      carried out" has not been fully or competently carried out.

      These problems can be resolved by each vessel having a thorough and specific trials programme instead of a general checklist, which demonstrates its capability in the intact and in various failed conditions. Such tests should provide :-

As an indication of appropriate DP proving trials, checks of the following could be made:

a)   In Harbour

    i.   Correct siting and mounting of all equipment and cabling.

    ii.   Correct wiring up of all power supplies, data cabling and equipment.

    iii.   Correct functioning of all equipment (including data input systems, computers, interfacing equipment, thruster units and power supplies) by electronic and functional testing.

    iv.   Effective shielding of all potential sources of electrical interference (including those which may only be used intermittently, e.g. telex).

    v.   Software checks and tuning.

    vi.   Correct functioning of all condition monitoring systems and alarms.

b)   At Sea

    i.   Correct functioning of all data input systems.

    ii.   Correct functioning of computers and interfaCING.

    iii.   correct functioning of power management system.

    iv.   Correct functioning of thrust units, including response times.

    v.   Optimum position-keeping performance by fine tuning of software.

    vi.   Measurement of position-keeping accuracy using independent means.

    vii.   Correct functioning of all automatic and manual changeover arrangements and procedures from primary to backup systems.

    viii.   Correct functioning of offset and heading change control.

    ix.   Satisfactory operation of DP system with bell running and then divers in water.

    x.   Position keeping in rough weather.

FIGURE 1

- a thorough test of all DP equipment
- a confirmation of the FMEA
- operator experience of failure modes

We are suggesting that every new vessel should undergo the above series of tests following the commissioning of all relevant equipment. Ideally the tests would follow on from the dp control system trials, often referred to as Customer Acceptance Tests (CAT).

Older vessels should be subjected to the same thorough tests and trials.

All vessels should then repeat the trials programme annually. The choice of an annual test is arbitrary. Most vessels have a slack period in the winter and prior to the first job can be convenient time for the vessel owner.

To develop a trials programme it is essential to produce an FMEA which detects possible failures of all DP related equipment and assesses the risk to the position keeping function of the vessel.

THE FMEA

4)   The analysis should set out to determine whether the level of redundancy aimed at by the designer has been achieved. In most cases non conformances with the original design aims are found and the analyst has to make recommendations for modifications to be carried out.

FIG.2 shows the typical systems and sub-systems relevant to DP vessel control which should be analysed. A few examples from a typical FMEA are shown below.

The FMEA has already defined the power generation capacity that should remain available after any single failure in that system. The example shown in FIG.3 demonstrates this vessel's "worst case failure" of the essential services to the diesel electric plant. There is no advice in the guidelines on how such a failure mode should be treated. A sudden loss of water would change the severity of the worst case single failure dramatically.

The power distribution system has in most cases determined the maximum number and configuration of thrusters which will remain on line after any single failure. FIG.4 shows a typical scenario where some thrusters fail due to loss of one voltage distribution and another goes down because a voltage distribution transformed from the first also fails. Most actual thruster failures (FIG.5) are restricted to one thruster and the effect of loss of a thruster is usually less than the severest power generation or electrical distribution failure modes.

Reference sensor failures are usually restricted to loss of one or two at a time with the power distribution being the main factor. See FIG.6. The DP control failure modes are well documented and single failures likely to cause loss of both control computers have had to be accepted due to the limitations of design. Hopefully the newer systems will prove to be truly redundant. The Guidelines still allow diving support

DP SYSTEM

Power Generation (10)

FO (11)
Diesel LO (12)
Alt LO (13)
LTFW (14)
HTFW (15)
S/W (16)
AIR (17)
FVC (18)
PM (19)

Power Distribution (20)

10Kv (21)
440V (22)
220V (23)
110V (24)
Automation SWBD (25)
Thruster 24V dc (26)
Various services, (27)
24V dc

Thruster Group (30)

Thruster (31)
Speed Control (32)
Azimuth Control (33)
FW Cooling (34)

Sensor/Ref System (40)

LWTW (41)
STW (42)
HPR (43)
Artemis (44)
Gyro (45)
VRS (46)
Wind (47)

DP Control (50)

ADP 503 (51)
ADP 311 (52)

UPS (60)

FIGURE 2

SYSTEM LEVEL 1:   DP SYSTEM (00)
SYSTEM LEVEL 2:   POWER GENERATION (10)
SYSTEM LEVEL 3:   LOW TEMPERATURE FRESH WATER COOLING SYSTEM (14)

FUNCTIONAL DESCRIPTION	14.
FAILURE MODE	02. Main circulating system loss of pressure.
FAILURE CAUSE	Electrical fault in ER substation or pump room MCC.
FAILURE EFFECT LOCAL	Both on-line pumps stop. Standby pump not available. Alarms.
FAILURE EFFECT END	No pressure in cooling water main to engine rooms on one side.
COMPENSATING PROVISIONS AND REDUNDANCY	Electrically driven LTFW pumps at affected engines may continue to recirculate water through engines and alternators. No immediate rise in temperature. In the event of total loss of pump room and local LTFW pumps, affected side of vessel can be cross connected manually to cooling water mains supplied from other pontoon pump room.
SEVERITY	Minor.
PROBABILITY	Low.
REMARKS ACTIONS	ER 440V switchboard failure affecting pump room LTFW pumps, local LTFW pumps and local HTFW pumps will eventually lead to shut down of all generators on one side of centre line bulkhead.

FIGURE 3

SYSTEM LEVEL 1: DP SYSTEM (00)
SYSTEM LEVEL 2: POWER DISTRIBUTION (20)
SYSTEM LEVEL 3: 440V SYSTEM (22)

FUNCTIONAL DESCRIPTION	22. To distribute power to motor control centres, propulsion power panels, UPS, taut wires and 220V switchboards.
FAILURE MODE	01. Loss of port ER substation.
FAILURE CAUSE	Short circuit, transformer breaker faults.
FAILURE EFFECT LOCAL	Transformer breakers open. Loss of port 220V switchboard. Loss of LTFW and HTFW pumps.
FAILURE EFFECT END	Half thrusters and generators lost.
COMPENSATING PROVISIONS AND REDUNDANCY	Before failure sufficient power on line to meet limiting requirements. Substation can be supplied from main high voltage starboard switchboard by manual changeover to other transformer. Other reference sensors on line. Port 110V automation on batteries.
SEVERITY	Major.
PROBABILITY	Low.
REMARKS ACTIONS	

FIGURE 4

SYSTEM LEVEL 1:   DP SYSTEM (00)
SYSTEM LEVEL 2:   THRUSTER GROUP (30)
SYSTEM LEVEL 3:   SPEED CONTROL (32)

FUNCTIONAL DESCRIPTION	32. To perform speed adjustment of motors. (In one direction for azimuth thrusters and 2 directions for tunnel thrusters.)
FAILURE MODE	01. SCR failure.
FAILURE CAUSE	Supply fault, internal electronic faults, firing control failure.
FAILURE EFFECT LOCAL	Alarms from electronic protection circuits and for feedback.
FAILURE EFFECT END	1 Thruster stops or is stopped by operator.
COMPENSATING PROVISIONS AND REDUNDANCY	Other thrusters remain on line.
SEVERITY	Negligible.
PROBABILITY	Low.
REMARKS ACTIONS	

FIGURE 5

SYSTEM LEVEL 1: DP SYSTEM (00)
SYSTEM LEVEL 2: SENSOR REFERENCE (40)
SYSTEM LEVEL 3: ARTEMIS (44)

FUNCTIONAL DESCRIPTION	44. To provide position reference to DP by microwave range/bearing Fix/mobile units.
FAILURE MODE	01. Loss of Artemis.
FAILURE CAUSE	Power failure. Out of range. Operator error. Blind sector. Mechanical/electrical faults.
FAILURE EFFECT LOCAL	Alarm on DP console.
FAILURE EFFECT END	Loss of one position reference.
COMPENSATING PROVISIONS AND REDUNDANCY	Other references available.
SEVERITY	Minor.
PROBABILITY	Medium.
REMARKS ACTIONS	

FIGURE 6

SYSTEM LEVEL 1: DP SYSTEM (00)
SYSTEM LEVEL 2: DP CONTROL (50)
SYSTEM LEVEL 3: ADP 503 (51)

FUNCTIONAL DESCRIPTION	51. To provide thruster group with signals computed from references in order to maintain position.
FAILURE MODE	01. A, B, C computer failure.
FAILURE CAUSE	Simultaneous faults internally or in power supply, fire or water through window.
FAILURE EFFECT LOCAL	Alarms for computers and power.
FAILURE EFFECT END	Loss of DP control at bridge.
COMPENSATING PROVISIONS AND REDUNDANCY	2 redundant computers, power supplies with battery back up. Switch over to ADP 311 computer system and assume control at back up DP room.
SEVERITY	Major.
PROBABILITY	Low.
REMARKS ACTIONS	

FIGURE 7

with a simple DP system and a joystick backup. The Norwegian Maritime Directorate have already tried to overcome the total loss of dual redundant systems from, for example fire and explosion on the after bridge by the introduction of the NMD consequence Class rules which specify a back up DP station. See FIG.7.

UPS systems are an area where the number of total failures are greater than expected and have left some DP systems inoperable for several days. Most vessel owners have learned the hard way that it pays to have 2 UPS units. The example shown in FIG.8 and 3 UPS. There are now at least 5 vessels with this set up and a new vessel and two rooms full of UPS. Therefore, having analysed and stated the capabilities and limitations of the DP system a set of vessel specific proving trials procedures can be written.

CONDUCT OF TRIALS

5)   i.   All relevant shipboard equipment must be fully operational. In particular, all propulsion units and their controls, both manual and automatic, all power generation equipment, all computer systems and all position reference systems must be fully functional, including their alarms, stand-by units, battery back-ups, shutdowns, trips, etc.

        ii.   All tests to be co-ordinated by the Trials Manager with approval from the OIM/Master with full regard to safety of navigation of the vessel. (The Trials Organisation is shown in Figure 2).

        iii.   All tests will be carried out on full DP in realistic environmental conditions or with some varying load on the system induced by movements of the vessel.

        iv.   During the trials, the vessel's staff will assist in recording alarms and failure locally. **Locally** means not only at the DP console, but at the ECR, computer room, thruster room, etc.

               During failure tests, the system must not be reinstated until the DP operators, ECR staff and surveyor are satisfied they understand the full effects of the failure and that all the information or indications to show what has occurred have been noted.

        v.   When reinstating systems after failure simulations, two people will check that circuit breakers have been reinstated. Only when everyone is satisfied the system has been reset and has stabilised, will the trials continue.

        vi.   If there are any doubts about a test, it will be repeated. If test results are unexpected then the test will also be repeated. It should be noted that seemingly small or spurious faults in DP control systems may be the first manifestations of a more serious problem.

SYSTEM LEVEL 1: DP SYSTEM (00)
SYSTEM LEVEL 2: UPS (60)
SYSTEM LEVEL 3:

FUNCTIONAL DESCRIPTION	60.
FAILURE MODE	02. Loss of output from UPS 1.
FAILURE CAUSE	Inverter failure.
FAILURE EFFECT LOCAL	Alarms.
FAILURE EFFECT END	Loss of Computer A, Console A, HPR tracking transducer, Gyro 2, one wind display, one Artemis supply, VRS 1.
COMPENSATING PROVISIONS AND REDUNDANCY	UPS 2 supplies Computers B, C, Console B, HPR screen, forward LWTW, surface TW, one wind display, one supply to Artemis, Gyro 1 on ship's supply. UPS 3 supplies aft LWTW.
SEVERITY	Major.
PROBABILITY	Low.
REMARKS ACTIONS	Alternative (bypass) supply changeover is manual and not synchronised.

FIGURE 8

vii.   Test will continue only when all those involved have been informed
       and (where unnecessary) suitable communications have been set up,
       e.g. DP console to thruster room.

viii. The test will not only prove hardware redundancy and DP capability
       after failures but also that the operators having the necessary
       training and experience to use the system and deal successfully
       with such failures.

The above shows a set of basic ground rules we have found useful in
practice. The first is of great importance and may involve considerable
work by manufacturers engineers to ensure all equipment is operational,
e.g. injection testing of high voltage breakers. Owners must have the
opportunity to have system designers onboard. The Vessel Superintendent
may be the ideal Trials Manager as the Master or Chief Engineer are
normally too heavily involved to manage all the aspects of such a set
of tests.

Realistic environmental conditions are essential at some point during
the trials but this is not usually a problem in the northern North Sea.
The recording of results is the area where most arguments occur. It is
of the utmost importance that the observers agree upon the results and
that the results are as predicted. If disagreement exists then repeat
tests should be carried out until all interested parties are satisfied.
Of course it may be that unexpected results have occurred as a fault
exists or the system is not in fact as assumed in the FMEA. We have yet
to find a vessel with a perfect set of "as built" drawings.

Incorrect reinstatement can cause disastrous failures in subsequent
tests. Indeed the desire to get on as quickly as possible with the test
programme can lead to even longer trial trying to find out what went
wrong. A vessel may even lose its credibility if several tests go
seriously wrong due to incorrect setting up. The trials are an
excellent opportunity for testing the effectiveness of the onboard
checklists used for initially setting up the DP system.

Owners already involved in this type of trial try to have on board as
many as possible of both sets of personnel as it provides an excellent
opportunity for training. The personnel who have carried out such
trials programmes have given a very positive response. All have stated
they had a better understanding of the vessel systems, capabilities and
limitations. Such trials should reduce the risk of a failure mode
becoming a serious incident i.e. reduce the operator error element.

FIG.9 is a page from a set of trials procedures. It shows 4 failure
modes on the diesel generator governor system. The switches to be
operated, fuses to be pulled, sequence of events and personnel carrying
out the tests are shown. Note that in addition to the rules for
"conduct of trials" a warning has been given about reinstating the
system. Should any observers question the validity of the tests then
reference to the FMEA will explain in its descriptive section exactly
how this particular subsystem operates.

FIG.10 shows the method used to record the results. These should be
filed in the descriptive trials report with the procedures. Therefore

a future auditor can gain confidence and a good understanding of the vessel's systems and their effectiveness by using the FMEA and trials reports.

FIGS.11 & 12 show another annual DP trials format which is being used on several vessels. The first items are to ensure that essential maintenance, with documentary evidence has been carried out on all relevant equipment. The trials procedures are thorough and space has been provided for recording results and making remarks. A fully descriptive set of procedures have also been developed which may be used to assist in reviewing the trials.

Method:

A.   Switch off 115V supply to electronic governor from L2. The switch is on the Amot Panel.

B.   Reinstate the 115V supply and remove the fuse from the 24V backup supply to the governor.

C.   Do not reinstate fuse (See B) and switch off 115V supply (See A). This will put governor on to 6V battery backup supply.

D.   Reinstate supplies to electronic governor.

NOTE: For the following tests 5.5 to 5.10, an electrician or mechanic will stand by the affected MCCs and record the breaker settings using the checklist. When the MCC power is reinstated he will ensure that the breakers and all associated machinery are returned to the status before the test. A second person (electrician or mechanic) will double check the breaker status using the checklist.

FIGURE 9

TEST NO. 5.4 C	DESCRIPTION: Loss of Power to one Governor

METHOD:

Switch off 24V dc power supply to governor at front of Amot panel on No. 2 Generator.

Generator online - 2, 3 and 5 at 3.2 MW each.

Thrusters online - 1, 4, 6 and 7

EXPECTED RESULTS:

<u>AT MCR:</u>

Online   Capacity   =   17MW
            Load   =   10MW
            Spinning Reserve   =   7MW

After switching off power supply the generators went to:

                             No. 2 =   5MW
                             No. 3 =   2MW
                             No. 5 =   2MW

No. 2 rose to full load over several seconds and gave an audible/visual alarm on the Amot panel for "Fuel Rack Overload".

RESULTS:

<u>MCR:</u>

STBD generators OK.
Thrusters 2, 4, 6, 9 and 10 stay in DP Em. Gen started. Port ER 440V substation repowered by auto changeover to 10,000/440 transformer fed from starboard 10kV bus. Thruster cooling automatic changeover valves open.

<u>503 Console:</u>

Failure of AFT,FWD and LWTW. UPS1, 2, and 3 loss of alternative supplies from emergency 220V switchboard.

<u>Navigation Console:</u>

Loss of hydraulics on thrusters 3, 5, 7 and 8. Loss of LO pump on No. 1 thruster.

COMMENTS:

Both LWTW should not have been lost. Aft LWTW recently changed from superstructure substation (starboard side) to emergency substation (port side).

WITNESSED BY:.......................................DATE:.../.../...

FIGURE 10

Annual DP Test Checklist

VESSEL NAME				DATE:		
SECTION 1 - REVIEW OF DOCUMENTATION - MAINTENANCE HISTORY						
1.1						
THRUSTERS	BT1	BT2	BT3	AZ1	AZ2	
MAINTENANCE RECORDS	YES	YES	YES	YES	YES	
OIL ANALYSIS	YES	YES	YES	YES	YES	
REMARKS: Last oil analysis taken week 10. Awaiting results.						
1.2						
GENERATORS		1	2	3	4	
MAINTENANCE RECORDS		YES	YES	YES	YES	
OIL ANALYSIS		YES	YES	YES	YES	
RUNNING HOURS						
AT PRESENT						
LAST ANNUAL TEST						
REMARKS: Oil sample as above.						
1.3						
SWITCHBOARD POWER MANAGEMENT						
SUBCONTRACTORS REPORT	DATE			REPORT NO.		
REMARKS: ABB started annual testing.						
To be completed week 13.						
1.4						
DP SENSORS						
TAUTWIRE		1	Y	2	Y	
HPR TD		PORT	Y	STARBOARD	Y	
VRS						

FIGURE 11

Annual DP Test Checklist

VESSEL NAME						DATE:			
3.4									
440V SYSTEM (BUS TIE OPEN)									
FAIL M1/P1 TRANSFORMER									
POSITION MAINTAINED?						YES		NO	
THRUSTERS STOP	BT1	Y	BT2	N	BT3	N	AZ1	Y	AZ2	N
UPS ALARM					YES			NO		
EMERGENCY GENERATOR START					YES			NO		
STBD TW ALARM					YES			NO		
STBD HPR RD FAIL					YES			NO		
110V ALARM G5					YES			NO		

REMARKS: Thrusters stop due to lack of hydraulics and 220V to ECU. AZ1 has standby hydraulic pump powered from P2 which starts. AZ1 can be restarted if 220V bus tie is closed. UPS and G5 charger fed from emergency switchboard. These will alarm and be restored when emergency generator starts. Cooling water pumps in stbd ER for engines and thrusters also fail but this has no immediate effect on positioning.

3.5									
440V SYSTEM (BUS TIE OPEN)									
FAIL M2/P2 TRANSFORMER									
POSITION MAINTAINED?						YES		NO	
THRUSTERS STOP	BT1	N	BT2	Y	BT3	Y	AZ1	N	AZ2	Y
UPS ALARM					YES			NO		
PORT TW ALARM					YES			NO		
PORT HPR RD FAIL					YES			NO		
110V ALARM G6					YES			NO		

REMARKS: As per 3.4. Port engine room affected.

3.6									
6kV SYSTEM (BUS TIE CLOSED)									
GENERATORS 1, 3 & 4 ON LINE. ALL THRUSTERS ON.									
TRIP GENERATOR 1 AND OPEN M1/M2 BUS TIE.									
THRUSTERS STOP	BT1	Y	BT2	N	BT3	N	AZ1	Y	AZ2	N
POSITION MAINTAINED?						YES		NO	

FIGURE 12

6)   MOBILISATION, NEW LOCATION AND PRE-WORK CHECKS

To fully integrate the testing of DP vessel the work cannot be restricted to just an annual full programme of trials. These trials have to be backed up with complementary trials for each situation, mobilisation, position change, bell change etc. Four levels are in operation and each is less onerous in scope and complexity but thorough enough to ensure equipment is still available and working. In addition the mobilisation tests will always have an element superimposed on them concerning the particular work to be undertaken.

It is this that clients and their representatives can concentrate on after reviewing the previously documented trials data onboard. Mobilisation tests should be job specific and to check equipment is working not an exercise on testing the whole design of the vessel. Similarly the tests for a new location should concentrate on checking items particular to that location and its restrictions as well as checking that other items are still configured as they should be.

The key to the practical success of such a system is the ease and relevance found by operators when they use it. The documents should remind them of the particulars idiosyncrasies of their vessel and hence contribute greatly to safety with a low nuisance factor.

7)   DP VESSEL OWNERS ASSOCIATION

Since writing the first sections of this paper the industry has moved on and we are now moving closer to "getting the DP act together". Essentially the vessels taking onboard the comprehensive trials philosophy are giving all their clients and their clients representatives a detailed vessel specific check list filled in from real trials and backed up by an independent FMEA. Owners are taking the initiative to stop surprises that cost them (and their clients sometimes) time and money.

For this to be effective it has to be thorough and acceptable to offshore operators i.e. it has to satisfy their statutory duty of care to, for example, the HSE. In this regard DPVOA has presented the scheme to the UKOOA Marine and Diving committees and is now in consultation in detail with the various interested parties.

The system also needs to be kept alive and relevant. This will be achieved by keeping a flow of technical information on DP failures from owners to all members. Advances or new permutations of DP are continuously in progress and there will always be lessons to be learned from the practical implementation of such advances.

The annual trials programme will need to be updated to reflect the increase in knowledge. This is another pillar of the DPVOA.

A third primary aim of DPVOA is to update the Guidelines and achieve an internationally agreed code. This work is well advanced and has the great advantage of separating the various types of DP Vessel. The minimum DP requirements are not the same for ROV survey work, offtake tankers and diving support.

The vision of DPVOA is that of risk qualification and the concept of acceptable risk. No single fault should cause a catastrophic failure ...... unless that fault is so unlikely and remote as to be not worthy of serious consideration. In other words if nobody anywhere has ever heard of it ever happening the risk must be acceptable.

SUBSEA SUPPORT VESSEL FOR THE NINETIES

JENS CHR. LINDAAS, KJARTAN VARTDAL, ARNFINN VIKA
Stolt-Nielsen Seaway A/S
P.O. Box 370
Stoltenberggt. 1
5501 Haugesund

ABSTRACT In this paper Seaway's development project "Underwater Support Vessel for the 1990'ies" is described. The vessel is being aimed for both the diving and the ROV/ROT market. The main characteristics of the vessel is presented, as well as the main topics being studied in the project. Further, the following systems have been described more in detail, which partly already are in operation on existing vessels:

- A new concept for hyperbaric diving system (patent pending).
  The concept consists of 8 vertical identical chamber cells configured in a circle, and a turret-based handling system for the bells.
- The DIVEROV concept.
  This diving assisting vehicle, based on a standard ROV, has now been in operation over one year. The DIVEROV is flexible in the sense that it also can function as an advanced ROV for stand-alone tasks.
- The module handling system now in operation on the DSV "Seaway Condor".
  This system incorporates both active and passive heave compensation for safe and efficient installation and retrieval of modules on subsea production systems.

1. Introduction

The development project "Subsea Support Vessel for the 1990'ies" was started in 1988 in order to coordinate our development efforts within several areas related to diving and ROV operations. The aim has been to provide results from the different development areas in such a form that they can be applied to specify Seaway's underwater support vessel and work procedures for the future. Some sub-projects have already resulted in systems and methods in operation on our existing vessels.

153

Volume 27: Subtech '91, 153–168.

Primary functions for the vessel have been defined as follows:
- Diving related to IMR, construction work and contingency, especially for deep water diving.
- ROV-operations related to the diving operations and intervention work where the ROV is used as a stand-alone vehicle.

Also module handling, well services and laying of cables and flexible flowlines have been defined as functions possible to adopt to the vessel.

2. Subsea Support Vessel - Main Characteristics

The following main characteristics/requirements have been defined for the vessel:
- Mono-hull vessel, approx. 100 x 20 m.
- Diving system and other equipment dimensioned for state of the art diving to 450 msw, and optimized for diving to 350 msw (continuous diving possible).
- Deck area min. 800 m².
- Lifting capacity min. 150 tons, active heave compensated.
  (We are challenging the crane manufacturers to produce a 250 tons crane with same weight and dimensions)
- Accommodation for min. 100 persons.
- Heli-deck.
- Two diving bells, also to be used simultaneously.
- Air diving to be arranged for.
- Work moon-pool min. 7 x 6 m.
- Two ROVs on board.
  One for observation purposes and one advanced work-ROV with diving assisting capabilities (DIVEROV, ref. Ch. 5).
- Weather criteria for operation of vessel: min. 40 knots wind and 1,5 knots current.
- Transit speed: min. 12 knots.
- Dynamic positioning system to fulfil requirements from authorities and clients as a minimum.
- Active stabilization/compensation of heave and rolling movements to be arranged for.

3. Main Topics Being Studied - Status of Project

Today, the status is as follows for the different sub-projects which have been defined into this project:

3.1 GENERAL ARRANGEMENT AND SPECIFICATION FOR A NEW MONOHULL VESSEL FOR UNDERWATER OPERATIONS.

In this sub-project the results from the other sub-projects are implemented into the design of a new monohull vessel.

A concept has been developed and is now being further detailed in order to prepare a specification for the vessel.

Today, the vessel measures 100 x 20 m, has 6 azimuth thrusters for optimal station keeping capabilities and a flush working deck of approx. 1800 m². Lifting capacity is 2 x 150 tons. The diving system consists of 8 chambers, each containing 3 divers. The two bells can be mated on the top of the chambers through utilizing a turret system. Diving capabilities are down to 450 m.s.w.

Two ROVs with dedicated handling systems will be implemented in the vessel. These are one work-ROV type for complicated work tasks and diving assistance (DIVEROV), and one observation ROV.

The design of the hull is innovative and has been optimilized for best possible motion characteristics when staying in a fixed position.

The creative new hull design and heave and roll movement compensation systems are configured to achieve work ability near to a semi submersible vessel.

3.2 CREW OPTIMIZING THROUGH IMPLEMENTATION OF NEW TECHNOLOGY AND MULTI FUNCTION POSITIONS

A report has been prepared pointing out different possibilities for optimizing the crew within different levels and areas.

3.3 MAKE THE CHAMBER CONTROL AND DIVING CONTROL MORE EFFECTIVE THROUGH POSSIBLE AUTOMIZATION OF FUNCTIONS AND ERGONOMIC STUDIES

A preliminary report has been prepared stating possible functions that can be automatically controlled, logged and documented. Further, a possible lay-out for the chamber control and the diving/ROV-control has been proposed. The study has been based on utilizing a modern computer based control system, similar to what is being used in the process industry. We have departed from mechanical connections and valves, and have replaced them with electronics. Still manual override from decentralized panels will be possible.

3.4 IMPROVE THE "LIVING STANDARD" AND ENVIRONMENTAL CONDITIONS IN THE HYPERBARIC CHAMBERS

A report has been prepared stating which factors are influencing the environmental conditions for the divers. Criteria to improve the "living standard" in the chamber system have been proposed.

On the basis of the new chamber system which has been proposed into our new vessel concept (ref. Ch. 4), these criteria will be used for further detailing of the chambers and the life support system.

3.5 IMPROVE THE TOTAL WORK EFFICIENCY THROUGH THE USE OF DIVERS AND ROV IN COMBINED OPERATIONS

This sub-project has been defined into the DIVEROV project, ref. presentation in Ch. 5. The DIVEROV has been in operation on the DSV "Seaway Harrier" since December 1989.It is now important to follow-up the operational experience, in order to be able to improve the concept.

3.6 WIDEN THE WEATHER WINDOW BY THE USE OF HEAVE COMPENSATED LIFTING SYSTEMS

A report has been prepared, describing the different technical solutions that can be implemented to obtain heave compensation on lifting systems.

The effect of these systems to widen the weather window has also been evaluated.

These results and of course our operational experience with already implemented systems (module handling system on the DSV "Seaway Condor", new 120 t crane on the DSV "Seaway Pelican" and new bell handling system on "Seaway Pelican") will be used to specify the lifting- and handling systems for the new vessel. The module handling system on "Seaway Condor" has been described more in detail in Ch. 6.

In fact, as we can see it today, the rolling of the vessel will in many cases be more critical than the heave. We have therefore started to look at possible means to counteract the rolling movements, and have some new ideas that now are evaluated.

3.7 IMPROVE EFFICIENCY BY IMPLEMENTING NEW COMMUNICATION SYSTEMS

In this on-going study possible means to improve communication on board the vessel and between the vessel and shore are evaluated. Promising ways of making the communication more efficient and less costly have already been identified.

3.8 IMPROVE EFFICIENCY FOR BRIDGE AND ENGINE CONTROL FUNCTIONS BY AUTOMIZATION OF FUNCTIONS AND UTILIZATION OF MULTI FUNCTION POSITIONS FOR THE PERSONNEL

A preliminary report has been prepared stating different possibilities to improve the overall efficiency. These results will be further evaluated, ending up in a specification adopted to the vessel concept.

3.9 ORGANIZING THE PERSONNEL ON BOARD THE SHIP AND SPECIFICATION OF EDUCATIONAL PROGRAMMES

As new technology will be implemented for many functions and multi function positions will be defined for the personnel, educational- and training programmes will be required.

3.10 FOLLOW-UP AND IMPLEMENTING RESULTS FROM OTHER ON-GOING RESEARCH AND DEVELOPMENT PROGRAMMES RELATED TO THE DIVERS PERSONAL OUTFIT AND EQUIPMENT, WORK PROCEDURES AND TOOLS

In this sub-project the aim is to implement results from other internal and external projects like HADES and FUDT into the specification of the vessel, equipment and work procedures.

4. Hyperbaric Diving System

A totally new concept for hyperbaric diving system has been developed.

The concept consists of 8 chamber cells, configured in a circle (ref. fig. 4.1a). Each chamber cell is a vertical cylinder, approx. 4,5 m high and 3,5 m in diameter (ref. fig. 4.1b). The chambers are divided into two compartments, one wet part for change, shower and toilet, and one dry part as combined day room and sleeping compartment. Normally 3 divers will stay in each of the cells. The diving bells are mated vertically on top of the chambers.

The handling system for the bells are attached to radial, movable structures, in such a manner that each of the two bells can be mated to any of the chambers. The bell is then positioned over one of the two moon-pools configured alongside the same circle as the chambers, and lowered into the sea.

Active heave compensation is integrated in the handling systems.

The diving system/bell handling system has been found patentable and a patent is now being applied for. The system is thus protected by international patent laws.

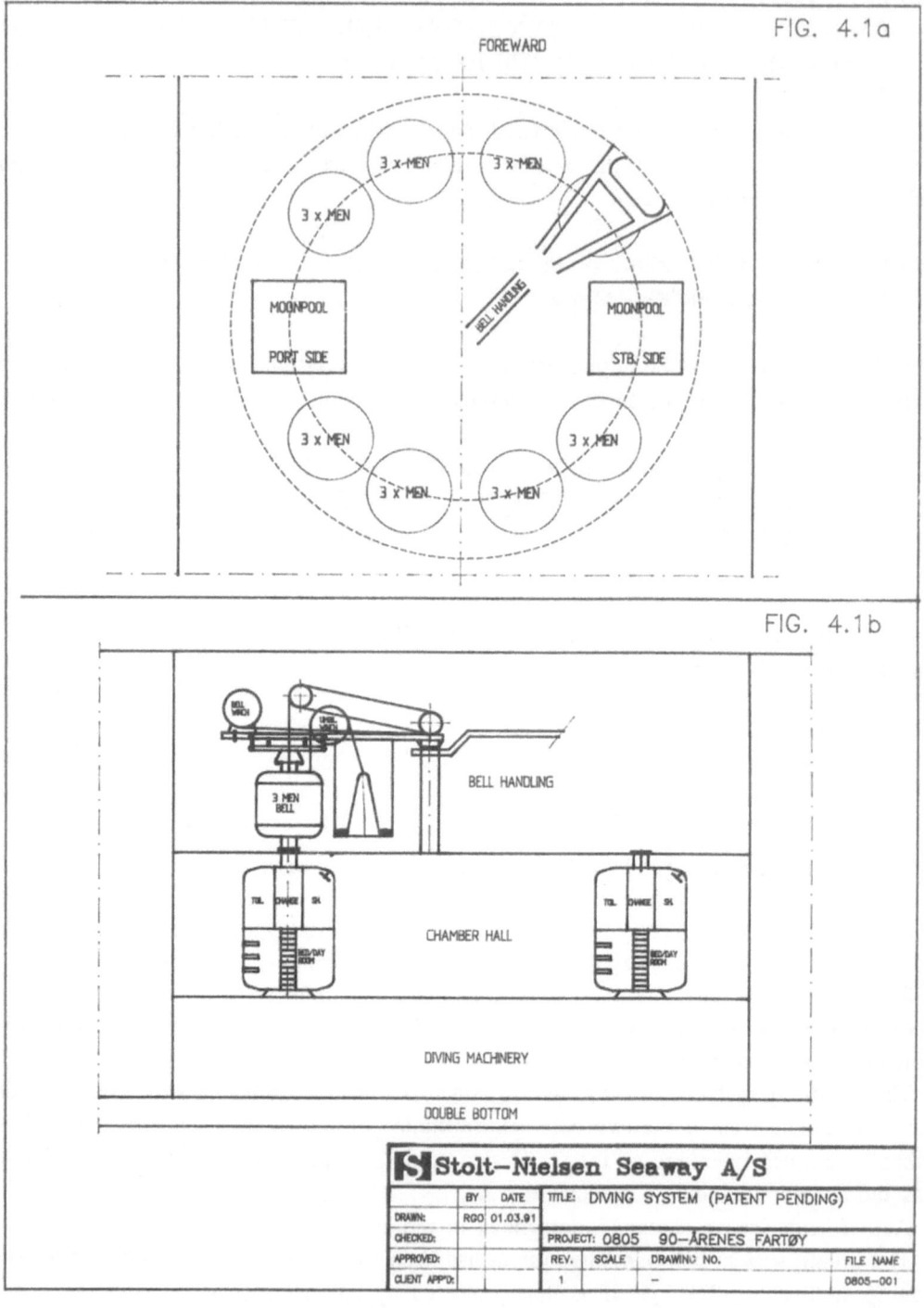

FIG. 4.1a

FIG. 4.1b

5. The DIVEROV Concept

5.1 GENERAL

The DIVEROV concept (ref. Fig. 5.1) was developed as an answer to an increasing concern about the cost effectiveness of diving operations. The DIVEROV is not just another advanced multipurpose ROV, nor is it a dedicated tool that can only be utilised while the diver is in the water. It has the possibilities of both alternatives, without the limitations. It can supply the diver with hydraulic or electrical power, carry tools and heavy equipment,
as well as offer flexible and stable support from which the diver can work. When there is no diver in the water, or assistance is not needed, it can perform work on its own utilising its advanced manipulators, multitude of video cameras or any of the optional modules developed. The concept is based on a well-known standard vehicle, the Triton from Perry Offshore Inc.

The DIVEROV concept has been developed in close contact with Phillips Petroleum Company, Norway, with the contract for diving services on the Ekofisk Field in mind. The water depth here is 70 m. The benefits to the operational efficiency will increase with increasing depth. The vehicle has now been in operation more than one year.

5.2 DESCRIPTION

The DIVEROV system is based on an upgraded (75 HP) Triton with a Tether Management System, TMS. The system is depth rated to 1000 m. Additionally it is equipped with:
- One 7-function Schilling master slave manipulator.
- One 7-function Kodiak manipulator.
- Mestotech 971 colour sonar.
- One Osprey Pan, Tilt, Rotate (PTR)-camera.
- One SIT camera.
- Three other cameras are mounted in accordance with varying requirements.
- Attachment arms with suction pads and hydraulic freeze joints. These offer the stability required for standalone manipulative operations as well as dive support. Usually one arm in combination with the vehicle's auto-park mode gives sufficient stability for most tasks.

The DIVEROV module consists of the following equipment:
- Secondary "dirty" hydraulic system, that is insensitive to seawater contamination of the hydraulic fluid. Hydraulic tools can therefore be connected/disconnected by the diver.
- MPI unit, complete with ink-container, UV-lamp and prods/coil capable of delivering up to 1500 A at 5V.

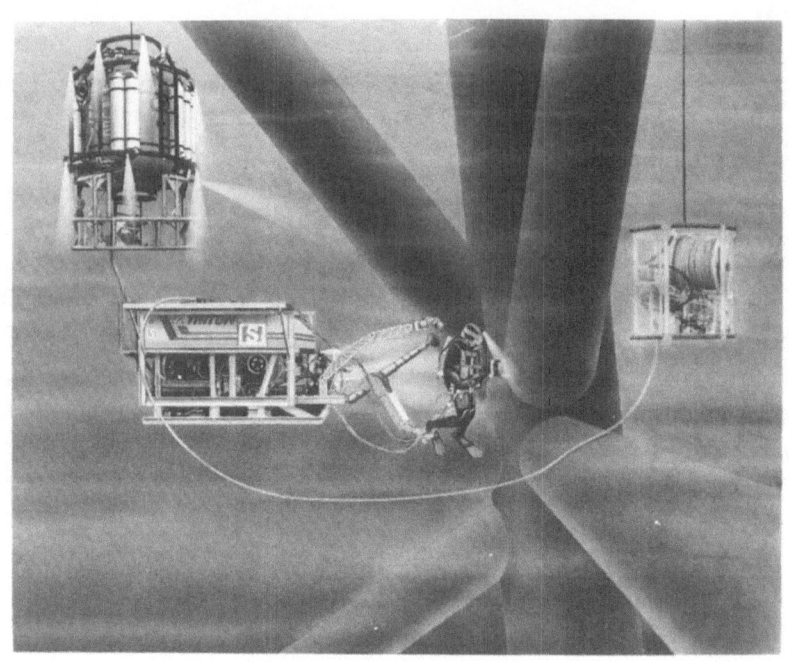

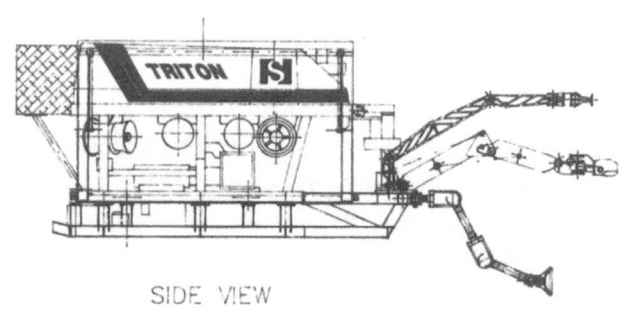

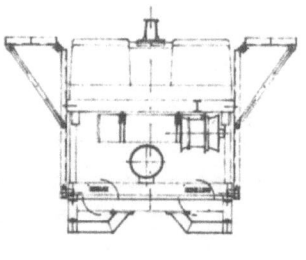

SIDE VIEW   FRONT VIEW

FIG 5.1
DIVEROV CONCEPT

- Interface and telemetry system for Eddy Current inspection equipment.
- Tool basket at the rear of the vehicle.
- Foldable work platforms on either side of the vehicle, that are utilised for direct support of the diver's work position.
- A manipulator held support device that can be accurately positioned on the diver's instructions, to offer him support in locations where access for the entire vehicle is restricted.
- Guards to prevent ingress of the diver's hands, feet or equipment into the vehicle's thrusters.

The electrical system is designed with the safety of the vehicle as well as the diver in mind. Insolating transformers and active line insulation monitoring with trip devices (Megacon) are utilised to comply with AODC "Code of Practice for the Safe Use of Electricity Under Water".

5.3 WORK TASKS IN "STANDALONE" MODE

INSPECTION
The DIVEROV is suitable for many inspection tasks on a "standalone" basis:
- It is well suited to carry out General Visual Inspection (GVI), however this is in most cases more cost-effectively carried out with a smaller vehicle, e.g. the Sea Hawk.
- Close Visual Inspection (CVI) generally require cleaning before inspection is carried out. With its wide range of cameras and mounting alternatives, it can solve most tasks. If accurate measurements are asked for, this can require special equipment, e.g. spotrange sonar/laser unit.
- Cleaning for CVI or non destructive testing (NDT) is time consuming and one of the areas where the benefits of the DIVEROV is obvious. It is capable of efficiently removing marine growth and protective coating with high pressure water jetting, gritblasting or with hydraulic brushes.
- Ultrasonic thickness measurements, Cathodic Potential (CP) measurements as well as Flooded Member Detection (FMD), are NDT techniques for which it is well suited.

MAINTENANCE, REPAIR AND CONSTRUCTION
The vehicle is well suited for work within these categories. Examples of tasks which can be executed in "standalone" mode:
- Dredging, removal of seabed material.
- Installation of sandbag supports and protection mats.
- Debris retrieval and clearance.
- Cutting of bolts and wire.

- Preparation work related to alignment frames and habitat prior to hyperbaric welding.
- Assistance during pull-in and connection of cables and flowlines.
- Intervention on subsea production systems (equipped with the necessary tool-packages).

5.4 WORK TASKS IN DIVEROV MODE

GENERAL TASKS
The following tasks are of general nature, common within inspection as well as repair, maintenance and construction work:
- Provide the diver with support from which he can carry out his work effectively. This is an important role, since significant amounts of dive time is spent while the diver is establishing a work position from which he can handle tools or equipment. While swimming freely, the diver can only exert a force of a few N.
- Supply light to the worksite and a picture of the working diver to the diving supervisor. This is one of the traditional dive support tasks for ROVs.
- Carry tools and equipment from the support vessel to the worksite. This can save time by eliminating the need for "downlines".
- Provide the diver with an outlet for hydraulic power whereby handling of hydraulic hoses from surface is eliminated.

The last two items also have a secondary effect in reducing the demand for manpower on deck on board the support vessel.

INSPECTION
The DIVEROV module has an integrated MPI unit, thereby providing the diver with outlet for magnetizing current, fluorescent ink as well as ultraviolet light right at the worksite. Recently, it also has been equipped with interface and telemetry system for Eddy Current inspection.

MAINTENANCE, REPAIR AND CONSTRUCTION
The DIVEROV can exert a force of up to 2000 N with thrusters alone. It will also be possible to add a variable buoyancy module giving the vehicle a lifting capability of nearly 4000 N. Lifting and pulling is necessary whenever an item is being installed, and this can easily be carried out with the DIVEROV.

5.5 EXPERIENCE FROM OPERATION

The vehicle has now been in operation over one year, and it has clearly demonstrated its potential.

The flexibility in using the vehicle either as a "standalone" vehicle or for diving assisting purposes, dependent on the task, has been proved to be very valuable in practical operations. An increase in total efficiency of 100% has been demonstrated in certain cases.

Initial problems related to pressure drop in the hydraulic system during certain power consuming operations have now been solved by upgrading the vehicle to 75 HP.

6. The Module Handling System on "Seaway Condor"

6.1 GENERAL

Seaway has together with Elf Aquitaine Norge A/S developed a module handling system for installation/recovery of subsea well modules. The handling system is designed to be operated on board the DSV "Seaway Condor", ref. Fig. 6.1 and 6.2.

The system has been successfully used on the East Frigg Field since 1989. In addition to module handling the system also has been utilised for cable laying and connection.

The system consists of horizontal and vertical module handling equipment, operated on the deck and through the working moonpool. It includes all the necessary equipment for the following operations:

- Longitudinal and transverse transportation of modules and tools on deck, without free lifting operations.
- Lifting of modules and tools through moonpool. During these operations the modules/tools are guided by a cursor system.
- Deployment of modules/tools in an actively heave compensated mode (motion compensation of a freely suspended load) or in uncompensated mode.
- Establishment of a low constant tension in the lift wire after landing of module/tool on template.
- Controlled increase in lift wire tension during the pick-up sequence, while the system compensates for vessel vertical movements.
- Retrieval of modules/tools in a passively compensated mode or a combination of active and passive compensation, where the passive part functions as a back-up shock absorber.

6.2 OPERATIONAL CRITERIA

Max. wave height for operations: 3 m
Max. pitch/roll for operations: +/- 5 deg.
Active compensator stroke selection. Load amplitude +/- 2.0 m
Passive compensator stroke selection. Load amplitude 1.12 m

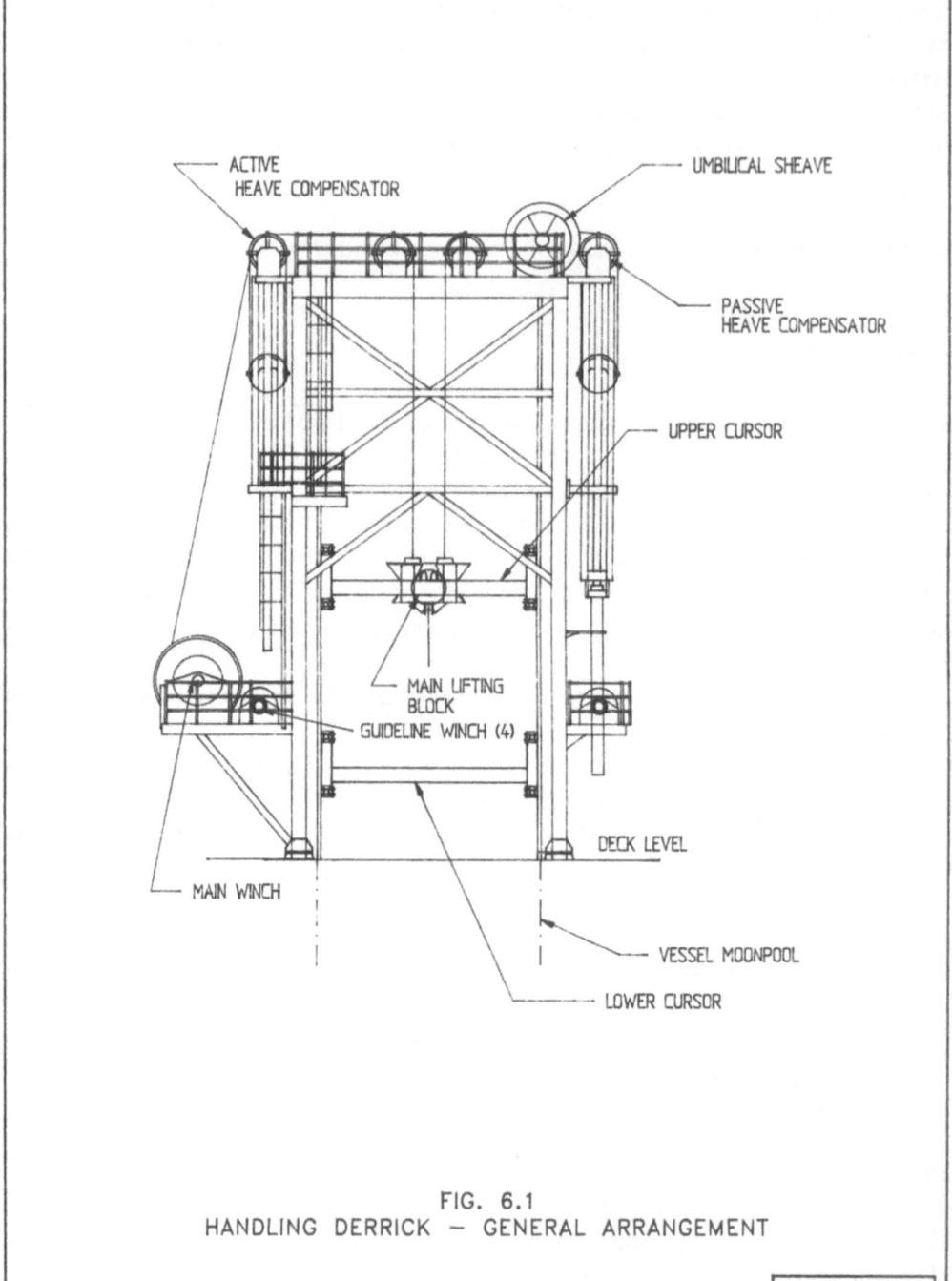

FIG. 6.1
HANDLING DERRICK — GENERAL ARRANGEMENT

CAD: INT\011-19\SYSD-F3

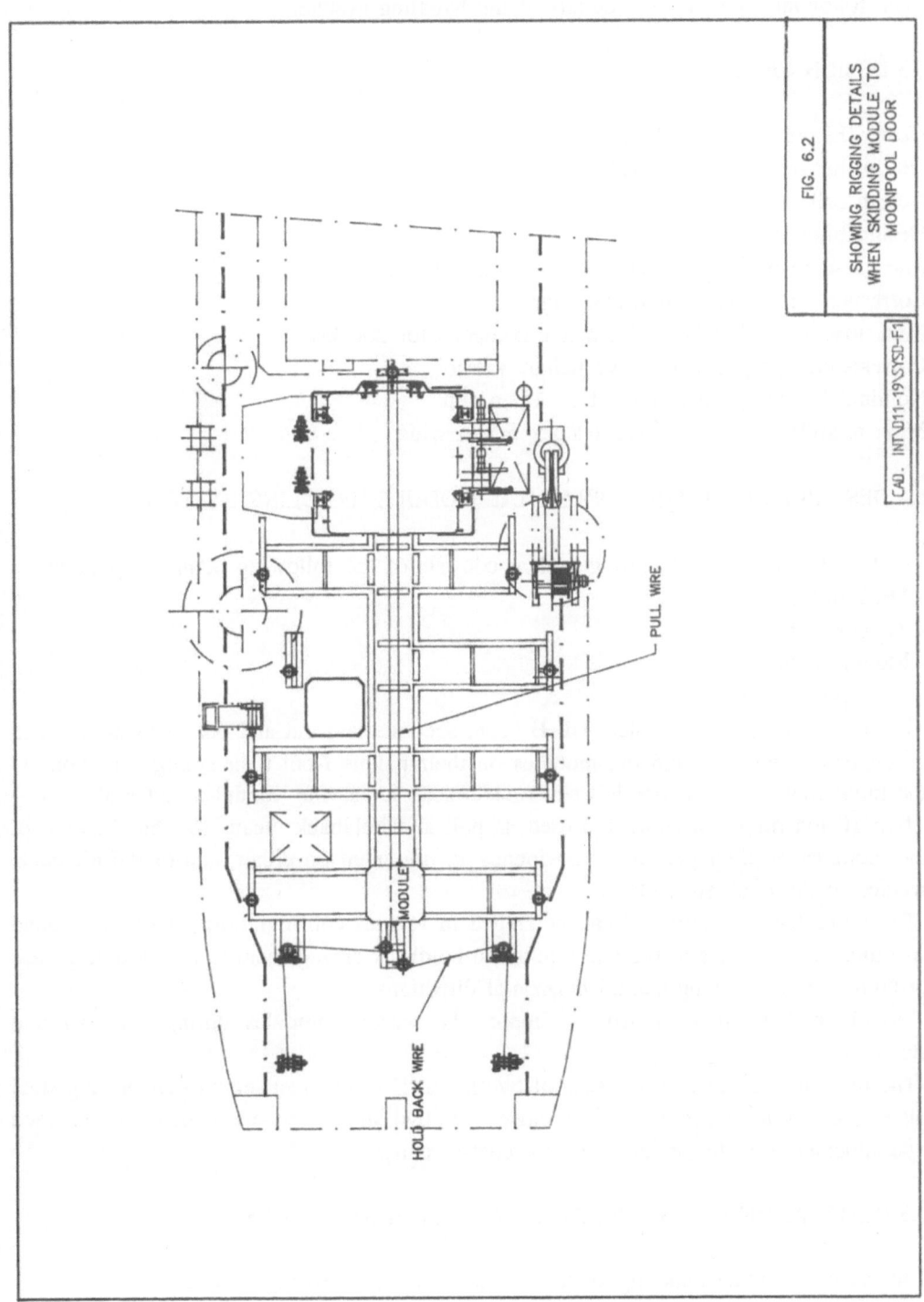

FIG. 6.2

SHOWING RIGGING DETAILS
WHEN SKIDDING MODULE TO
MOONPOOL DOOR

CAD. INT\011-19\SYSD-F1

The following gives design details of the handling system.

6.3 DESIGN CRITERIA

Design heave velocity 1.5 m/s
Design heave acceleration 1.7 m/s2
Design roll +/-4 deg.
Design pitch +/- 4 deg.
Max. load to be handled with passive compensator 550kN
Corresponding significant wave height 2.0 m
Max. load to be handled with active compensator 200 kN
Corresponding significant wave height 3.0 m
Nominal lifting speed at max. load 15 m/min
Max. module size in moonpool 5.5 x 5.5 m wide

6.4 DESCRIPTION OF THE HORIZONTAL MODULE HANDLING SYSTEM

The horizontal module handling system consists of the following main components:
- Deck rail system
- Tugger winches
- Module pallets
- Moonpool doors

The deck rails consist of steel HE-B beam sections laid out and bolted to the deck in a network system, to skid the modules on their pallets from their storage positions to the moonpool. The network has seven sidetracks for storing modules and tools.

Two 10 ton tugger winches are used as pull and holdback means for the skidding of the modules on their pallets. The winches are mounted on either side of the aft deck, welded to the deck supports.

The wires lead forward and can be rigged in various configurations, over and around specified roller guides to provide a pull and holdback arrangement to the module pallets in both lateral and longitudinal horizontal directions.

Module pallets are designed to support the various modules during skidding and storage.

The main moonpool door consists of two main 'I' beam steel sections supporting steel plate. The door is secured when closed by two door stoppers which are operated hydraulically from the moonpool door control console.

6.5 DESCRIPTION OF THE VERTICAL MODULE HANDLING SYSTEM

The vertical module handling system consists of the following main components.
-    Main winch

- Heave compensation system
- Guideline system
- Umbilical handling system
- Cursors
- Control and power pack containers

The main lifting winch has a single looped block with SWL of 65 tonnes and a depth capability in a double wire mode of 180 meters. The system can be used in a single wire mode with a SWL of 32.5 tonnes and a depth capability of 380 metres.

The heave compensation system consists of one active and one passive compensator. The two systems are completely independent of each other.

The active compensator alone is used for stabilising a freely suspended load. The load shall be between limits 5 tons and 20 tons. This compensator is used only during the last phase of a deployment, when landing the load.

The passive compensator is in effect a taut-wire device, in which the wire tension may be varied between 20kN and 200kN. During deployment the passive compensator is "dead", i.e. in its extreme outer position, corresponding to a low system pressure. When the load is situated on the template, the passive compensator maintains a low or moderate wire tension. Prior to a lift-off operation the system pressure may be increased, by pumping gas into the accumulator, to pre-tension the wire before lift-off is initiated.

The guideline system consists of four guideline compensators, four winches, a common accumulator section and a common power pack for the four winches.

The guideline compensator will maintain the tension between an upper and a lower limit of 1-5 tons as the vessel heaves.

A hydraulically powered friction sheave is mounted on the derrick top platform for umbilical handling. This system is designed to keep a constant tension in the umbilical during deployment or retrieval. The constant tension function is obtained whether the module is stationary at the seafloor or is being deployed/retrieved.

The cursor system consists of two cursor frames (upper & lower) and four vertical guide rails running down the derrick corner beams from the top platform down to the bottom of the moonpool.

The upper cursor function is to stabilize the main lifting block as it is moved up and down in the derrick and moonpool, and to prevent it from being affected by vessel and marine environment movements.

The lower cursor main function is to stabilise tools and modules connected to the main block as they hang in the derrick and pass through the moonpool.

The Control and Power Pack Containers contain all control and power facilities to operate the module handling system. The handling system can thus easily be mobilised and demobilised on the vessel according to needs.

7. Conclusive Remarks

Generally speaking, the project "Subsea Support Vessel for the 1990'ies" help Seaway to maintain the position of being in the forefront in underwater operations. Not only will we be ready to build a new vessel for future requirements when the opportunity calls for it, but we are also in a position to implement improvements on our existing vessels.

DIVING AND UNDERWATER SERVICES - AN OVERVIEW

JOHN WESTWOOD
Douglas-Westwood Associates
Whitebeams
Pett Bottom
Canterbury CT4 6EH
England

ABSTRACT.   In 1982 the diving and underwater services business became a billion dollar industry, a position it has never since regained. For the next four years sales declined, the crash of '86 only being part of this process. The period since 1987 has however seen an increase in revenues, with the industry probably turning over about $850 million in 1990.   The aim of this paper is to give an overview of the commercial development of the North Sea diving and underwater service industry and consider the evolution of the technologies now in use.   It concludes by commenting on the future business prospects for the industry set against a scenario of 'post Gulf war' oil prices.

1. Introduction

Over the past 27 years the North Sea has graduated from an outpost of the US offshore industry that employed a few divers to the world's largest market for subsea technology. But from the beginning a number of inter-linked factors have predominated:

- the short commercial lifetime of some technologies, an example being manned and diver lockout submersibles.

- the continual tendency for gross over-estimation of potential markets and a lack of awareness of alternative technologies, together leading to some commercial disasters.

- the recurring problem of over-supply, in virtually every case a new technology has reached gross (often 100%) over supply within 5 years of its commercial acceptance.

- the impact of diver replacement technology.   Between 1981 and 1985 the growth rate in ROV activity was ten times that in diving.

Volume 27: Subtech '91, 169–179.

- the considerable investment into non-productive 'R&D'.
The channelling of much needed funds away from much needed
product development into high profile 'blue sky'
projects.

The investment in hardware has been enormous. Over the
period some 94 dive support vessels have been built
worldwide, plus 22 MSV's and 17 submersible/ROV support
vessels, over 1,200 ROV's have been built, together with
scores of manned submersibles.

The oil price crash of 1985/6 at last forced the North Sea
oil companies into properly examining their costs and
defining their real need for underwater services. In the
past 5 years the offshore oil industry has matured, the
realisation having dawned that the industry is about
producing oil, not offering services, and the diving
companies have rationalised and begun their transformation
into underwater engineering contractors. What does the
future hold for these new groupings?

2. The Market - Forecasts And Reality

2.1 THE FORECASTS

> "Forecasting is always difficult,
> particularly if its about the future....
> so if you must do it, do it frequently".

Figure 1 shows the sales of the worldwide diving &
underwater services industry, together with some forecasts.
Figure 2 attempts to adjust these for inflation, to bring
everything to 1990 money values. It can be seen that in
most cases, forecasts made in the mid 1980's overestimated
the size of the 1990 business by 100%! Why?

Probably the greatest error made by forecasters is that
their forecasts of tomorrow are all too often based on
today's way of doing things. Secondly, it is easy to forget
the production of a forecast can in itself change the
future. For example, the forecasting of vast expenditures
in underwater inspection & maintenance caused the oil
companies to examine how the need for such expenditure could
be 'designed out' of future installations. Thirdly, the
real costs of many products and services should <u>reduce</u> with
time, not increase. This is the experience of many
industries.

2.2 OVERSUPPLY

The timescale from introduction of a new technology to its
gross oversupply has in most cases been around 5 years, or
sometimes less. In a number of cases the situation has been
exacerbated by a market downturn.

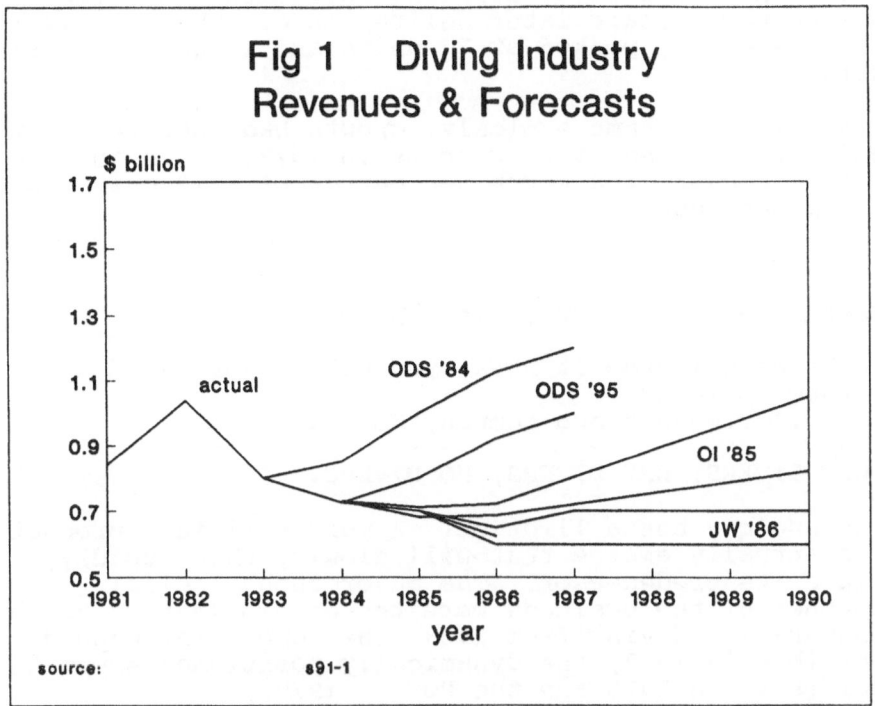

Fig 1   Diving Industry
Revenues & Forecasts

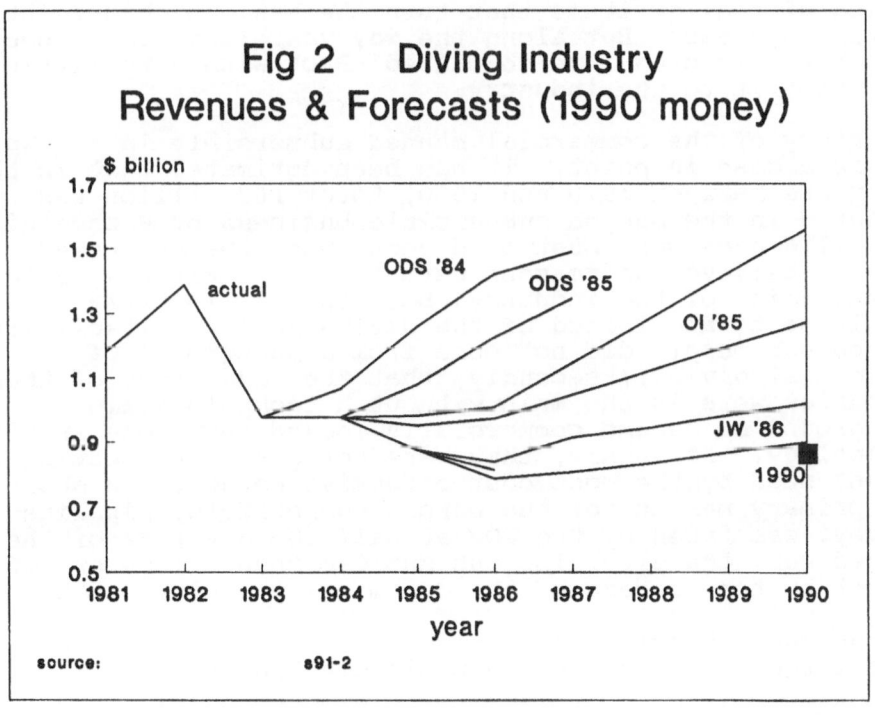

Fig 2   Diving Industry
Revenues & Forecasts (1990 money)

The first ROV support of drilling operations occurred in
1980, but three years later deliveries of ROVs, mainly for
drilling support, peaked at 77 units and market saturation
occurred.

Previously, this same explosive growth had occurred in the
manned sub business, from 2 units in 1973, to 22 in 1978,
when in the event the North Sea market of 1979 only had work
for 9 manned subs.

3. Technology - Blue Sky, Black Hole.

"We have looked at ROVs, you will never get
them to work"
........manned sub company MD. 1978.

3.1 WET DIVERS, DRY DIVERS, NO DIVERS.

Any technology has a lifetime. A more efficient product
will eventually emerge that will slowly, then rapidly,
displace its predecessor. The major technological
milestones of the business were the development of heliox
and saturation diving from 1963, the commercial manned
submersible in 1972, the dynamically positioned support
vessel (DSV) in 1975 and the ROV in 1979.
In addition there has been the continuing development of the
myriad of smaller items that turn the basic machines into
working systems. But along the way vast financial resources
have been directed into so called "R&D" with very little
real benefit to the industry.

The story of the commercial manned submersible in the North
Sea is a case in point. It has been estimated that in the
ten years between 1970 and 1980, "over £60 million was
invested in the manned submersible business by 6 companies
and all 6 have gone bust or discontinued the business".
Why? I believe the reasons were twofold. Firstly, a lack
of knowledge of the industry, both the oil industry
requirements and indeed of the diving business itself (the
manned sub people did not come from a background of
commercial diving); secondly, that the management of these
companies were in the main without a technological
background so became commercially locked in to one specific
technology. So in all, they were not geared to meeting the
client need by the most cost-effective method. In short,
the primary market for the manned submersible, pipeline
survey, was taken by the ROV at half the day rate of the
manned sub; the diver lockout sub was then driven out of the
market by the success of the dynamically positioned DSV.

Autonomous submarines have also died a death, after having
very large sums invested in their development.

3.2 RESEARCH AND DEVELOPMENT.

So we see a continuing interplay between the market place
and R&D. Over the years a great deal of oil company,
service company and government money has been put into R&D.
In any review one must question the value of much of this
investment.

It is interesting that many of the quantum leaps in
technology did not realy come from the diving companies of
the day where in most cases the 'plumbing' thinking was in
command. The commercial manned submersible however, had its
origins in ocean research and naval submarine builders.
Similarly, commercial ROVs came from US navy developments
and parallel civil work in the UK; the diving companies DSVs
relied upon the positioning technology first developed to
position drill ships. The diving companies were (with some
exceptions) notoriously slow to take up the ROV, seeing it
as a competitor, rather than another tool. So it was others
who received the lions share of the R&D funding.

Much government funding has been absorbed by the defence
contractors and other large groups, mainly due to the fact
that they have the organisation necessary to apply for the
funds and it is easier for government to deal in a few large
units of funding than many small ones. The large defence
contractors have been notably unsuccessful in the commercial
underwater technology business. Again, much of this is
associated with a lack of knowledge of the market and what I
will term the 'private venture' syndrome, (private venture
is the defense industry expression for the very unusual
experience of funding their own product development). But
the success of such groups in 'R&D fund mining' has been a
factor causing a virtual 'funding famine' amongst the many
small companies that supply much of the industry's
specialised high technology.

Most importantly, these other organisations, such as the
defence contractors and in their day the manned sub
operators, did not have the diving companies knowledge of
customer needs or the opportunity to apply alternative
technology to routine diving jobs. So it was a then diving
company, Sub Sea Offshore, who opened up the largest market,
drill support, to ROVs with operations on a BP drilling rig.

3.3 BRIBERY AND BROWNIE POINTS.

No review of the last 25 years could be complete without a
comment on the 'Brownie Point' system which has 'encouraged'
much oil company money into work on diverless subsea
production. As discussed later, the UK will have many
subsea completions in future years, but virtually all will
be within diver depths. In a recent straw poll we did not
find a single oil company with plans to install a diverless
subsea production system in the UK sector. Good news
perhaps for the UK diving companies, but bad news for export

prospects of exporting technology, as tomorrow's
developments in deeper foreign waters will be designed for
installation and operation without underwater
intervention.

What will the North Sea companies have to offer for these
deep water field developments? The industry constantly
seeks proven low-cost technology and although a number of
experimental ideas have been researched in the North Sea,
deep water diverless systems will be developed and proven in
other areas such as the USA and Brazil. To assure a share
of future markets, what is needed is not just the spend on
R&D, but more government efforts into getting prototype
hardware into the field, where it can be proven prior to
international marketing.

3.4   TECHNOLOGY

The extent of technical progress can be judged from a recent
Subsea International advert claiming 'more than 3,000 hours
ROV bottom time, more than half at depths below 1,500 ft'.
As part of the HYDRA programme, Comex are planning a record
manned dive to 2,300 ft. Important as depth may be, the real
challenge for cost-effectiveness lies in shallow water where
we have seen a proliferation of low cost ROVs, and Perry
building a tracked vehicle for use on shallow water
pipelines in the Gulf of Mexico.

In subsea production concepts, cost effectiveness and
simplicity have replaced the grand projects conceived before
the post oil price industry shake-out. Modular systems, in
which whole units are changed have given way to the idea of
replacing individual high risk items by ROV or ROT. Again,
this will eventually result in intervention technology that
is cost-effective in diver depths.

So what next, where will the industry go in the next 25
years?

4. The Future.

The factors that will shape future business can be regarded
in three groups:

        THE MACRO FACTORS

        THE GROWTH MARKETS

        THE CORPORATE RESPONSE

4.1   THE MACRO FACTORS

OIL PRICES - much more can be said on this subject, but to
us here today the most important factor is that the industry
is contemplating a future with very little real increase in
oil prices, perhaps in real money terms $20 might prevail
until the end of the century.

SAFETY - beyond the individual problem solving approach we
must question the very need to design offshore installations
for permanent manning. Much has been said on the subject of
offshore safety, but one question remains - is it reasonable
to expect people to have to work in an environment that has
the proven potential for absolute catastrophe, if we can
develop the technology to run offshore oil and gas
production from the shore? The way ahead must be in
combination of subsea developments and unmanned platforms,
delivering products to existing infrastructure or direct to
the shore.

FRONTIER GIANTS - It is logical that the potential for
another 'giant' field find must be greater in the deep water
frontier areas of the world than in the highly explored
shallow water areas. Earlier this year, BP & Shell
confirmed a major find on their Mars prospect in the Gulf of
Mexico. Situated in 3,100 ft of water, 130 miles offshore,
BP say it is potentially the largest prospect it has
participated in since the discovery of its major North Sea
fields.

EXTENDING THE DISTANCE - Overall, more than 70% of the North
Sea's discoveries are within 35km of existing
infrastructure. Current satellite developments extend up to
15 or so km, so a relatively small increase in flow
distances could allow economic access to a lot of oil &
gas. The enabling technologies of Multiphase Flow and
Subsea Separation are making steady progress.   The
implications of this are that the day is approaching when
the virtual elimination of the offshore platform can be
contemplated. A number of multiphase pumps are on trial
onshore or topsides in Italy, the Netherlands, the UK,
Sarawak and other locations, and in Europe alone at least 20
oil companies are funding developments of subsea multiphase
pumps. Subsea multiphase pumping is moving closer to
reality and very soon we should see experimental pumps
installed on subsea wells where the risk of interruption to
a significant cash flow is low. The major requirement is
now to solve the problems of powering them, both in terms of
the electric motors and the costs of the very long high
power cables.

SUBSEA COST REDUCTION - The stripping away of the frills is
reducing the costs of subsea developments and making them
more competitive with platforms. There are about 200 subsea
wells in the North Sea which accounts for about half of the
world's producers. The OSO forecast a further 350 over the

next 10 years giving a market of £350 million. They also
reported as stating that because supply just about matches
demand, the major manufacturers would seem to have an
assured future where they 'do not compete on technical merit
but in terms of approach, putting the pieces together to
come up with the most efficient solution'. So we see
developing a set of standard subsea building blocks, with
the design effort being in their configuration.

It has been estimated that in 1991 money Shell's 9 well UMC
development would cost £1 billion, whereas a comparable
modern diverless deepwater development would cost £125
million. And the costs differences between 'diver assist'
and 'diverless' will continue to reduce. (A Mentor
Engineering comparison between a diver assist and a
diverless development gave about a 25% difference in 1990).

Subsea development costs will continue to see some reduction
both in terms of capital and operating costs. CAPEX will
benefit from:

- 'catalogue engineering', as mentioned above, more use of
  off-the-shelf products and less special one-off designs

- standardisation, which will reduce hardware costs

- reductions in drilling costs as automation of the drilling
  process increases efficiency by reducing rig time per
hole.

OPEX will fall as the greater use of standardised components
gives higher reliability and reduces the logistical problems
and allows subsea field maintenance to be contracted out on
a 'turnkey' basis.

GROWTH IN GAS - Europe could face a substantial shortfall
in gas supplies by the end of this decade as the emphasis on
natural gas continues. Political and environmental concerns
have made gas the fuel of the future. Currently some 16% of
Western Europe's energy supplies come from natural gas, but
demand will rise sharply over the next decade. There are a
number of possible suppliers, including Norway, where a 15
year supply contract for 2bn cu m per year contract has been
signed between the Norwegian Gas Marketing Consortium and
the UK's National Power.
Then there is gas from Libya, Tunisia and Algeria - a new
pipeline is under study from Morocco to Spain for completion
in 1995. The Soviet Union is another supplier but with its
own problems.

Further afield, and much longer term, there is the much
discussed idea of the TransAsean system, a $10 billion,
5,000 mile gas pipeline linking producers and markets in the
countries of Thailand, Indonesia, Malaysia, Brunei,

Singapore and the Philippines. Feasibility studies are being
conducted by a Franco-Italian consortium.
In all a bright longer term future for pipelines.

4.2   THE GROWING MARKETS

A GROWTH IN SUBSEA PRODUCTION   - the Subsea Production
Yearbook review of worldwide subsea field development
prospects identifies 16 for 1991, 33 for 1992 and 40 for
1993, with overall, Petrobras having nearly twice as many
prospects as any other operator.

DECOMMISSIONING - North Sea platform decommissioning is seen
by some as the next boom market. CNWWM recently gave the
period 2005 to 2015 as the period when the main wave would
occur and estimated costs for decommissioning of all current
and planned fields as £4.7 billion in 1991 money. But does
anyone really believe that the oil companies are going to
spend such vast sums of money to recover a load of scrap
iron in a situation when there is no cash flow against which
to offset these costs?   The platforms may well be removed,
but methods will be developed that will effect
decommissioning at much lower costs than those currently
forecast. Again a situation will occur that the forecasts
of great expenditure will change the future.

BOOMING PIPELINE MARKET - as discussed above, the next few
years will see a booming market and there may also be a
replacement market coming, nearly 1,300km of North Sea line
will be between 11-20 years old by 1995.

SMALL MARKETS - these constantly emerge. For example, 25% of
the US National Bridge inventory have no design or as built
plans which can be used to determine susceptibility to
scour. A Federal five year mandate on underwater bridge
inspection programmes is expected to nett millions of
dollars annually for engineering and diving companies and
the ADC has instituted a training course. This year the US
National Cooperative Highway Research Programme went out for
proposals to develop functional equipment to determine
subsurface bridge foundation characteristics.

INTERNATIONAL - the North Sea market will eventually decline
and of all areas, we expect the Far East to have the most
rapid growth.

4.3   THE CORPORATE RESPONSE

Since the mid '80s a number of mergers have seen the
consolidation of the worldwide supply of subsea wellhead
equipment into four major companies, Cameron, Vetco Grey,
FMC and National. Since then there has been a period of
fairly rapid growth for the underwater contractors caused by
an increase in subsea developments, plus the ESV programs,
plus platform refurbishment work.

The major diving companies are now well through their
metamorphosis into underwater contactors, offering complete
packages of turn-key services; engineering, divers, ROVs and
the diving support vessels. The technological impact has
been that oil companies 'scopes of work' have slowly been
geared towards the turn-key resources of these major
contractors. (So we may be reaching a situation where the
small independent diving vessel owners may slowly, but
surely, be forced out of the market).

Formation of large corporate groups is likely to be based
either on installation vessel capacity or corporate muscle,
for example, the evolution of Stenna and Rockwater, the
development of Northern Offshore. (It is interesting to note
that the three North Sea cable burial specialists are all
parts of major corporations, NOS is Cable & Wireless; UDI is
GEC Marconi; and BT Marine is British Telecom). Meanwhile
the Norwegians will probably continue to confuse the rest of
us with their continual rounds of corporate marriages and
divorces.

Most importantly the EPIC contracts are being announced:.
- As part of a $40m turn-key contract, Comex 'engineered,
procured, installed, commissioned and delivered a field
onstream to Total Indonesia (Tuna Tambora)'.

- Agip's award of the first major (£30-35m) engineering,
procurement, construction and installation contract to
Cameron Offshore Engineering for the seven wells of the Toni
field.

The oil companies will also change, with smaller in-house
engineering groups acting in a management role, contracting
out the engineering.

5. Conclusions

My own experience of the past 25 years has shown that in the
underwater business:

1. The starting point is to first define your market.

2. Then develop an understanding of the forces driving that
   market. Then, most importantly,

3. DEVELOP YOUR NEXT MARKET .....

In the final analysis, the underwater industry has shown
itself capable of adapting to new challenges. The manned
submersible business of the early 1970's led to commercial
ROV's of the 1980's, then the platform inspection vehicles
of the 1990's. Some of my own original team are now
planning the decommissioning of nuclear power stations using
their experience of remote operations. Now Oceaneering have

a company 'Oceaneering Space Systems'.   Comex have been
awarded several contracts by the European Space Agency to
train astronauts for extra vehicular activities.

"And then develop your next market"........

ADVANCES IN ROV TOOLING
FOR SUBSEA CABLE REPAIR AND MAINTENANCE

AUTHOR : I D BONNON
BT (MARINE) LIMITED

ABSTRACT. Recent years have seen significant advances in ROV tooling techniques for the successful repair and maintenance of subsea cables. This paper outlines the developments which have been made, with particular reference to the advances which have been made to burial systems which are currently in use today. These burial systems allow cables to be rapidly buried, either during a repair process, or in post lay burial mode. The primary system is the use of water jetting systems. These ROVs can work in water Depths down to some 2,000m, and cut trenches upto 1m deep in soil strengths of upto 600 kpa, which is virtually a soft rock. This allows for cables to be successfully buried in the seabed, providing adequate protection from hostile seabed activity, such as fishing trawler boards or dragging anchors.

The development of the advanced burial systems has gone hand in hand with new technology developments in sensing and positioning equipment, which is allowing for precise and efficient repairs and maintenance of subsea cables. Subsea machines are becoming more reliable and durable and can be rapidly mobilised to undertake maintenance operations in as little as a 24 hour period. The proven success that the repair type ROVs have had of late has set up a new generation of rapid post lay burial machines. These machines bury surface laid cables and pipes by use of jetting tools or mechanical excavators. The advances are saving cable and pipe owners and users significant sums of money in terms of zero outage time and the resulting loss of production or communication downtime.

1   Introduction

Virtually all of today's telecommunication and offshore umbilical and power cables will be buried wherever there is a risk of damage occurring. Submarine cable ploughs have been developed which sufficiently bury cables to a minimum depth of 600mm. Figure 1 shows a typical BT (Marine) Limited Plough and Figure 2 shows how a wedge of soil is lifted by the plough and the cable deposited at the base of the trench, before the wedge is then lowered to rest on the

181

BT(MARINE)LTD   PLOUGH **1**

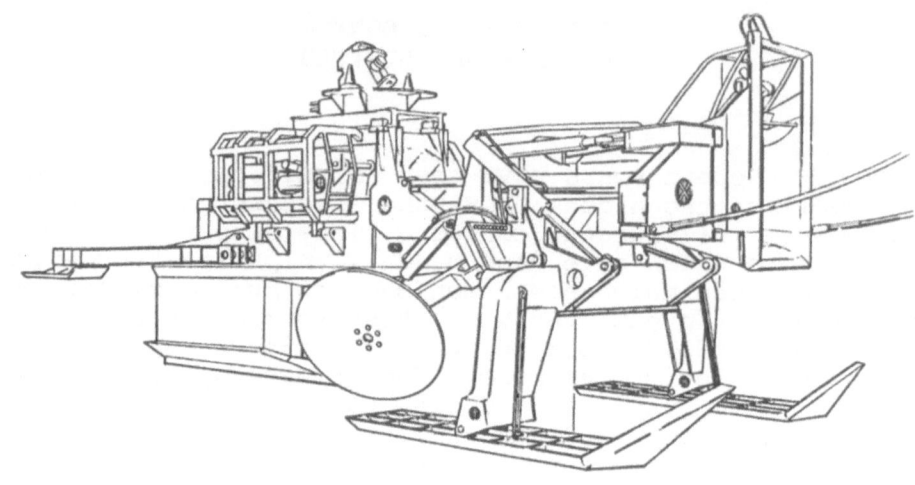

FIG   **1**

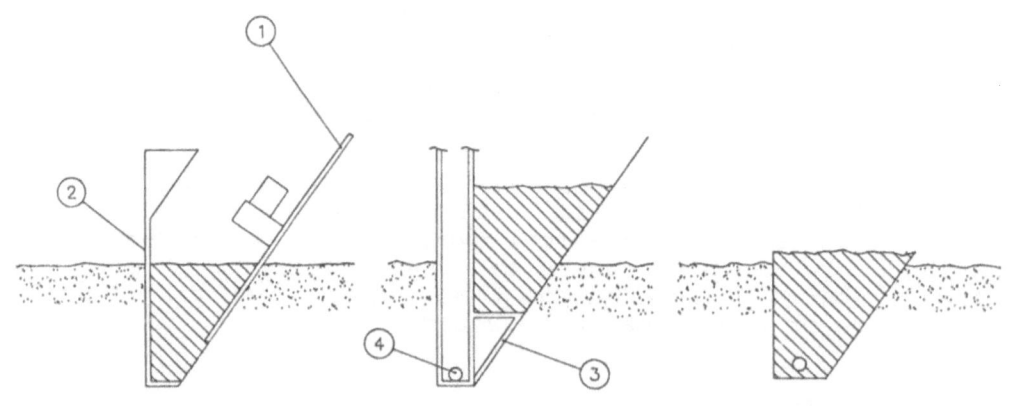

1.   A TRAPEZOIDAL PRISM OF SOIL IS CUT BY THE FREELY ROTATING DISC CUTTER (1) INCLINED AT 35 DEGREES FROM THE VERTICAL, AND THE SHARP VERTICAL KNIFE (2)

2.   THE PRISM OR WEDGE OF SOIL SO DEFINED IS GENTLY LIFTED FROM BELOW BY A RAMP OR HORIZONTAL SHARE, (3) RISING ALONG THE SLOPING SURFACE CUT BY THE DISC, AND THE CABLE (4) IS PLACED UNDERNEATH IT.

3.   THE WEDGE OF THE SEABED THEN RESTS ON THE CABLE.

FIG. 2

cable. Whilst the risk to a cable is minimised once it has been buried, there is still a chance of damage occurring, particularly from an engaged ship's anchor which can penetrate upto several metres. The type of repair which will then have to be undertaken will usually depend on the depth of water where the fault has occurred. As well as faults occurring from dragging anchors, it is possible other types of faults can occur. There are sites around the world in the deep ocean floor where volcanic activity takes place, which can result in damage to cables. Earthquakes and other forms of seismic activity can also damage cables. In addition, if a cable is laid on a very rugged rock surface on the seabed, there is always the risk of the cable chafing. Latent manufacturing defects can also be another cause of failure.

As previously noted, submarine cables are generally buried for long lengths by towed plough systems. However, the use of post lay burial devices, has of late, become increasingly popular. This is due to their produce friendly handling nature and low cost. The devices are also used for inspection and repair of subsea installed cables and umbilicals, plus the burial of small diameter pipe systems.

2   How ROVs Undertake Subsea Repairs

In order to undertake a successful repair on a buried section of subsea cable, an ROV system would adopt the following procedure to complete the work. Figures 3 to 8 show a typical repair sequence for the SCARAB III repair and maintenance vehicle. The procedure is outlined in more detail below.

2.1   CABLE DETECTION

Following deployment from a cableship, the ROV would locate the cable, usually by detecting a tone placed on the cable. Cable detection can now be undertaken with cable tracking in both AC and DC modes. In addition, new systems allow for cables to be detected in Passive mode. Providing there is typically $25mm^2$ of steel in the cross-section of the cable, usually derived from the armouring, then a cable can be detected. Obviously pipes are relatively easy to locate and track.

To detect the fault the ROV tracks along the cable until a fault location is detected. This will be obvious in the perceived broken signal strength on the cable locator system. In addition, it may also be possible to use the

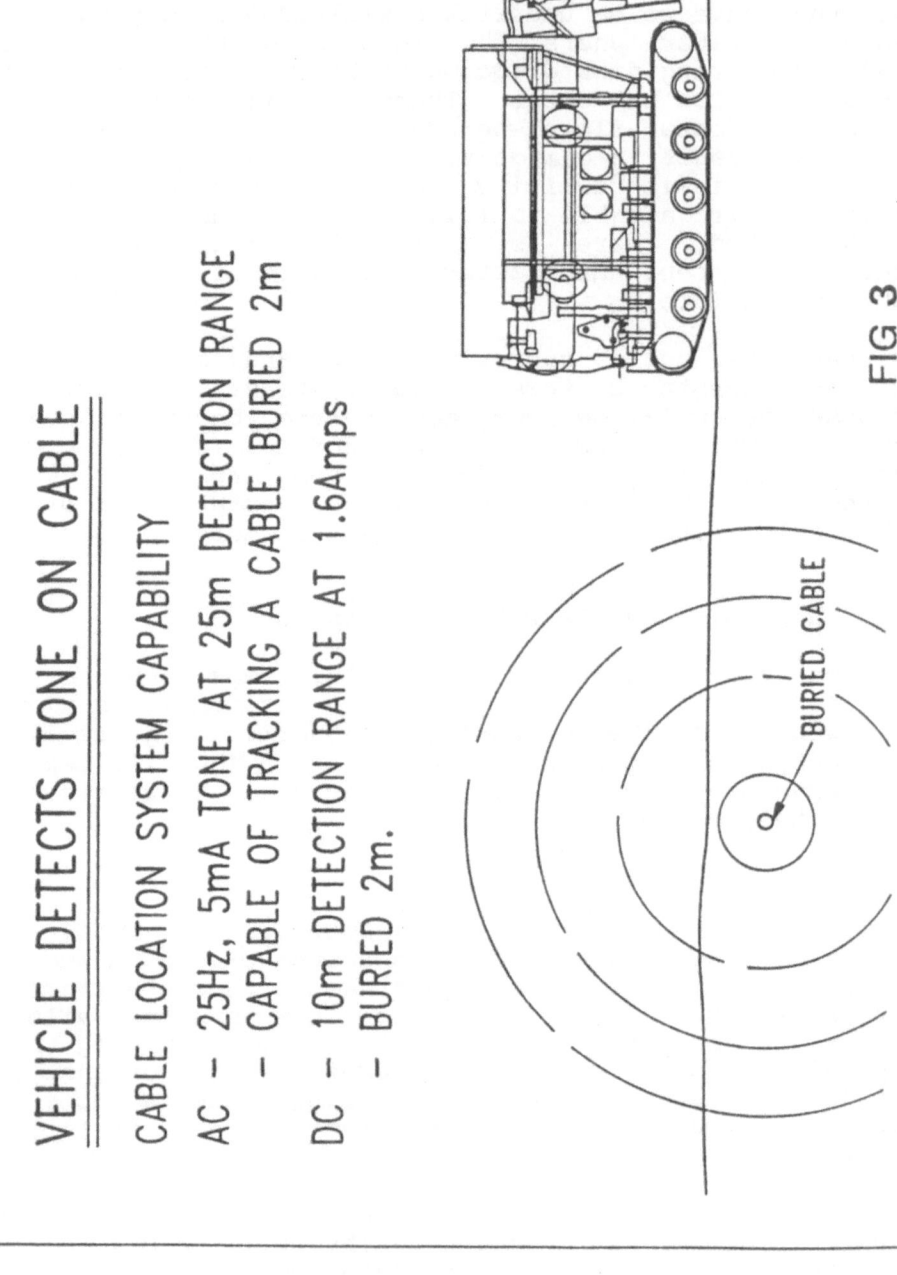

VEHICLE DETECTS TONE ON CABLE

CABLE LOCATION SYSTEM CAPABILITY

AC — 25Hz, 5mA TONE AT 25m DETECTION RANGE
   — CAPABLE OF TRACKING A CABLE BURIED 2m

DC — 10m DETECTION RANGE AT 1.6Amps
   — BURIED 2m.

BURIED CABLE

FIG 3

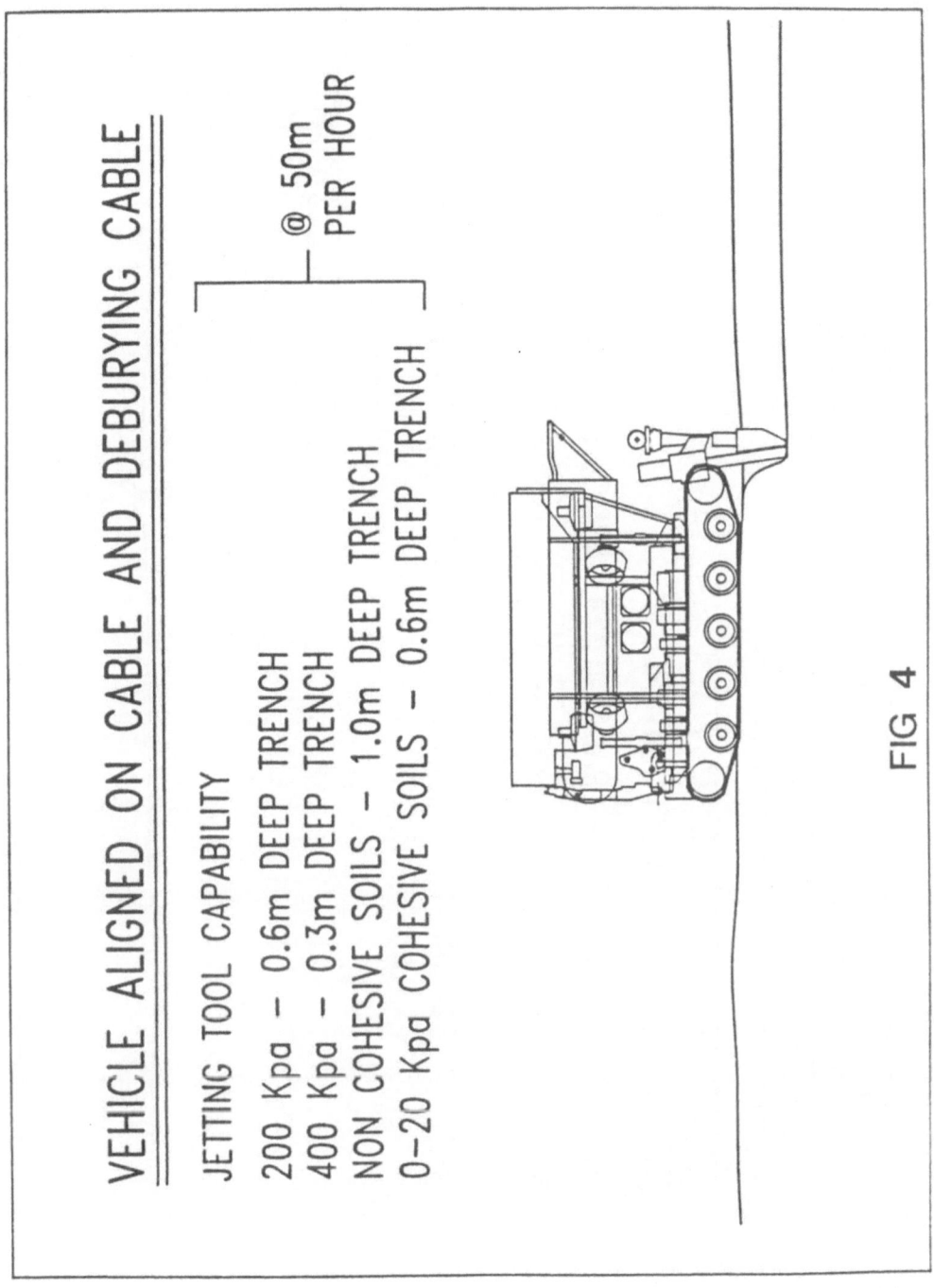

VEHICLE ALIGNED ON CABLE AND DEBURYING CABLE

JETTING TOOL CAPABILITY

200 Kpa – 0.6m DEEP TRENCH
400 Kpa – 0.3m DEEP TRENCH
NON COHESIVE SOILS – 1.0m DEEP TRENCH
0–20 Kpa COHESIVE SOILS – 0.6m DEEP TRENCH

@ 50m PER HOUR

FIG 4

VEHICLE APPROACHES TRENCH & DEPLOYS CABLE

CUTTER & CABLE GRIPPER

TOOLS ARE DEPLOYED BY 2 OFF SEL TA9
SEVEN FUNCTION MANIPULATORS.

GRIPPER CAPABILITY – RANGE OF UNARMOURED AND ARMOURED
CABLES UP TO 110mm DIA.
23000 Kg LIFT.

CUTTER CAPABILITY – ALL TELECOMS CABLES TO 110mm DIA.

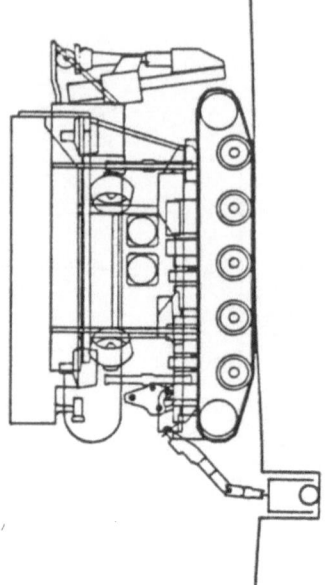

FIG 5

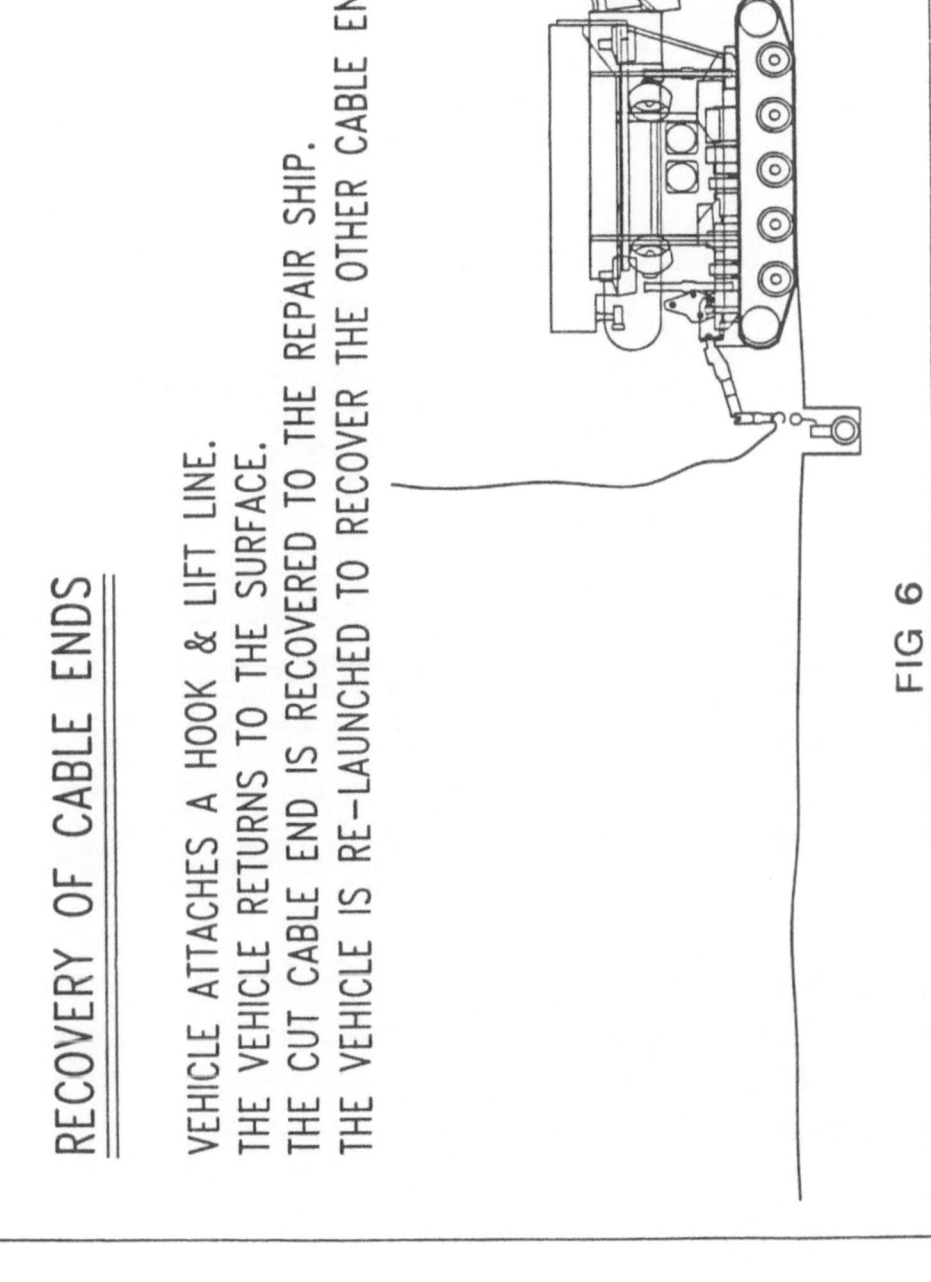

RECOVERY OF CABLE ENDS

VEHICLE ATTACHES A HOOK & LIFT LINE.
THE VEHICLE RETURNS TO THE SURFACE.
THE CUT CABLE END IS RECOVERED TO THE REPAIR SHIP.
THE VEHICLE IS RE-LAUNCHED TO RECOVER THE OTHER CABLE END.

FIG 6

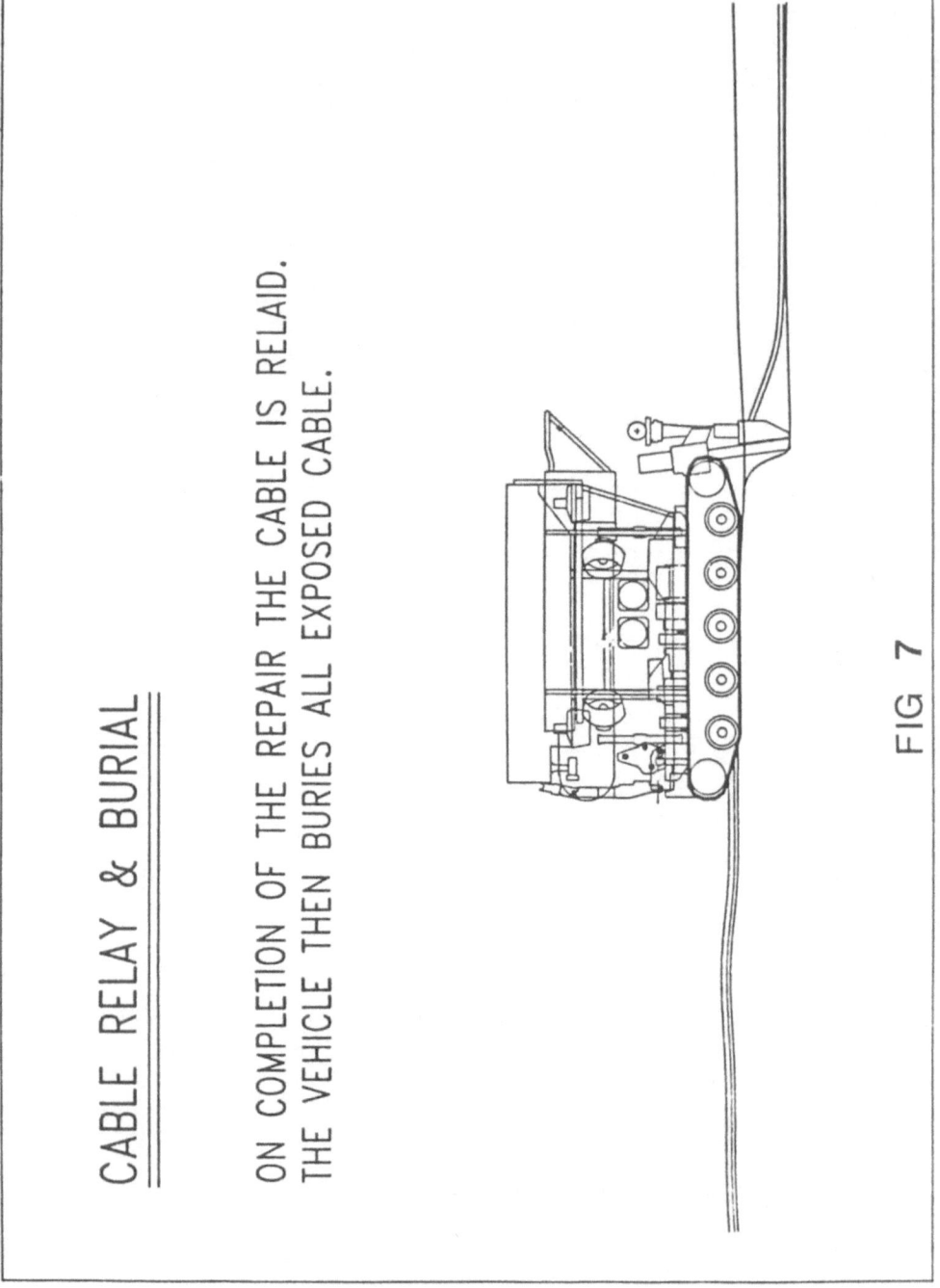

CABLE RELAY & BURIAL

ON COMPLETION OF THE REPAIR THE CABLE IS RELAID.
THE VEHICLE THEN BURIES ALL EXPOSED CABLE.

FIG 7

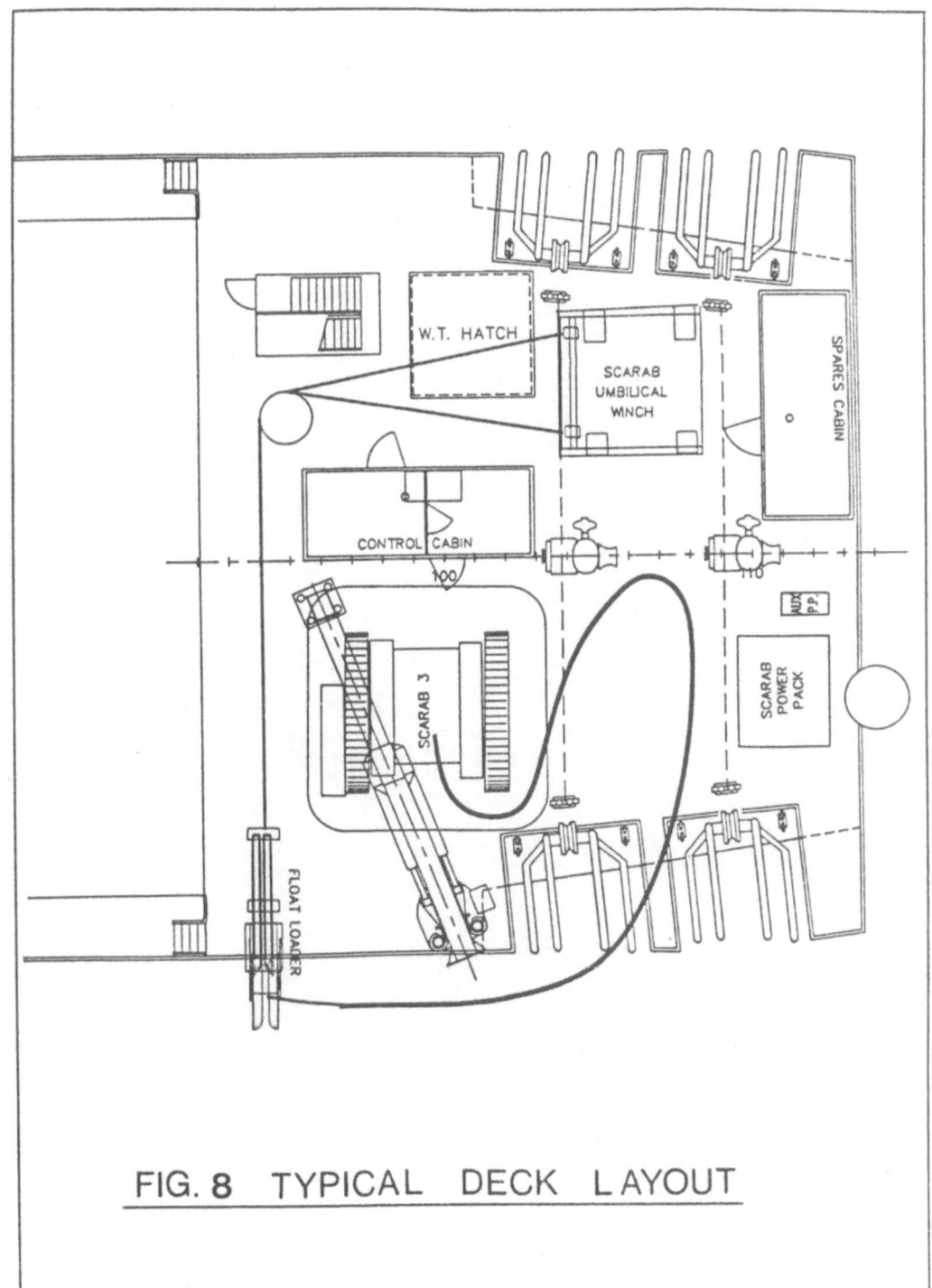

FIG. 8   TYPICAL   DECK   LAYOUT

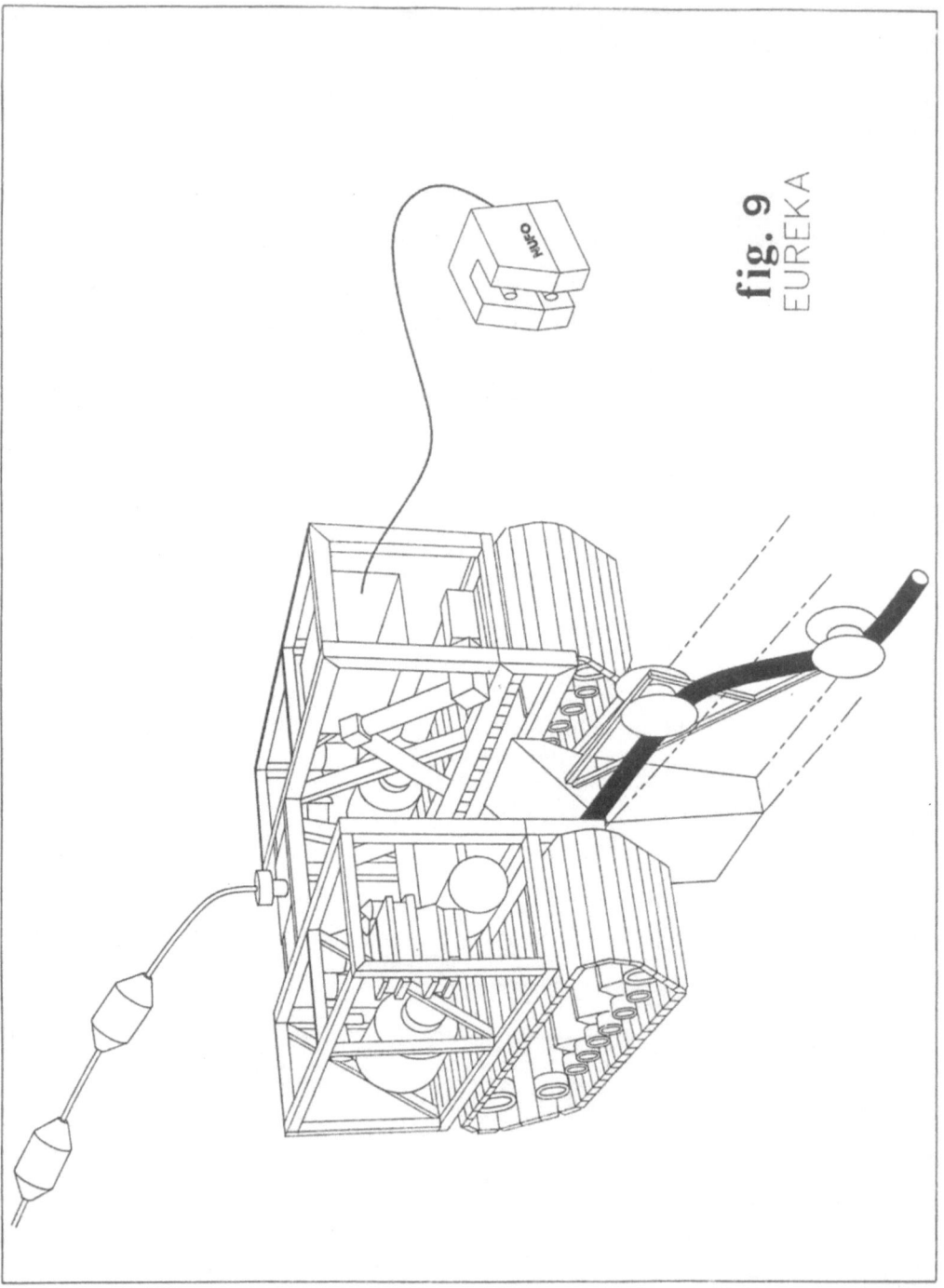

fig. 9
EUREKA

surveillance equipment on the vehicle to look for evidence of anchor or trawler scars by using the cameras or sonar equipment.

2.2   DEBURIAL OF CABLE

Once a fault has been detected, the ROV will then proceed to debury the cable in the vicinity of repair, typically using the jetting tools fitted to the vehicle. Burial and deburial can be undertaken in soils with strengths upto 600 kpa, which is virtually a soft rock. It may be necessary to excavate along the cable route at certain distances, away from the fault location, as to ensure that when the cable ends are recovered to the cableship there are no excessive strains induced into any optical fibres or conductors within the cable.

2.3   CABLE RECOVERY

With the cable exposed, the manipulators on the front of the vehicle would usually deploy a stored cable cutter. This would then make a single cut in the cable. The manipulator arm then typically replaces the cutter into it's storage position and removes the cable gripper. The gripper would then be attached to one end of the cable. It is then possible to attach a lift line from the cableship to the gripper and then one end of the cable is recovered to the cableship. This process is also repeated for the other end of the cable.

2.4   CABLE REPAIR

With both ends of the cable now onboard the cableship, testing and jointing to repair a bight of cable can be undertaken. Both ends of the cable are tested to check for faults in both directions, and once a faulty section has been identified and cut out, a repair bight is spliced into the cable system. Once this process is completed, the repair bight is carefully lowered to the seabed.

2.5   CABLE BURIAL

Obviously the repaired section of cable now lies as a surface bight of cable on the seabed. If a fault has occurred previously at this location, then the risk of repeated faults is very high. Therefore, every effort is made to ensure that the cable is reburied on an expedient basis. This is achieved by redeploying the ROV system, which once again locates itself on the cable using it's locator system, and then proceeds to bury the section of repaired cable.

2.6   TIMESCALES

Most ships are capable of mobilising a repair spread as
outlined above in a period of time less than 24 hours. Once
at sea, providing weather parameters are reasonable, a
complete operation as described above can be undertaken in a
24-36 hour period. Within the last year a very successful
repair operation was undertaken on a power cable in the
North Sea using a technique very similar to the one outlined
above using the BT (Marine) Limited Trencher vehicle.

3   Applications of Post Lay Burial

Following the successful development and application of the
specialist ROV intervention tools, a new generation of post
lay burial type devices have evolved which specialise in
both post lay burial and also still retain a repair
capability. These devices have used the advanced burial
technology which has evolved over the years in the ROV
systems and now complements primary plough installation
methods.
Post lay burial can be particularly advantageous in the
following areas :

- Offshore umbilical installation
- River and Estuary crossings
- Low cost installations
- Shore Ends
- Non Ploughed areas

One of the most difficult technological problems of late,
whilst installing offshore umbilicals, has been the ability
to use a burial device capable of working from the J tube of
one platform, right up to the J tube of another platform.
If ploughed systems are used, there is always a section of
cable which has to be surface laid in the close proximity of
the J tube, which then has to be buried either by the use of
divers, or alternatively, covered with expensive matting. A
post lay burial device which has the capability to work
close into platforms safely and efficiently, obviously
minimises the cost of the umbilical installation. Then,
providing the rate of progress of the burial tool is of
sufficient speed whereby it can be competitive in relation
to ploughing operations then the overall installation
becomes very attractive.

There are an increasing number of cable installations where
the requirement is to cross a river, an estuary, or lake,
which ensures that the shortest route between two points on
land is obtained to minimise the cost of an overall cable
installation. Therefore, with a device which is capable of

burying cable from a beach, through the water and up the beach on the other side, in one simple, quick and efficient operation, provides a significant benefit in terms of installation costs.

Worldwide, with the production of new fibre optic communications systems which are of low cost and of a dependable nature, has led to pressures for low cost installations. This allows subsea cable systems to be competitive with land based microwave systems. This occurs particularly when making coastal loops, or inter-island communication links. One favoured procedure therefore is to use the relatively expensive cableship spread to lay the cable and then complete the installation with a much cheaper vessel, deploying a post lay burial device, which can then successfully bury all sections of cable, including the lengths of cable which may be laid through the surf zone and up the beach. A unit capable of doing this operation is of great benefit to the cable installation process.
In addition, if the whole spread can be mobilised on land on one side of the crossing, the device can then work it's way across the span of water, and once the installation is complete, disengage itself from the cable and then return to the original shore station.

The burial of cable shore ends has always been a difficult process. Normal procedure has been to use the towed ploughs in as close as inshore as they can get, and to also bury the land based portion of cable with civils type equipment. It was then possible on certain occasions to revisit the site with a trenching device and attempt to bury the section of cable between low tide mark and start of ploughing operations. One limitation would often be that the trenching machines would have to be submersed in water, therefore again there would tend to be a length of surface laid cable, from the low tide point down to the minimum operating depth of the trenching machine. A device which was capable of working from the completion or start point of ploughing, right through the shore end area, which would consist of subsea, surf zone and then up the beach, has obvious benefits. Especially, if the operation can be totally conducted off a cheap vessel such as a barge, rather than an expensive cableship, or DP vessel.

During a normal ploughing operation, the plough has to be recovered at certain points along the route. This will typically occur either where there has been a need to replace the wear points on the plough, or alternatively, where the plough has to be recovered due to bad weather or any other operational restrictions. In these areas there may be evidence of significant hostile seabed activity, ie trawler boards etc. Therefore, it is essential that these sections of cable be buried to afford the overall cable system the maximum level of protection. The solution for this problem is to use a post lay burial device, again off a potentially cheaper vessel than a cableship, in order to complete the operation. The device can therefore survey the route and bury any areas of cable which are exposed on the surface.

It also may be possible to have a totally integrated spread whereby the post lay burial device accompanies the plough system to do an operation in it's totality. This would mean that the plough could be used to it's fullest extent during the cable installation from point A to point B, and on the return of the vessel from point B to point A, the post lay burial device could be deployed to bury any surface sections of cable, so that the complete cable system can be installed in the most shortest possible timescale. This option may be slightly more expensive by tying up the expensive cableship for certain periods of time to do post lay burial operations. However, the mobilisation of a second vessel is negated.

4   Post Lay Burial Machines

The BT(M) Eureka system, as illustrated in Figure 9, is a good example of a purpose built post lay burial device, which satisfies all the anticipated applications for post lay burial. The system has a hybrid jetting/plough tool which allows the product, which may be cable or small pipeline, to be buried safely and at a very efficient rate of progress. In event of a total power failure the Eureka system can be lifted clear of the product by the dedicated surface support handling system, without damaging or putting any force onto the product. The tool utilises jetting power to effectively cut into very stiff seabeds, ie clays upto 200 kpa and then uses the 5 tonne tractive force which is generated from the subsea tractor unit, to effectively remove a wedge of soil and to place the cable or small pipeline in a trench. Then, there will be certain amount of tendency for natural backfilling which will then afford the product maximum post burial protection. The other novel concept of the Eureka system is that it can also be directly interfaced to a mechanical cutting device. This therefore allows efficient work to be undertaken on land where a water supply may be a problem, and supplements the jetting tool so that all environmental conditions can be coped with.

It will be possible to handle the Eureka system both on

land, through the surf zone, and down to water depths of upto 1,000m.

Eureka consists of a subsea tractor unit controlled from a host ship by means of an umbilical tether. It is also capable of undertaking sustained subsea burial operations for distances typically upto 20km before recovery for maintenance.

The system comes with it's complete dedicated handling system, which allows it to be mobilised rapidly upon any suitable vessel of opportunity. It has all the necessary sensing and control features, which allow it to undertake sustained operations subsea. Eureka will also be expected to find subsea cable, both buried and surface laid, uncover buried cable, bury and rebury cable and to also perform visual, electronic and acoustic inspection of all types of surface laid cables and pipes. Eureka also has the capability of taking pertinent visual and data records during any operation. It's typical progress rates are as follows :

        Tracking speed with no burial
        being undertaken   3,000m/hr

        Burial speeds -
        Jetting tool deployed
        to 1m depth, product engaged

        Sands   500m/hr
        Gravels   400m/hr
         Clays upto 20 kpa   350m/hr
        Clays upto 50 kpa   300m/hr
        Clays upto 100 kpa   100m/hr
        Clays upto 200 kpa   50m/hr

The system is also versatile and flexible, such that it can be adapted for other future uses.

5   Conclusions

The advances which have been made in ROV tooling for subsea cable repairs and maintenance has yielded significant benefits for the cable owners. ROV intervention for repairs and preventative maintenance has ensured that cable systems, once installed, are achieving high degrees of serviceability and reliability. The advances in burial technology which were achieved in developing ROVs for repair and maintenance has now been onpassed to the post lay burial type ROVs. These burial systems are now offering a new service to cable owners, in as much as that the installation method allows low cost and reliable installations which can challenge more traditional land based communication methods.   The key to the success of these post lay burial ROVs in the future will almost certainly depend on their durability and reliability. In the next few years where the systems are extensively used in both arduous conditions and for greater lengths of time in more harsh environmental conditions, they will have to achieve a track record which impresses the owners of the cable systems.   This record will not only have to demonstrate that cable systems can be buried, but also that they must be buried to the prescribed depth of burial and at a rate of installation commensurate with the installation budgets for the cable systems. The technological advances have now been implemented and their successful usage and application is awaited.

THE APPLICATION OF WORK CLASS ROV'S

An overview of present capability and one Companys view on the way forward.

George S. Robertson
Sonsub North Sea Ltd
18 Farburn Terrace
Dyce
Aberdeen   AB2   0DR

ABSTRACT

Sonsub North Sea Ltd is a part of the Sonsub group of companies which has been involved in the application of remote systems technology to the subsea environment for a number of years.

Primarily, Sonsub has been supporting the various requirements of the Offshore Oil and Gas Exploration and Production industry.   The use of robotics and other remote systems technology to successfully complete work tasks in the subsea environment, has presented challenges that have required an innovative approach to the use of telerobotics.

Remotely Operated Vehicles (ROV's) are a form of robotics that have been utilised in the Offshore Oil and Gas Industry since the early 1970's. However, until recently, their use has been restricted to observation and light manipulator work, utilising a variety of rate feed, spatially correspondent or tactile feedback manipulator systems.

The Challenge remains to develop systems that have the ability to carry out human equivalent tasks presently performed by divers, thus enabling personnel to be removed from this hazardous environment.   There are numerous difficulties to be overcome in meeting this challenge and this paper addresses a particular philosophy and methodology adopted by one company to service the needs of its customers.

The Problems

Taking telerobotics and remote handling technology into the subsea environment presents its own unique problems. The most obvious problem is ensuring that all electric, electronic and mechanical components are suitably protected against seawater and the pressures encountered when working at depth.

Other less obvious, but equally critical problems related to the operating environment are as follows.

Volume 27: Subtech '91, 197–204.
© 1991 *Society for Underwater Technology*.

Sea Surface Conditions

This affects the launch and recovery of the relevant ROV systems and can cause considerable delay to operations whilst waiting for weather conditions within the system's launch and recovery limits. If work is taking place from a floating vessel rather than a fixed offshore structure, unfavourable surface conditions can also have a dramatic effect on the station-keeping ability of the vessel. The need to carry out repair and maintenance to quite sophisticated electronic equipment on the deck of a heaving vessel in a severe storm, can challenge the skill and patience of the most qualified and experienced Technician.

Ocean Currents

The business end of an ROV system is basically a vehicle which is able to "fly" through water on the end of a control tether or umbilical carrying all power, signal and data conductors.

An ROV is flown in a similar fashion to a helicopter, in that it is free to move in all planes and has the ability to hover. The ROV is subject to prevailing ocean currents as a helicopter is subject to prevailing winds. One major difference is that the helicopter pilot has the advantage of "presence" and a feel for the conditions affecting the machine.

The ROV pilot does not have this added dimension and can only fly and operate manipulators by vision with limited assistance from tactile sensors. It is this lack of direct "presence" or "feel" that can make the performance of manipulative tasks extremely difficult.

Currents also induce considerable drag on tethers and umbilicals and this can have a detrimental effect on vehicle performance and stability, particularly in hover mode.

Water Turbidity

More often than not, vision is the only sensory feedback available to the ROV pilot. It is not unusual for suspended particulate matter to restrict visibility to less than one metre, which can have a severely detrimental effect on operations and can cause a temporary cessation of activity. Turbidity can be a particular problem when working at, or near the seabed, where thruster wash can throw up dense clouds of unconsolidated material, restricting visibility to zero.

Manipulator Dexterity

The dexterity, reach and strength of ROV manipulator systems has generally been one of the major limiting factors governing the ability of an ROV to accomplish a given task.   Much reliance has been placed on robotic manipulators to carry out tasks that would sometimes be difficult, if not impossible, for the human arm to accomplish.

The Solutions

The aforementioned problems are highlighted as an indication only of the complexities facing the operator who wishes to accomplish meaningful work tasks successfully in the subsea environment utilising telerobotics.   Years of experience has seen the Underwater ROV Industry evolve along a slightly different path than that being taken by the more traditional Land-Based Robotics Industry which has had the advantage of operating in a more controlled environment.

One of the more challenging problems that manufacturers and operators of ROV systems have had to overcome is keeping the ROV itself stable in mid-water positions so that manipulator tasks can be successfully performed.

The approaches to solving this problem have been varied, and range from sophisticated software-controlled manipulator management systems which tolerate a reasonable degree of vehicle movement, but retain the manipulator end effector in a stable position, through to simple mechanical clamps which secure the ROV to some point on the fixed structure on which work is to be performed.

To date, one of the most successful solutions to this stability problem has been the combination of a vehicle dynamic positioning (D.P.) system and an articulated suction attachment device.

The articulated suction attachment device consists of a rubber suction pad on the end of a hydraulically lockable arm fitted to the required position on the ROV. The ROV approaches the structure or assembly to be worked on, is positioned at the work location and the attachment device is placed on a convenient point. A pump is then activated which evacuates water from behind the pad and the pressure differential secures the pad to the surface. The arm is then locked into position, giving the accelerometers and inclinometers built into the vehicle's D.P. operating system, a firm point of reference. The accelerometers and inclinometers sense any vehicle movement that may be induced by ocean currents and respond by automatically instructing the vehicle's thrusters to counter the movement.

This method of "soft coupling" the vehicle to the work site greatly reduces the risk of damage to either the work site or the vehicle, and offers the ability to "break away" from the work site quickly and easily, should this be necessary. This type of system also offers maximum flexibility of operation, as the attachment device requires only a very small area onto which it can attach the suction pad, which has a reasonable tolerance to marine growth that may exist at the work site. Therefore, no specially designed docking or attachment points are necessarily required to be pre-installed or retrofitted to the work site.

A major contributor to vehicle instability is induced drag and vortex vibration on vehicle control tether/umbilicals.

A vehicle which is deployed from the surface via a control umbilical, can have hundreds, if not thousands, of feet of umbilical subject to ocean currents from various directions.

The induced drag over several hundred feet of umbilical can be quite high, and a vehicle utilises much of its available power just overcoming this drag factor. Obviously, this drag also makes it extremely difficult to maintain a stable course or position.

This problem has been largely overcome by the use of Tether Management Systems (TMS), which are basically a submersible winch assembly to which the ROV is docked and lowered to the desired working depth. Once at the working depth, the vehicle is released from the TMS, the winch begins to pay out the neutrally buoyant flying tether and the vehicles proceeds to its work location. This means the vehicle is only required to pull neutrally buoyant tether horizontally, thus avoiding any umbilical drag. In this configuration, it is normal for a vehicle to have a horizontal excursion range of up to 600 feet.

Once at the work site, the critical sensory feedback is vision. Recent years having seen a marked improvement in camera technology for underwater use. The introduction of CCD cameras has greatly improved operational reliability, and image enhancement technology is offering greater ability to operate in extremely low light levels conditions.

Present generation large work class ROV's are able to carry numerous sensors such as magnetic anomaly detectors, sub-bottom profilers, side scan and sector scan sonars, salinity and temperature sensors, current meters and other systems considered as "passive" peripheral equipment.

The real "work" that an ROV is expected to perform is typically carried out by the manipulator systems. Numerous manipulator types are available, ranging from basic rate controlled 3 function "grabbers", to extremely dextrous, almost human equivalent 9 function spatially correspondent arms complete with force feedback facility.

The type of manipulator selected depends on the work the ROV is expected to perform. For general construction/maintenance type work, it is usual to select a combination of one 5 function and one 7 function manipulator. The 5 function is used to carry out heavy duty work tasks, such as handling and applying impact wrenches, rotary wire brushes, cut-off wheels, rigging equipment etc.

The 7 function is used where more fine control and dexterity is required. Typically, this would be to handle close-up inspection cameras, placement of sensors such as ultrasonic inspection and cathodic potential measurement probes, placement and alignment of measuring devices etc.

The Work

The general acceptance of the latest generation ROV systems as being truly work capable, is resulting in an escalating demand for ROV's to accomplish increasingly more difficult and sophisticated tasks.

These demands for greater capability are giving cause for re-examination of the role of ROV systems in these more complex tasks. As tasks become more complex and demanding, the ROV Operator is faced with two choices:

A. **Continue to increase the capability of ROV systems to enable them to meet the challenge of increasing tasks complexity.**

This approach would require the introduction of increasingly complex subsystems into ROV design to enable the vehicle to carry out tasks requiring a high degree of strength, stability and manipulator dexterity.

The design of such sub-systems would have to take into account the operating environment and address the more practical requirements of work site access and ease of handling, which place limits on vehicle size. As vehicles become more complex in order to successfully accomplish these more demanding tasks, they necessarily become larger, requiring more power to drive and control them. Given that there are very fee of the repetitive tasks that are ideally suited to the application of robotics, a very real danger exists that by attempting to continually increase the capability of ROV's to suit what essentially is an everchanging scope of work, an escalating spiral of costly development is embarked upon that will eventually commercially strangle the Industry.

B.   **Consider the ROV system itself as being only a prime mover or delivery system and a submersible power and control unit.**

The adaption of such a philosophy provides for a more focused development programme to be followed for what we traditionally recognise as ROV's.

The result should be a new generation of robust, reliable, powerful systems that are compact, manoeuvrable and easily maintained in the field. They no longer have to be "all things to all people", but must simply have the capability to transport, place and, if required, provide controlled power to tooling modules that are tasks specific. Such an ROV will still have the capability of performing tasks by use of its manipulators and add-on tools such as impact wrenches, cable cutters etc., but the more complex tasks requiring high degrees of stability, precision positioning, high strength or other capabilities beyond that normally expected of an ROV, will be carried out by task specific tooling modules deployed by the ROV. Once these modules have been positioned, the ROV is then free to act in a support role providing assistance and monitoring where required.

It is this approach that is now beginning to be adopted by not only the ROV Operators, but also the larger end-user companies who see cost-effective solutions to some of the more complex challenges they face, particularly in deeper waters.

Two case studies are worth considering to fully illustrate the advantages of remote intervention utilising task specific tooling modules.

The first case related to an inspection system designed to achieve; double wall, single image radiography, full weld volume ultrasonic scanning and an approved method of magnetic particle inspection in water depths in excess of 80 msw. A dedicated, platform deployed, inspection module was designed, built and commissioned in favour of a diver or ROV-based system for the following reasons:

- Support logistics, deck space and load induced on the production platform were significantly reduced using an ROV deployed inspection module.

- A module approach proved to be more cost-effective than a diver base alternative.

- The module's integrated power and control system provided the means to achieve precision tracking and positioning of inspection sensors and sources, including ultrasonic probes and radiography sources.

- The system had built-in datum reference encoders which provide position confirmation data and accurate and repeatable inspection techniques.

- The module approach enabled the ROV to be relocated after the inspection module had been positioned, to undertake site preparation for subsequent inspections. This also allowed the ROV to be recovered to the surface without interrupting the inspection process.

- The module may be positioned by a variety of work class ROV's, thus providing the client with a broader range of opportunity for future inspections.

- A dedicated inspection module, with purpose supplied communication umbilical, enables dedicated data communication, from probe arrays for example, to be transmitted to surface processing equipment without the inherent limitations of available conductors in the ROV umbilical.

- Less restrictions on tasks performed above water at the time of inspection, results from using a remote system over the diver alternative.

The second case study, relates to a module designed, built and commissioned to support a deep water construction task which necessitated the bleed-down of trapped air, removal of blanking caps and installation of diffuser nozzles at 232 underwater sites. To meet this requirement, a dedicated work module, positioned by the ROV, was selected in favour of possible alternative means, for the following reasons:

- The task specific module provides power, control and spatial work envelopes that could not be effectively incorporated within the space frame of a typical work class ROV.

-       Position feedback encoders located within the
        module, enable the operator to perform the requisite
        tasks in black water conditions.

-       After positioning the module on site, the ROV is
        able to detach and undertake additional support
        tasks, thus increasing overall system productivity.

-       The dedicated module allows repetitive, heavy duty
        subsea tasks; in this case the installation of eight
        nozzles per mission, to be performed without
        immediate surface support.

-       The described system provides a less hazardous
        alternative to the diver option which, in this case,
        was of particular importance.

The change in emphasis from over-engineering the ROV to
the use of the ROV as a delivery and support system and
more engineering being directed into the task modules,
has given the more technically competent ROV operating
companies the opportunity to develop a specialised
engineering expertise related to the adaption of remote
systems technology.

There remains little doubt that this approach is the way of
the future and we will begin to see some radical changes in
the way tasks are accomplished by the use of remote handling,
particularly in the underwater environment.

R.O.V. PERSONNEL TRAINING PAST, PRESENT & FUTURE

C. W. BELL
Wray Castle College
Ambleside
Cumbria
LA22 0JB

Abstract

It is now accepted, within the R.O.V. Industry, that a comprehensive training programme with recognised standards is required. The benefits of this can be summarised as follows:

1. Enhanced Safety
2. Improved Operational Efficiency
3. Improved Quality of Service
4. Improved Job Satisfaction and Retention of Personnel

This paper examines the progress made since a paper was presented to the Subtech '89 Conference on behalf of the A.O.D.C. called manning levels and training standards for R.O.V. operations. The proposed Induction and C.S.W.I.P 3.3U Courses are referred to and a comprehensive programme of training modules is outlined as a possible means of extending the training available to meet future needs.

In particular an argument for competence based training in line with the governments proposals for N.V.Q.'s (National Vocational Qualifications) is put forward, to replace the currently popular knowledge based system.

This paper also emphasises the importance of assessment and the need to broaden the entrance requirement to ensure that companies have the opportunity to employ a balance of technical ability with offshore and underwater experience.

The paper concludes by suggesting that a comprehensive and complete programme of competence based assessed training modules be developed and that the cost of this will be minimal compared with the benefits to be realised, as

Volume 27: Subtech '91, 205–216.

personnel are often prepared to fund their own training to aid career progression and even if the funds were to come from an increase in day rate paid by operators this would amount to less than £5 per person per day.

Wray Castle College is commited to developing this training programme in a conscientious and professional manner but requires that the industry supports this development by recognising standards and the benefits that will be achieved.

1.   Introduction

It is now accepted, within the R.O.V. Industry, that a comprehensive training programme with recognised standards is required.

This paper examines past and present training requirements and facilities and offers a future training programme. Elements of this programme are already established and recognised by the industry, others, presently under development could become so in due course.

Courses presently recognised by the Industry are knowledge based but it is envisaged that future training programmes will be competence based in line with the Governments recommendation for National Vocational Qualifications (N.V.Q.'s).

The benefits to the industry of further development and recognition of this training programme can be summarised as follows:

1.   Enhanced Safety
2.   Improved Operational Efficiency
3.   Improved Quality of Service and Results
4.   Improved Job Satisfaction and Retention of Personnel

The scope of this paper is limited to R.O.V. operating personnel and does not consider the training requirements of clients reps, managers or other associated personnel.

2.   Background

A paper was presented to the Subtech '89 Conference by Brian Redden and Herb Newbury on behalf of the A.O.D.C. (The International Association of Underwater Engineering Contractors) R.O.V. Personnel Training Standards Institute entitled: "Manning Levels and Training Standards for R.O.V. Operations". This paper laid the foundations for minimum industry entrance requirements and a Pilot/Technician Induction Course.

We at Wray Castle were working on a similar Induction Course at that time with financial assistance from the Government's Training Agency. Since then we have held eleven Induction Courses and trained over eighty people.

In May 1991 the A.O.D.C. published proposed minimum entrance requirements and objectives for the basic induction course that are similar in most respects to those implemented by ourselves.

A moderate degree of success with our induction courses has led us to consider the future training requirements of the industry and to write this paper.

3.   Past Training

It is fair to say that in the past, no industry recognised training scheme was available for R.O.V. Personnel. This is not to say, that Companies were not doing the training but rather that training standards were not widely recognised. Operational Training was gained simply by personnel being included in an operational team and being introduced to the techniques and systems supplied for that contract.

Operators would occasionally commit vessels and R.O.V.'s for training during the winter months usually in sheltered waters. This training being carried out away from oilfield installations concentrated on technical training and operation of specialised systems and techniques.

Companies were also in some cases, arranging brief 'in-house' induction courses and technical courses were sometimes being given by equipment manufacturers.

Training was therefore rather 'Ad Hoc' and lacked any recognised standards or entry qualifications.

4.   Present Training

The A.O.D.C. published proposals for a basic Induction Course in May 1991 which has been widely circulated and when finalised wll be recognised. These include the Course Objectives, Course Management and entrance qualifications. This Course is essentially knowledge based and is recommended to take at least 36 hours to complete. This Course lays down an excellent foundation for a complete R.O.V. training programme but, it is appropriate that we make some comments at this stage in the course development in the light of our experience of running similar courses over the last two years.

3.1   PAST TRAINING FACILITIES

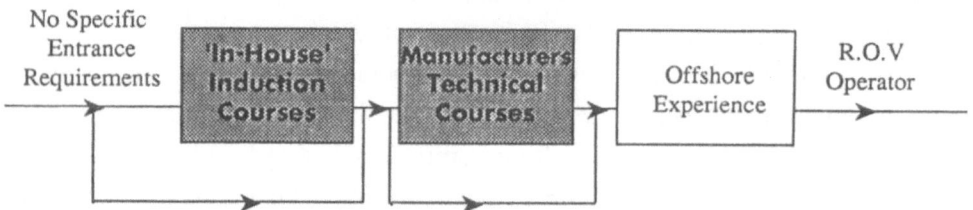

   INDUSTRY RECOGNISED TRAINING

   TRAINING AVAILABLE BUT NOT RECOGNISED

FIGURE 1

The objectives are based on trainees gaining a knowledge or outline knowledge of the subject areas. The government is presently providing the impetus for Colleges of Further and Higher Education to implement the N.V.Q. (National Vocational Qualification) System administered by the N.C.V.Q. (National Council for Vocational Qualifications). The implications of this are far reaching but it is evident that the N.V.Q. will be competence based and candidates will be assessed on the Competent performance of work related activities.

The Pilot/Technician induction course is doubtless a vocational qualification and as such may develop along national guidelines for N.V.Q.'s and be competence based.

As an example of what this will mean we can look at the 5th objective of the Seamanship Section of the proposed A.O.D.C. Induction Course which states that he candidate must have an outline knowledge of the use and care of lifting equipment.

Because it is required that the candidate have an outline knowledge only this objective can be achieved through lectures or reading the manual and this knowledge is not assessed.

For this objective to be achieved according to the criteria for N.V.Q.'s the candidate would be required to be assessed as competent in the care and use of this equipment, which would involve carrying out actual lifting exercises in simulated working conditions to the level of competence specified in the assessment criteria. This principle can be extended to all essential areas of training where competence is required. As a further example competence can be achieved in the basic piloting skills of maintaining headings, maintaining visibility, umbilical awareness etc, in sheltered water with a basic R.O.V. prior to type training, in a similar manner to the training of aircraft pilots. We have found this to be extremely successful because once a trainee has mastered the basic piloting skills and gained confidence, he can quickly convert to more advanced types of R.O.V.

In order to accommodate future requirements of N.V.Q.'s we recommend that candidates who complete the induction course at Wray Castle then undertake a second week of skills training in order that candidates reach minimum levels of competence in areas that are essential to safe and efficient R.O.V. operations. To assist employers and candidates we also offer assessment of the knowledge and skills that are required to achieve these levels of competence. There can be no doubt that it is better for a pilot/technician to undertake his first tour of duty offshore if he is competent and is known to be able to act in a safe manner and be a useful team member than if he is not. It is thus envisaged that as the provision of training facilities develop R.O.V. training will become competence based.

The proposed entrance requirements ask that candidates for the Induction Course have a minimum of City & Guilds in a relevant technical subject (U.K.) or equivalent in non-U.K. areas.
AND

Minimum of 2 years experience in the appropriate discipline which should have been obtained within the previous 3 years of employment.
OR

An academic qualification (minimum HNC) or equivalent in a relevant technical subject.

These requirements exclude people with relevant offshore experience alone, for example divers who have worked alongside R.O.V. crews. In our experience to date we have found that divers have been able to offer a great deal to employers as R.O.V. crew members after successfully completing our induction training. Firstly they have a good sense of safety and knowledge of offshore working practices and are therefore extremely useful on deck during launch and recovery operations, secondly their experience of actually undertaking underwater tasks enable them to become excellent pilots with a good feel for situations and knowledge of the tasks being performed.

It is our belief that divers trained in the basic technical and piloting skills are a useful component of a balanced operational team.

We have found that where a course has been attended by candidates with a high level of technical qualifications alone that the operational aspects of the training have often been performed in a less than satisfactory manner. We believe therefore that a balanced team must consist of both technical ability and relevant offshore experience. To illustrate this point we can give as an example a diving company operating in Saudi Arabia that has retrained its divers and technicians through our course to operate an R.O.V. and are operating it very successfully without the need to employ any specifically R.O.V. experienced personnel. This approach has given this Company the advantage of being able to undertake jobs requiring both diving and R.O.V. operations with just one crew and take advantage of the cost savings. Balancing R.O.V. teams with suitably re-trained divers and technical personnel can therefore have advantages and experienced divers should not be excluded from the Induction Course solely because of the lack of recognised technical qualifications, particularly if they attend the course under Company sponsorship. This approach has the additional advantage of unifying the industry.

For reference, I include a list of the areas of knowledge covered by the proposed A.O.D.C induction course:

1.   Background
2.   Safety
3.   Survival Training and First Aid
4.   Seamanship
5.   R.O.V. Operations
6.   R.O.V. Systems

Full details of the Course objectives and requirements are available from the A.O.D.C.

At the time of writing this paper full details have not been finalised regarding the proposed courses and examinations for platform inspection pilots (C.S.W.I.P. 3.3U) and pipeline inspection (C.S.W.I.P.) 3.3U(P) but it is believed that they will become available in the near future, and will form an important part of the training programme.

At the time of writing, no other proposed recognised training exists although 'in-house' and manufacturers technical courses are common place but don't generally have recognised standards.

4.1   PRESENT TRAINING FACILITIES

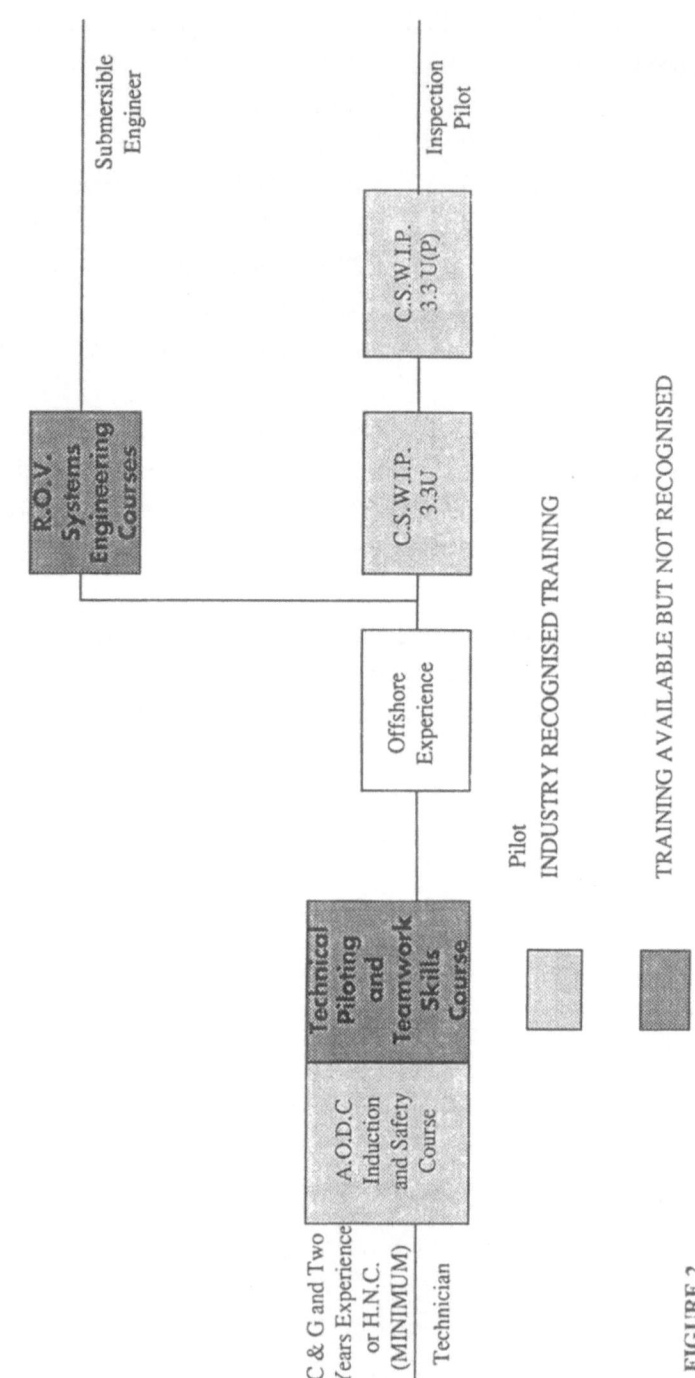

FIGURE 2

5. Future Training

Training that will meet the industries future needs will consist of a Comprehensive Structure of Modules. There will be a requirement for modules of training to be available in each of the following areas:

1. Pilot/Technician Induction
2. Systems Engineering
3. Inspection
4. Supervisors
5. Client Appreciation
6. Specialised Training Modules. As more specialised systems and techniques are introduced to the industry.

It is envisaged that facilities will develop that will offer these modules of training and that the levels of competence achieved and assessed will be recognised. We at Wray Castle are certainly commited to this structure of training and expect that it will gain wide industry support and recognition.

5.1 FUTURE TRAINING MODULES

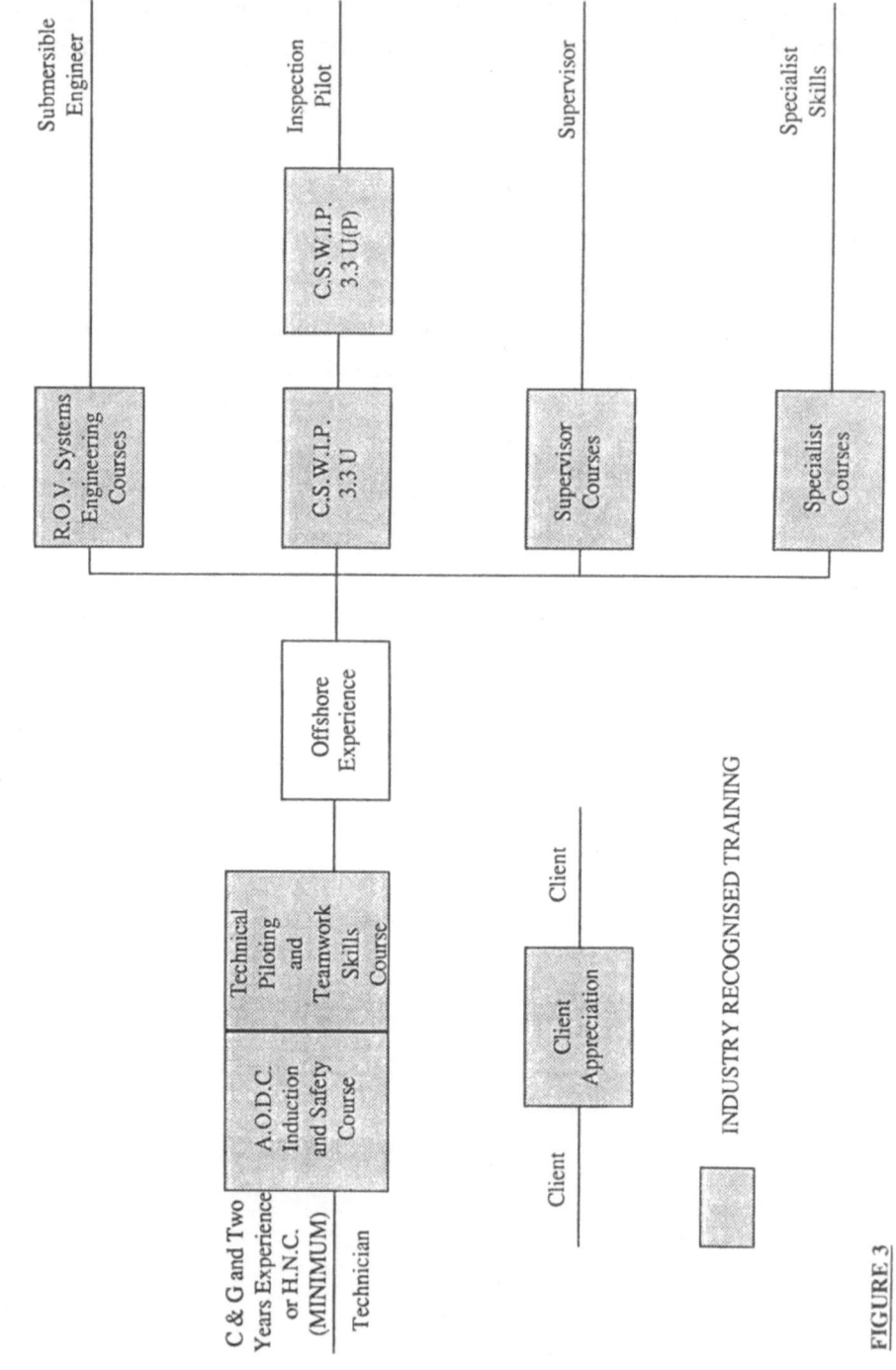

FIGURE 3

6.   Funding

It is obvious that there is a great deal of work required to establish this comprehensive training programme and facilities. However, an excellent start has been made by the introduction of the A.O.D.C. Induction Course and C.S.W.I.P Inspection Courses.

If the Courses were to be fully funded by an increase in day rate paid by the operators we estimate that the day rate would be increased by £5 per person per day. If these funds were passed on to the training organisations, then there would be sufficient to allow for the development of the full training programme.

It has been found, however, that in many cases personnel are willing to fund courses themselves in order to improve their Career prospects. Therefore, funding for this future training provision could come from a combination of small increases in day rates for personnel, and from the personnel themselves.

7.   Conclusions

The R.O.V. Industry now recognises the advantages of establishing and recognising a comprehensive programme of competence based training modules.

Development of training facilities is becoming established, and the will exists to fully develop these to meet future training requirements.

This training programme can be fully established at a very minimum cost to operating and contracting companies but with a very worthwhile benefit in terms of operational safety, downtime, efficiency and quality of results. The main requirement now is for operating and contracting companies to recognise these modules and to support their development.

8.   References

1.   A.O.D.C. 057 (1991) ' Basic Induction Course for R.O.V. Personnel'
2.   Redden Brian G and Newbury Herb 1988 'Manning Levels and Training Standards for R.O.V. Operations' in SUBTECH 89 Conference Proceedings.

3.   Further Education Unit (1988) N.C.V.Q. and its implications.
4.   Bell C (1991) Wray Castle College training Modules and Entrance requirements. Ambleside, Cumbria.

Part 3
Equipment Development

THE FUDT-PROJECT
(RESEARCH AND DEVELOPMENT PROJECT IN DIVING TECHNOLOGY)

RUNE BRÅTHEN,
Statoil,
P.O.Box 620,
N-4001 Stavanger,
Norway.

CATO HORDNES,
Norsk Hydro,
P.O.Box 646,
N-5001 Bergen,
Norway.

ABSTRACT. The paper gives a background, organisation etc of the norwegian R&D project in diving technology, which is a multi company project.

1. Introduction

For at least the last 15 years, the use of manned underwater operations versus the use of remotely operated vehicles has been discussed. Throughout this period, it has been proven that diver assisted operations are still needed and will be needed for an indefinite period. Development has, however demonstrated that divers now are more often supporting unmanned operations, rather than the opposite, as it was some years ago.

Detailed evaluations must be carried out, in order to select the most cost effective sub sea intervention method, in connection with the development of offshore oil and gas fields. However, even if unmanned intervention methods are to be used, it is still necessary to qualify the manned intervention method, ie, the use of divers. This is due came into force 01.01.91, explicitly states that the oil companies must qualify the use of divers, if it cannot be documented that it is unlikely that divers are to be used in the oilfield´s lifetime.

Currently, the development plan for the norwegian continental shelf, includes several fields located in water depths greater than 180 meters, which is the defined by the Norwegian Petroleum Directorate (NPD) as "deep diving". Hence the volume of deep diving on the norwegian continental shelf will increase in future, whereas, only a very few operational dives have previously been carried out to these depths. Further, the pipelines from offshore fields may have to cross the norwegian trench, therefore, intervention methods down to 360 meters may need to be qualified, although the offshore installations themselves are located in relatively shallow waters.

2. History

Previously, Statoil financed several diving related research and development projects. These projects were not satisfactorily coordinated. Consequently, the central diving unit in Statoil started a project which was named: "FUDT" (Forsknings- og

219

utviklingsprosjekt i dykketeknologi = Research and development project in diving technology).

The intention of this project is to coordinate the existing projects; to improve the internal management; to improve the selection process of R&D-projects and to improve the utilisation of the resources.

In 1989, Norsk Hydro and NPD joined the project with Saga Petroleum entering the cooperative effort in 1990.

3. Organisation

The financing bodies (Statoil, Norsk Hydro, Saga Petroleum and NPD) are represented on the board of FUDT. (A sub project may in addition be supported by other companies and institutions such as: BP Norway, Stolt-Nielsen Seaway A/S, the norwegian research foundations: NTNF and NAVF, The Norwegian Oil Industry Association: OLF etc.)

NUTEC is defined as the main cooperative/liason institution, but other institutions such as SINTEF, the University of Bergen, various diving contractors and development companies, etc, have been given separate work packages.

4. Background

In spite of world-wide efforts and relatively large R&D-projects in Norway such as Statpipe, Troll (DSIS) and the Oseberg Transportation Project etc, satisfactory solutions to central and other important areas still do not exist. Areas of concern, at the start of the FUDT-project were: the long term health effects of diving, decompression profiles, divers individual equipment, bail-out, communications equipment, toxicology, bacteriology, etc.

There has been a lack of implementation of the results of some of these projects, which may be explained by their short-term character. These projects did not support the long term research requirements and the continuous activity, which is necessary, in order to meet the "challenges" and provide solutions.

Also, short-term projects are relatively expensive. A project with a long term profile in areas with high priority, will have a much better cost-efficiency. (The average cost for one of these short term projects could have financed the total FUDT-project for more than 10 years!)

5. Objective

The objective of the FUDT-project is to develop the necessary competence in diving technology, to ensure that future manned underwater operations can be carried out in a safe and cost-effective manner.

To accomplish this, the project:

- defines the central problem areas in relation to future manned underwater operations,
- will carry through each sub-project in accordance with an established priority list,
- will contribute to the stability and construction of relevant and competent R&D units in Norway, where strategic competence can be accumulated for solving future tasks,
- seeks an effective communication between the relevant professional environments within operational diving and the R&D institutions, in order for the industry to define their problems for the R&D personnel and to ensure implementation of the results.

The duration of the project is estimated at 5 - 7 years.

6. Implementation

Previous R&D activities in diving technology have often assigned the implementation of the results a low priority. In the FUDT-project, it is emphasised that implementation of the results is mandatory. It is assumed that the sub-projects will accordingly, present proposals for implementation of the results.

Elements in FUDT´s policy to start and speed up such an implementation process have been:

- annual conferences for the industry (previously held only in the norwegian language),
- induction of operational diving personnel in the sub-projects (where possible),
- publishing of project results in scientific and popular science magazines, including trade publications,
- establishment of a "Reference Commitee".

An essential element, to ensure implementation of the results, is to give the industry an opportunity to have influence on the selection of sub projects. Furthermore, the industry must be able to monitor, and have some influence on the various sub-projects.

A "Reference Commitee", with representatives from the diving contractors, the diver´s trade union, the State Diving School and other relevant R&D units has been established. The objectives of this commitee are to propose a ranking of the project proposals received by FUDT. The commitee is of vital importance; to ensure that only relevant activities are in progress. Furthermore, the commitee assists in the implementation of the results.

The "Reference Commitee" is organised and chaired by NUTEC.

7. Sub-projects

The FUDT-project is divided into three main areas:

- technological and operational matters,
- working environment and health effects,
- implementation, organisation and education.

Within each of these areas, separate and specific sub-projects are defined, in accordance with the overall strategy.

FUDT´s sub-projects in 1991 are as follows:
- respiration (incl silent bubbles and oxygen toxicity)
- toxicology (contamination by trace substances, etc.)
- bacteriology (pseudomonas and its factor, disinfection etc.)
- HPNS (pilot study)
- decompression (correlation of doppler data with dive and medical parameters, complement activation, etc.)
- helmet design
- breathing equipment
- monitoring equipment
- development of microphone
- diving bell ergonomy
- 3 basic research sub-projects (oxygen effects, effect of pressure on cells, thromboembolism)
- FUDT-coordination, seminar and implementation.

8. Results

The utilitarian value of the FUDT-project can be reflected by the project´s goal: "safe and cost effective diving operations down to 400 meters". A delay in the development of a deepwater offshore oil field would create enormous economic consequenses. Future operations, especially the deep diving operations, will be able to increase efficiency, to a large extent, with only marginal improvements of equipment.

Some results from the FUDT-project thus far:

- determination of long term neurological effects of diving, including correlations to possible causal relations,
- determination of long term effects of diving on lung function parameters with indications of the causal relations,
- development and integration of the "NUTEC" unscrambler into the Stocktronic communication system, which gives an intelligibility of 75% down to 400 meters,
- development of a breathing system which will meet all requirements in NPD/DEn´s new guidelines for breathing equipment down to 400 meters, both for the primary and secondary system, and with a bail-out duration of more than 15 minutes,
- development of emergency lock-out diving suits acceptable to 400 meters,
- development of improved electrically heated ergonomically designed survival suits with improved emergency scrubbers,
- detailed knowledge of pseudomonas aeruginosa, disinfection procedures and agents which have lead to a large reduction in occurrence.

HYPERBARIC PIPELINE WELDING BEYOND 600msw: A CONCEPTUAL PROPOSAL

J.F. Dos Santos, P. Szelagowski, H. Manzenrieder, H.-G. Schafstall
GKSS Research Centre GmbH
Institut für Anlagentechnik
Max-Planck-Str., D-2054 Geesthacht
Federal Republic of Germany.

ABSTRACT. Economical and political circumstances have been continuously pushing exploration and production of mineral resources. However, for much of the world, the best potential for major new oil and gas reserves lies beneath 450msw. Eventual welding repair operations at such depths can only be carried out by fully automatic systems, part of a diverless underwater working station. Such station should feature: flexibility inside the working cell, self-sufficiency and reduced distances for data transfer and proximity between control units and their executive counterparts. To achieve these objectives a repair station consisting of a submersible, alignement frames and a purpose built habitat has been proposed. This study presents a conceptual description of this repair station, its main elements, working procedures and development requirements to realize it.

1. INTRODUCTION

Since the number and age of offshore installations raises every year, there is an increasing need for repair & maintenance systems and procedures. Although underwater pipeline repair technology has been of concern to companies for a long time, the actual depth range in which repair work could be performed is still more or less behind the depths in which pipelines are laid.

Taking the Mediterranean Sea as an example, more than 80% of its area lies on depths exceeding 200m and more than 50% at 1000m. In addition to that an increasing number of new fields are being discovered in depths beyond 500m in the Mediterranean [Tassini, 1987] and at the Campos Basin in offshore Brazil [Freire, 1989]. Operators hesitate to exploit these resources without contingency procedures for pipeline repair and maintenance since an eventual pipeline failure in deep waters could have dramatic consequences to the environment and production.

Although successful hyperbaric manual welding trials have been executed in 600m simulated water depth [Jansen et al., 1987], 180m appears to be the limit for offshore

223

manned intervention, at least in the Norwegian Sector of the North Sea [Delauze, 1989]. It seems evident that pipeline repair concepts for water depths greater than 300m - 400m must focus on robotics and remote sensing if repair intervention is going to catch up the capabilities that already exist in the drilling, production and transportation phases of the offshore petroleum industry.

The two different diverless underwater repair methods generally considered for deep water repairs are mechanical connectors and hyperbaric welding. Various repair systems based on mechanical connectors have been designed, such as the TITUS system [Lerique and Tangeland, 1989]. Mechanical connectors are believed to offer slightly better pre-conditions for automation than hyperbaric welding and as a matter of fact, some of the available systems have already undergone offshore trials. However, they also present some disadvantages such as the requirement for accurate metrology of the contact surfaces, the probable need of dummy nodes and the lack of mechanical strength. Moreover, depending on the system, the effort for pre and post repair activities offers no additional time benefit [Alexander and Quin, 1984].

Currently, hyperbaric welding is still the best known repair method either as a manual operation (carried out by welder-divers) or as diver assisted mechanized welding. Manual pipeline welding has been successfully qualified at 450m simulated water depth in accordance to British Standard 4515 and API 1004 [Delauze, 1989]. Mechanized orbital welding systems are presently successfully operated down to 450m depth [Mecklenburg et al., 1991] using the gas tungsten arc welding (GTAW) and the gas metal arc welding (GMAW) processes. More relevant to repair in deep waters are the satisfactory results obtained with a robotic station down to 1100 msw [Dos Santos et al., 1991a] using the GMAW process. The high technological level and reliability achieved in underwater welding have already encouraged the development of a system using an orbital welding set-up assisted by a manipulator which can form the basis for a future diverless repair spread [Rougier, 1991].

The growing potential of hyperbaric welding associated with the recent developments in robotics for underwater work suggested an additional alternative for a diverless repair station. This repair station, consisting basically of robots and manipulators to prepare the pipe and to weld the spool piece, would be assisted, controlled and monitored by devices currently used in underwater work such as commercial manned submersibles and ROV's. This station should also be self-sufficient, or in other words, be capable of operation independently from surface vessels or platforms.

This work presents a conceptual description of such diverless repair station, its main sub-systems and some aspects on the repair procedure.

2. HYPERBARIC DRY WELDING PIPELINE REPAIR

2.1. ESSENTIAL ELEMENTS OF A PIPELINE REPAIR PROCEDURE

The performance of a pipeline repair underwater could be in practical terms, divided in

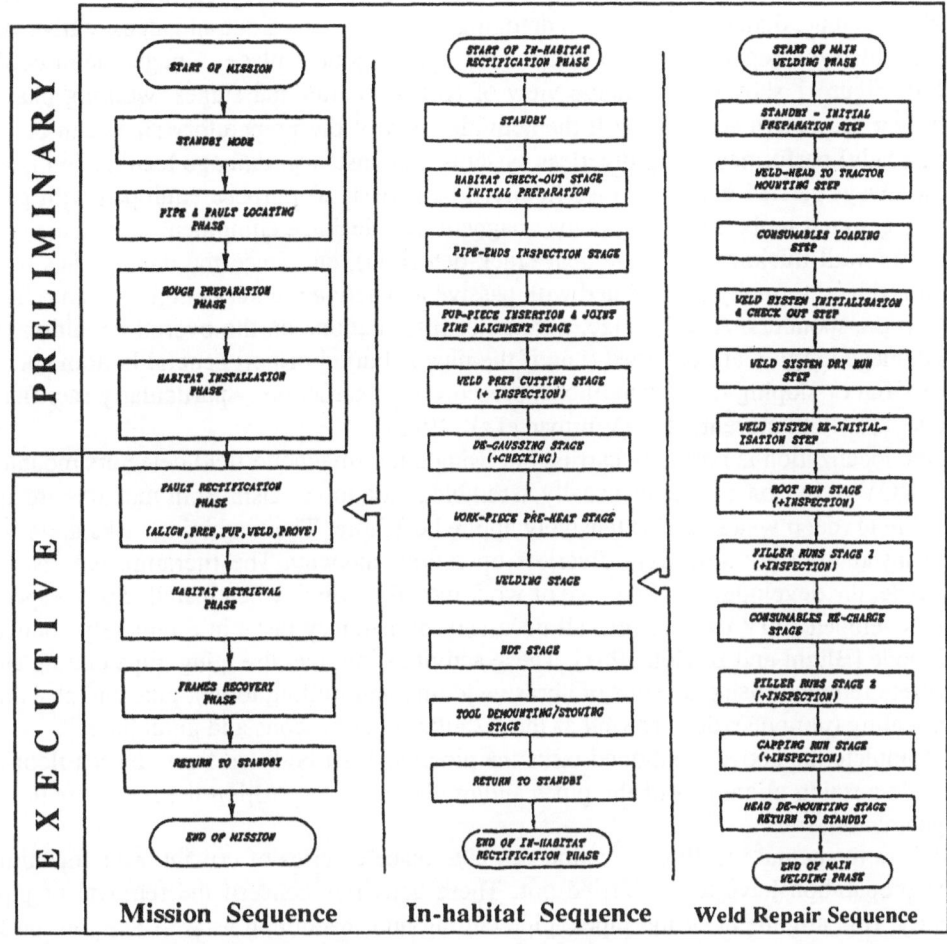

Figure 1. Underwater welding repair procedure. Exemple of a repair carried out using an orbital system [Knagenhjelm, 1985].

two main stages: a preliminary one in which the basic preparatory operations (i.e.: location of the pipe, damage assessment, dredging, etc.) are carried out and an executive stage in which the repair operation and related activities take place [Knagenhjelm et al., 1985]. Figure 1 shows a schematic view of typical operational stages, working phases and their respective activities. All the activities part of the preparatory stage can be (or are already) performed using diverless systems. For instance, damage localization is almost always carried out using pigs (tethered or tetherless) purpose equipped with geometric gauging tools, camera systems, magnetic flux leakage equipment, ultrasonic sensors (for wall thickness measurement crack detection), etc. Once the damage has been localized it is normally demarcated with passive and active markers ranging from reflectors to transponders. At this stage, a range of informations on the overall conditions at the damaged region are obtained (i.e.: if the pipe is buried or not, general bottom conditions - flat or sloping -, soil conditions, presence of obstructions - particularly sacrificial anodes -, bottom current, etc.) [Wittman et al., 1985].

Pipe localization is normally carried out using either magnetic field detectors mounted on a ROV or by sonar systems usually assembled on a towed fish. Both methods are applicable to deep waters and extensively applied offshore. The remaining tasks up to habitat installation are nowadays diverless operations anyway. The literature reports, for example, the development of a range of work modules which when installed on a especially developed ROV can perform all necessary preparatory tasks in a remotely controlled mode [Blight and Baylot, 1987]. These activities involve: dredging, pipe cutting and concrete removal (using a series of abrasive jetting and milling tools), pipe end cleaning and sealing (plug insertion) as well as the installation of beacons and guidelines.

To complete the above mentioned tasks the pipeline must be lifted from the sea floor to provide a rough alignment of the pipes and/or clearance to position the habitat with the alignement frames.

Before the actual welding operation can take place a series of crucial tasks regarding pipe preparation have to be carried out. These activities concern the removal of pipe ovality, fine alignment of the pipes, pipe cutting and machining as well as spool piece installation, alignement and support. The mechanized (diver assisted) pipeline repair (orbital welding) systems currently used offshore are able to perform most of these activities in a remotely controlled mode [Håbrekke and Knagenhjelm, 1989]. Most of these systems are provided with a split ring cage equipped with independently controlled hydraulic rams, each fitted with a sensor providing a position feedback signal. The "rerounding" operation is then software controlled. The joint preparation is conventionally carried out using lathes or milling modules mounted on the rotating face plate of the pipe clamp [Blight and Baylot, 1987].

The installation, alignment and support of the spool piece is normally performed by external support arms mounted on the pipe clamps or through a number of internal support systems. These internal systems can support the weight of the spool piece and provide a flexible "joint" between the spool piece and the pipe ends. More advanced systems can reform the pipe end, align the pup, support it during welding and eventually house a gamma source for post weld single wall radiography. In both cases, the pigs are "flushed out" of the line after welding by the pipe cleaning pigs [Blight and Baylot, 1987].

The welding operation itself involves two additional related activities: pre-heating (and interpass temperature control) and demagnetising of the joint. The pre-heat treatment is carried out using either induction or resistance heating, temperature being controlled by thermocouples welded to the spool piece. To reduce arc blow caused by remanent magnetism hall effect probes are installed on the weld head to provide readings of magnetism in the groove. These values are then used to calculate the current supplied to demagnetizing coils set at each side of the weld [Delauze, 1989].

The operational orbital welding systems presently available for pipeline repair on the sea bed apply exclusively the Gas Tungsten Arc Welding (GTAW) process with cold wire addition. This process is the most suited to root pass welding, offers the possibility of autogenous welding and enables sophisticated stop/start sequences. Apart from that the deposited weld metal is not affected by the effects of pressure on oxidation-deoxidation reactions meeting therefore, high toughness requirements. However, the GTAW process is characterised by slow welding speeds and low deposition rates.

Once the butt welds have been completed and tested the habitat is retrieved and the lifting frames are recovered.

2.2 DIVERLESS PIPELINE REPAIR: ORBITAL WELDING SYSTEM

A general overview of the literature and the present industrial practice offshore indicate that most of the tasks up to the preparation of the joint can be or are already performed without diver assistance. Although it could be argued that some of the equipment and techniques used in the preparatory stage (see Figure 1) are still in the early stages of automatisation, the joint preparation activities and particularly the welding operation itself present problems for automation.

A possible solution for the problems posed by full automation in hyperbaric welding has been proposed by Comex Services, Marseille with the THOR-2 system [Blight and Baylot, 1987; Rougier, 1991]. For this system it was necessary to design the habitat and the sub-sea module such that the equipment can be taken out and put back into storage and installed on the pipes by remotely controlled carriages and a manipulating arm. This manipulating arm is an 8-axis polar robot mounted in the roof of the sub-sea module and can be used to both install and service different equipment in the habitat.

The system consists of two task oriented set of components: pipe machining and pipe welding sub-systems. The former is composed by a tool orbital clamp which is primarily used to remove pipe ovality but also provides a rotating face plate which carries either the machine tools or the welding heads. The second element in this sub-system is a pipe machining tool which performs the following tasks: sawing, counterboring and bevelling of the pipe ends [Rougier, 1991].

The pipe welding sub-system consists of two welding heads which can be installed on the rotating face plate of the tool orbital clamp, welding power sources, pre-heating and demagnetizing elements, a laser groove tracking and the weld control system. The latter uses a 383 Intel processor with sub-sea and surface control stations. It has been built for simultaneous hot, fill and cap passes with weld heads [Rougier, 1991].

The THOR-2 has now been tested in a simulation facility and test pools, and is about to undergo an offshore trial in the North Sea. Although the systems has not yet simulated a diverless repair it forms the basis for a future deep water tie-in/repair spread [Rougier, 1991].

2.3 DIVERLESS PIPELINE REPAIR: ROBOTIC WELDING SYSTEM

A fully automatic robotic station could be an alternative to the manipulator assisted orbital welding system. Compared to orbital systems, robots present some distinct advantages, such as:

- handling more than one tool (i.e.:, welding torch, machining tools, grinding devices, NDE equipment, etc.),

- operating instruments and control panels (input/output user friendly interfaces),

- welding other than circumferential joints

- capability of carrying out tasks in regions other than the joint area.

As a matter of fact, during the late 1970's and early 1980's the possibility of using robot arms to completely replace the welder-diver was studied [Delauze, 1989]. At that point in time the robot configurations available were not particularly ideal for pipe welding. In addition to that the performance of some robot components (i.e.: actuators, path-measuring systems - encoders -, etc.) were significantly influenced by pressure.

Nevertheless, the accelerated development of the robot technology in the mid eighties encouraged the GKSS Forschungszentrum Geesthacht to set up a major research project to implement the use of robots in underwater work particularly for hyperbaric welding. The initial objective of this research programme was to modify and test an industrial robot for hyperbaric dry work down to 120bar [Aust and Szelagowski, 1985]. The device in question has six degrees of freedom with a load capacity of 15kg. The actuators are permanent initiated brushless rotary DC motors with integrated incremental positioning sensors. For outside pressure compensation all robot internal cavities have been drilled. To limit the access of dirt and dust into the robot bronze filters have been mounted on the pressure compensation apertures. The limit switches which define the end and reference position of each axis have also been modified for the envisaged working pressure (120 bar) [Aust and Domann, 1989].

The second development stage involved the further modification of this industrial robot to withstand immediate contact with sea water down to 1100msw, allowing therefore, its direct application in wet environment [Gustmann, et al., 1991].

Parallel to the development work in the robotic side, a R&D programme on welding technology aiming at water depths beyond 600msw was set up to define basic aspects of a welding procedure, such as: shielding gas flow requirements [Dos Santos et. al., 1988], filler metal/shielding gas combinations producing satisfactory weldment mechanical properties and optimization of welding parameters [Dos Santos et al., 1991a]. In addition to that key hardware components such as arc observation cameras [Dos Santos et. al., 1991b] and a seam tracking system [Dos Santos et. al., 1989] capable of working in the range 60 bar to 110 bar, were also developed.

The Gas Metal Arc Welding (GMAW) process (with solid and flux cored wires) was selected as the most promising process for application in deep waters, since there are still some serious doubts on the stability of GTAW arcs above 60 bar [Crane and Richardson, 1990]. The filler wire and shielding gases employed resulted in weld metal mechanical properties complying with the requirements of standards and guidelines commonly used for underwater welding work (i.e.: 107J Charpy impact energy at -20°C in 1000msw) [Dos Santos, et. al., 1991a]. A systematic evaluation of the effects of welding parameters on arc stability using an advanced power source led to a substantial control over the welding process which allowed positional welding down to 110bar [Dos Santos, et al., 1990].

The weldment properties and process control achieved, associated with the satisfactory performance and reliability of the main components and peripheral devices, demonstrated a good potential for an underwater repair station, based on the extensive use of robots and manipulators, for the pipe preparation and welding phases. The conceptual description of such an underwater repair station is presented below.

3. DIVERLESS HYPERBARIC REPAIR STATION

3.1. TASK DESCRIPTION

The repair operations are supposed to be carried out on buried or partially buried pipelines positioned on the sea bed and fully independent from surface support (of a vessel or platform). The task in question is to exchange a damaged pipeline section by a spool piece and/or hot tapping of a valve. Either financial and/or technical reasons indicate that this task shall be carried out "in situ" (hyperbaric conditions) and in dry environment.

The demands imposed by greater depths require dedicated technological solutions. For instance, the energy supply should be positioned closely to the operation place to reduce losses. As a matter of fact, this is a crucial factor to guarantee the self-sufficiency of the repair station. Furthermore, it is essential to keep the distances for data transfer as short as possible to guarantee effective data transmission.

To cope with these demands a diverless repair station has been proposed consisting basically of the following three main components (Figure 2):
- a submersible
- a habitat
- alignement frames ("H" or "A" frame)

The following items describe these three main components as well as their operational

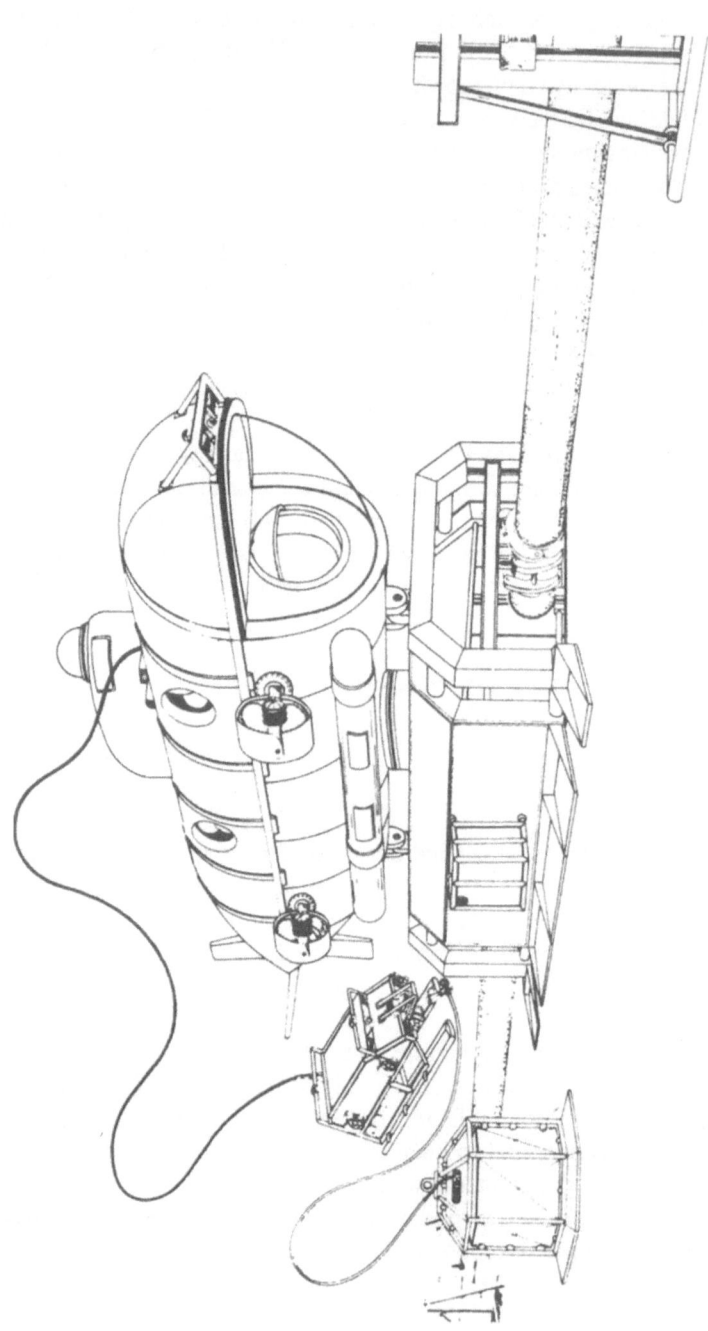

Figure 2. Schematic view of a diverless repair station consisting of a submersible, a purpose built habitat and alignement frames. Also pictured are a sub-sea energy generator and a ROV supporting the operation on the sea bed.

principles.

3.2. COMPONENTS OF THE REPAIR STATION

3.2.1. *Submersible.* The submersible is designed as a one-bar-system suitable for the depths in question. The necessary personnel consists of two crews: one to handle the boat and one to perform the repair operation itself. The boat carries all propulsion and manoeuvring systems as well as all the necessary connections to measure and control the required station components and sub-systems.

The submersible should be equipped with a flange with hydraulic clamps to connect the boat with the habitat In this flange all necessary connectors and penetrators for the components installed in the habitat shall be housed. This guarantees dry docking for all sensitive control and monitoring cables.

Manipulators are installed outside the submersible as well as docking arrangements for ROV's. The manipulators shall support the installation of the alignement frames and the habitat. Additionally, all control valves (installed outside), connectors or switches can be actuated through these devices.

The submersible shall carry one or two garages for ROV's. These ROV's will not only support the installation of the different components of the station but shall also be used in the preparation of the work site. Furthermore, they shall actuate valves, connectors, etc., mounted outside the vessel. These vehicles shall also be used for general site observation during the execution of the repair.

3.2.2. *Alignment Frames.* The alignment frames shall be lowered using guidelines. The exact positioning of the frames over the defective pipe section should be supported by the submersible.The frame system consists of at least two sections equipped with horizontally and vertically adjustable clamps to lift the pipeline and to perform positional adjustments. The clamps can be hydraulically or pneumatically operated and can perform the required adjustments to correct the pipeline position vertically and horizontally.

The frames have to be fitted with an autonomous pneumatic or hydraulic system for the preliminary installation stages.

3.2.3. *Habitat.* The habitat is a square shaped steel modular construction in which the bottom can be closed through sliding segments after installation over the pipeline (closed bottom type). The side walls are designed with cut outs which surround the pipeline. A tight fit should be obtained through hydraulic operated covers. Such a construction guarantees the possibility to keep the internal top side of the habitat dry during the transfer from the surface to the sea bottom.

Further features of the habitat are listed below (Figure 3):

a) on the top of the habitat there is a docking system for the submersible. Here all penetrators and connectors for control and monitoring tasks are mounted. These penetrators are closed by covers and therefore protected against water. After the docking

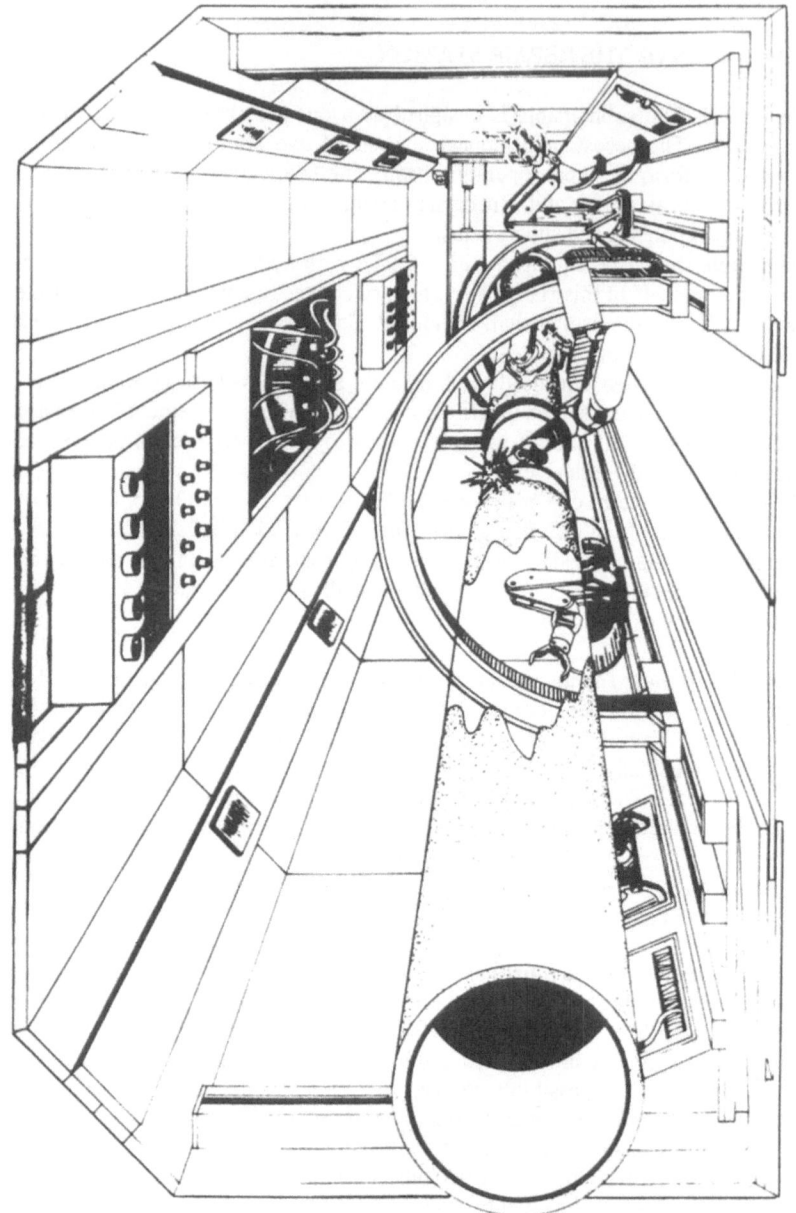

Figure 3. Schematic view inside the purpose built habitat. Pictured are the two robots mounted on semi-circular gantries and the two manipulators. On the habitat's cieling there are two sets of tool magazines and the connection flange between submersible and habitat (control and monitoring cables also pictured).

procedure has been concluded, the remaining water in the flange area shall be evacuated by compressed gas. After the pressure compensation between flange cavern and the submersible (one bar), the connectors are fitted at atmospheric conditions.

b) on the top of the habitat an additional connecting flange is installed for the shielding gas supply, necessary for the welding operations. The shielding gas storage can be mounted on a ROV. This ROV shall dock onto the habitat and therefore supply the required shielding gas. This procedure guarantees the exchange or renewal of gas supply according to the operational requirements (i.e., welding process employed) without interference with the remaining station components. Gas tanks, installed outside the submersible should supply the main chamber gas for the habitat.

Inside the habitat the following systems are installed:

a) two redundant robots are operated in the habitat. They are individually mounted on 180° semi-circular gantries in such a way that every position in the habitat can be reached. The robots are able to perform all necessary tasks (e.g., plugging, groove preparation, welding, testing, coating etc). Both robots are remotely controlled by a central computer on board of the submersible. They can be independently or simultaneously operated.

b) at the bottom region of the side walls there are rail mounted manipulators, which assist the robots (in tasks such as, removal of the defective pipe, inserting the spool piece, preheating, etc.)

c) special tools are stored in compartments (magazines) in the action radius of the robots or manipulators.

d) the spool piece or tap valve, pre-fabricated on the surface, is also located inside the habitat.

3.3. REPAIR PROCEDURE

Once the pipe has been located and the defective section has been demarcated, the subsequent execution of the repair can be divided in three main phases:

a) installation of the alignment frames
b) concrete removal
c) Installation of the habitat
d) pipeline repair
e) demobilisation of the repair station.

In the following items these phases are described in detail.

3.3.1. *Launching and Positioning of the Alignment Frames.* The frames should be launched by means of guide wires, which have been anchored by ROV's launched from the submersible close to the work site. The accurate positioning of the frames is controlled and supported by the submersible. Once the alignment frames are in position the sub-sea energy supply should be launched.

The clamps of the alignement frames are activated by a ROV. In this way the pipes are lifted and kept in a stable position.

3.3.2. *Removal of Concrete Cover.* The concrete removal (and eventual cutting of the steel mesh) should be carried out using the water jet process (probably with additional abrasives) and performed by one of the ROV's assembled at the submersible. In this way a better control and observation of the whole operation is achieved.

3.3.3. *Launching and Positioning of the Habitat.* The habitat is launched and positioned by the submersible (Figure 2). During the launching, the upper part of the habitat is kept dry by means of gas inflation from the bail-out system outside the submersible or habitat. Once the work station has been completely installed on the sea bottom the connection with an underwater energy supply (i.e.: closed-cycle diesel system) can also be established. Form fitted covers close the remaining side openings (between pipeline and habitat wall) allowing therefore the evacuation of the remaining water from the habitat. Before the repair work can be started, shielding gas containers (ROV transported) are launched and connected to the habitat.

3.3.4. *Performing the Repair.* The repair itself and related activities are to be carried out by the two robots and the two manipulators. For each individual task (machining, welding, testing, etc.) the robots will use conventional equipment and accessories (milling tools, welding torches, ultrasonic heads, etc.) fixed on magazines mounted on the habitat ceiling (see Figure 2). The same applies for the manipulators which according to the task (support of the damaged section during cutting, pre-heating, demagnetization, etc.) should use appropriate grippers (four-fingered grips, inductive heating pads, demagnetizing coils, etc.). The following activities are foreseen:
(a) cutting and machining
The defective pipe segment shall be cut using machining tools operated by the robots. In the process of this operation the groove preparation shall be machined on the pipe ends in order to receive the spool piece (prepared on the surface).
The cut-out (damaged) pipe piece should be supported by the manipulators provided with grippers. Once the cutting process has been concluded, the defective pipe section is disposed with the assistance of the manipulators. At this stage the pipe should be reformed either by an internal support system or hydraulic hammers actuating from inside the pipe.
(c) Insertion of the spool piece.
After the defective pipe segment has been cut off, the available pre-fabricated spool piece should be transported by a robot and accurately positioned by the manipulators. After that the spool piece is tack welded and the manipulators are free for other operations (pre-heat, demagnetization, etc).
(e) Welding
Once the spool piece has been inserted and tack welded the preheating pads shall be hold by the manipulators, in such a way that the robot arms have easy access to the grooves. Preferably the whole joint should be filled using one single welding process. However, the use of a non-consumable process (GTAW or Plasma welding) with cold wire addition might be required for the root and hot passes. In any case, regardless of the process

used, for both passes a manual override of weaving parameters and wire feed rate should be available. The welding torch module shall be provided with a camera system for the observation of the arc, the weld bead and the weld gap. The adjustment of the torch position, welding and weaving parameters is carried out by an adaptive control unit. For this a seam tracking system is included in the welding head.

(f) Testing

The quality of the weld should be tested by US devices during the welding operation. Measurements and results achieved for each layer are recorded in a central processing unit. Data monitoring and recording should allow the precise location of a defect or fault in the eventual need of a repair weld.

(g) Corrosion protection

Corrosion protection procedures should follow immediately after the conclusion of the welding operation. Different possibilities are available, although the spray coating using a multi-component paint will be preferably applied due to of its simplicity.

3.4. DEMOBILIZATION OF THE DIVERLESS UNDERWATER REPAIR STATION

Once the repair operation has been completed the habitat should be brought to the surface. The recovery of alignement frames, underwater power pack and any other additional device should be carried out and observed from the submersible.

4. RESEARCH AND DEVELOPMENT REQUIREMENTS

The technology required for the successful development of a diverless pipeline repair system viewing an eventual application at greater water depths (beyond 500msw) is not yet available. However, a cost analysis of a single repair shows that the costs and risks involved are so high as to justify a R&D programme to develop this technology.

The environmental and physical conditions expected at the working depths in question associated with the technological requirements of a diverless repair operation demands a dedicated vessel purpose built for the task. The data transfer rates, response characteristics and pressure sensitivity of some key components (i.e.: welding power source and respective controller, I/O units, etc.) in current diver-assisted repair spreads, forces their installation in pressure housings mounted directly onto the habitat or in its immediate proximity on the sea bed. With increasing depth (and therefore, data transfer distances) and complexity of the systems involved it is expected that the amount of components requiring protection against the environment and proximity to the repair region is also going to increase. More important than that are the eventual consequences to the repair operation in the case of either mal-functioning of I/O units, interfaces or controllers or faulty data transfer. Although the use of a dedicated vessel (submersible) has a critical influence on the final cost of the repair system, it is believed to be a decisive element in the successful execution of a deep water repair.

The development of free swimming remote operated vehicles has seen an explosive growth rate in the past few years. As a result of that ROV's with extended capabilities, increased size and horse power are now available. These vehicles can nowadays be used to:

(a) position and manipulate work tool packages,

(b) to function as a source of hydraulic power to operate those work packages,

(c) provide remote sensing and measurement capabilities,

(d) to provide direct visual inspection and scenario observation.

Therefore, the ROV technology present available can fulfil at a competitive price, the requirements of the diverless repair station described in this work.

The supply of electrical energy is essential for sub-sea working stations operating on an autonomous basis. A comparison among the different surface independent energy generators indicates that the closed-cycle diesel are the most suitable systems. In addition to that, closed-cycle diesel engines largely comprise proven elements and its level of development may be regarded as so high that no difficulties are to be expected in adapting to sub-sea operation with carbon dioxide as carrier. The increasing loss of efficiency at greater water depth due to the need for an exhaust-gas compressor can be considerably reduced by intercooling [Gietzelt, 1987].

The coordination among the four main components (two robots and two manipulators) inside the repair station is a more complex problem. However, the growing industrial demand for automation in assembly and/or construction tasks has created the need for programming and simulation systems for work cells in which two or more robots work in combination [Freund et al. 1990]. In addition to that the geometric and dimensional features of the elements involved in a pipeline repair are well known and have a fixed position inside the habitat. Thus, although the coordination of the four components involves some development work it can be realised with the software tools presently available. As far as peripheral devices are concerned, most of the tools and handling systems (e.g.: machining tools, sensors, welding torches, tool magazines, etc.) are either available or can be easily adapted for the envisaged tasks.

The welding operation poses one important question. It is presently not quite clear which process could be employed for the root and hot passes. It has been suggested [Richardson et al., 1990] that the GTA welding process would probably reach its practical application limit by about 60 bar. One alternative is the plasma welding process. However, although there is evidence that this process could provide stable operation above 60bar [Richardson, 1989], the welding torch is apparently still too big for narrow gap preparations. A second alternative is the GMAW with solid wire. This process has already been successfully used to perform root runs in the flat position down to 500m using both short circuit transfer and pulsed current techniques. Although preliminary studies on process stability above 60bar (using short circuit transfer) have shown encouraging results [Dos Santos, et al., 1990], the actual capabilities of the process to bridge root gaps at the pressure range in question have not yet been investigated. In case this process proves to be adequate for root runs above 60bar, the chemical formulation of the available wires would probably have to be optimized for the this pressure range.

5. CONCLUDING REMARKS

The general concept for a diverless repair station as described in this work has been based on the current state of the art in underwater technology. Three fundamental aspects have been taken into account :
• flexibility inside the working cell (hence, the extensive use of robots and manipulators),
• self-sufficiency (hence, the provision of a surface independent energy supply and the installation of all control and monitoring devices in the submersible), and,
• reduced distances for data transfer and proximity between control units and their executive counterparts (hence, the need for a submersible).

The equipments, systems and technologies required are in its overwhelming majority already available. As a matter of fact, most of the foreseen problems are expected to be found in the combination of all different components in one working station rather than in the development of individual elements or sub-systems.

However, three development areas should be addressed:
• the coordination among robots and manipulators for all the individual tasks,
• the realization of a closed-cycle diesel energy generator, and,
• the development of satisfactory procedures, processes and techniques for root and hot pass welding at the envisaged working depth.

Important to notice is that regardless of the repair concept (be it with manipulator assisted orbital system or fully robotic) the three topics mentioned above must be further pursued if repairs are to be carried out beyond 600m at all. However, it must be also emphasized that these topics are currently being investigated at different institutions worldwide. For instance, the requirements concerning welding technology for such depths are currently part of the long term R&D programme at the GKSS Forschungszentrum. It is therefore valid to assume that such a diverless, self-sufficient repair station might be a reality in the near future.

Another positive aspect in the development and realisation of this repair station is that the test facilities for the main element of this station - the habitat - already exist. The features (dimensions and pressure rating) of the main working chamber A1 of the GUSI complex (GKSS Underwater Simulation Plant) are sufficient to install a mock-up of the habitat and simulate repairs down to 1100m.

6. REFERENCES

Aust, E. and P. Szelagowski (1985) "Schweißroboter im Unterwassereinsatz" in GKSS Extern Report Nr.: 85/E/52, Geesthacht.
Aust, E. and H. Domann (1989) "Development of a positioning sensor for a subsea robot", in N.M. Patrikalakis, J.S. Chung and M.J. Morgan (eds.), Proceedings of the Eighth International Conference on Offshore Mechanics and Arctic Engineering, ASME, The Hague, Volume VI, pp. 129-138.
Alexander, F.J. and R. Quin (1984) "Emergency pipeline repair" in European Seminar

on Offshore Oil and Gas Pipeline Technology, Birmingham, pp.2-19.

Blight, J. and M. Baylot (1987) "THOR-2" diverless welding system in D.N. Waller (ed.), Proceedings of the Second International Conference on Developments in Automated and Robotic Welding, The Welding Institute, Paper 12, London.

Crane, A.D. and I.M. Richardson (1990) "Power source requirements for underwater welding", Draft Report Prepared for the IIW Select Committee on Underwater Welding, Kiew, 39 pp..

Delauze, H.G. (1989) "Welding in difficult conditions for the offshore and nuclear industries" in International Institute of Welding (ed.), Pergamon Press, Proceedings of an International Conference on "Welding Under Extreme Conditions", Helsinki, pp.1-17.

Dos Santos, J.F., P. Szelagowski, H.-G. Schafstall and D. Hensel (1988) "Mechanical and metallurgical properties of robotic underwater welds performed within a depth range of 100msw to 1100msw" in M.M. Salama, R. Denys, H.C. Rhee and J.Y. Koo (eds.), Proceedings of the Seventh International Conference on Offshore Mechanics and Arctic Engineering, ASME, Houston, Volume III, pp. 51-62.

Dos Santos, J.F., K.D. Kober, P. Szelagowski, G.F. Schultheiß and J. Thomas-Jeromin (1989) "Results of the application of a seam tracking system in robotic hyperbaric root pass welding" in International Institute of Welding (ed.), Pergamon Press, Proceedings of an International Conference on Welding Under Extreme Conditions, Helsinki, pp. 225-238.

Dos Santos, J.F. , P. Szelagowski, H.-G- Schafstall and A. Dobernowski (1990) "Preliminary investigations on the effect of short circuit variables on metal transfer above 60bar abs." in Proceedings of the 22nd Offshore Technology Conference, Houston, Volume I, pp. 127-134.

Dos Santos, J.F., P. Szelagowski and H.-G. Schafstall (1991a) "Hyperbaric flux cored arc welding in the depth range between 600msw and 1100msw: process behaviour and weld metal properties" in American Welding Society (ed.), Proceedings of an International Conference on Underwater Welding, New Orleans, pp. 120-134.

Dos Santos J.F., H. Manzenrieder, G. Reinsch, P. Szelagowski and H.-G. Schafstall (1991b) "Considerations on arc observation systems for hyperbaric welding" in Proceedings of the 1991 Offshore Mechanics and Arctic Engineering Conference, Stavanger, Paper 91-846.

Freire, W (1989) "Campos basin deepwater giant fields" in Proceedings of the 21st Offshore Technology Conference, Houston, Volume II, pp. 117-126.

Freund, E., H. Heck, K. Kreft and Chr. Mauve (1990) "OSIRIS - Ein objektorientiertes System zur impliziten Roboterprogrammierung und Simulation", Robotersysteme 6, 185-192.

Gietzelt, M. (1987) "Underwater energy supply" in H. Bianchi (ed.), GKSS, Proceedings of the PETROBRÁS/GKSS Workshop on Underwater Technology, Rio de Janeiro, Paper 12, 18 pp..

Gustmann, M., H.-R. Niemann, E. Aust and G.F. Schultheiß (1991) "Modifizierter Industrieroboter geht in die Unterwasser-Erprobung", Robotersysteme 7, pp. 37-40.

Håbrekke, T. and H.O. Knagenhjelm (1989) "Hyperbaric mechanised TIG-welding for underwater tie-in and repair operations, using the IMT (Integrated Modular Tool)" in International Institute of Welding (ed.), Pergamon Press, Proceedings of an International Conference on Welding Under Extreme Conditions, Helsinki, pp. 247-259.

Jansen, B., P. Loebel, H.-G. Schafstall, P. Szelagowski (1987) "Manuelles und mechanisches Verbindungsschweißen in hyperbarer Umgebung" in GKSS Extern Report Nr.: 87/E/31, Geesthacht.

Knagenhjelm, H.O., A.W. Morris, C.A. Pinches, G.S. Bellis and K. Gjermundsen (1985) "The development of a mechanised welding system for deep waters" in P.T. Houldcraft (ed.), The Welding Institute, Proceedings of the First International Conference on Advanced Welding Systems, London, pp.449-459.

Lerique, M.P. and T.G. Tangeland (1989) "Tie-in tool for underwater systems: TITUS" in Proceedings of the 21st Offshore Technology Conference, Houston, Volume II, pp. 331-336.

Mecklenburg, J, P. Szelagowski, J.F. dos Santos and H.-G. Schafstall (1991) "Development of an orbital GMAW procedure for the use in hyperbaric atmosphere" in H.-G. Schfstall, D. Seeliger and G.F. Schultheiß (eds.), GKSS, Proceedings of the 3rd International Symposium on Underwater Technology, Geesthacht, pp. 23/1-23/12.

Richardson, I.M. (1989) "Some characteristics of the constricted GTA (Plasma) arc at elevated pressures" in International Institute of Welding (ed.), Pergamon Press, Proceedings of an International Conference on Welding Under Extreme Conditions, Helsinki, pp. 215-224.

Rougier, R. (1991) "Automatic hyperbaric welding with THOR-2" in Proceedings of the 1991 Offshore Mechanics and Arctic Engineering Conference, Stavanger, Paper 91-843.

Tassini, P. (1987) "The Mediterranean - deep waters findings and developments" in Sixth (1987) International Symposium on Offshore Mechanics and Arctic Engineering, Panel I - Offshore Outlook, Houston.

Wittman, R.H., B. Thomsom and F. Nanni (1985). An optimum method for the repair of pipelines in deep water. In: International Conference on Deep Offshore Technology, Rome, 21 pp..

ROV OPERATED ELECTRICAL CONNECTOR

D B PYE
TRONIC ELECTRONIC SERVICES
SANDSIDE ROAD
ULVERSTON
CUMBRIA
LA12 9EF

ABSTRACT

The recent trend toward Diverless Intervention in the Offshore Industry has resulted in the development of remotely operated variants of a number of basic items of equipment.

TRONIC LTD in co-operation with SUB SEA OFFSHORE are currently engaged in the manufacture and test of an ROV Mateable connector based on existing technology where possible.

The significant criteria addressed during the programme include tolerance to mis-alignments in various planes, rough handling, ease of assembly, and repeatable electrical properties to name but a few.

The project is seen by the participants as on-going and a number of improvements have already been identified as being desirable in the finish product.

1.0 Introduction

This paper is intended to provide an over-view of the progress to date in the development of an ROV mateable connector. The programme is still in its early stages and, whilst significant features have still to be proven by long-term testing, recent tests have established the basic soundness of the design. The project began in December 1990 following initial discussions with Sub-Sea Offshore. Whilst the basic requirement for diverless systems has been long established and was the stimulus behind the replacement of manned submersibles with ROVs in the late 1970s, it is only within the past 5 years that the design of conductive couplers has progressed to a stage where sub-sea mating is achievable with consistency.

Whilst sub-sea mateable connectors existed prior to 5 years ago, in general they were unable to provide respectable and repeatable results or lacked the

241

considerable robustness required to withstand handling by
ROV. Additionally, none of the existing connectors were
designed with 20 year+ immersion in mind, being either
Neoprene moulded or constructed from unsuitable materials.

The design, therefore, needed to bring together the features
of several basic connector types, and to introduce the
robustness and tolerance to mis-handling inevitable with
remote connection.

In addition to the improvements achieved in connector
design, certain other parallel developments had to take
place to allow the requirement to be addressed. The most
significant of these is the establishment of versatile
handling systems, manipulators, tools etc, to allow the
precise connection and disconnection of couplers (both
hydraulic and electrical). Also of great importance to the
project is the development by various ROV manufacturers of
docking systems which can reduce the necessary dexterity of
manipulation and increase the mating force available.

Two separate approaches to remote connection are to be
considered:

1   Provision of 'Stab Plates' to enable connection to be
       made concurrent with equipment installation subsea.

2   Provision of fixed connectors on one half of the
       structure with the mating being carried out by ROV
       following installation of the remainder of the
       equipment.

Whilst both systems have their merits and limitations, it
was considered that, notwithstanding the necessity for 'Stab
Plates', it was the ROV mateable connector that provided the
most versatility.

The perceived limitation of the 'Stab Plate' system is that,
as it forms part of the structure being assembled, without
careful design the coupling can be subject to high, non
axial, mating forces and approach rates. Additionally,
replacement and maintenance is difficult and disconnection
involves removal of part of the sub-sea structure.

These therefore, were the elements which led to the

initiation of the project now underway. It is important to
note, however, that the development of 'Stab Plate'
connectors has proceeded in parallel and several adaptation
of standard products have been installed and commissioned.

2.0 Specification

2.1 Overall Requirements

> Our initial approach was to utilise as many
> existing and well-proven concepts as possible and
> to combine these with the features necessary to
> facilitate remote handling. These additional
> requirements are listed below:

2.1.1 Tolerance to axial mis-alignment

2.1.2 Tolerance to radial mis-alignment

2.1.3 Ease of handling by existing tooling

2.1.4 Acceptable IR figures with minimum mating force

2.2 Detailed Specification

> The following are the significant requirements of
> the product as initially specified.

2.2.1   Minimum 7-circuit, 10 Amp per circuit,
           250 V operating.

2.2.2   Depth rating: 3,000 ft

2.2.3   Mating cycles: 50 at rated depth

2.2.4   Service life: 20 years

2.2.5   Mating force: Minimum possible to
           achieve necessary IR

2.2.6   Minimum IR (wet mated).   500 MΩ

2.2.7   Field Assemblable, pressure compensated

Mating/De-mating to be achieved by a simple push/pull
action. Initially, certain components of the connector were
designed to suit specified handling equipment, in the long
term it would be preferable for tooling adaptors to be
designed to allow the connector to be standardised.

3.0 Product Description

design was initially based around the Tronic K3 7-pin
'Seawater Environment' connector, upwards of 3,000 of which
have been in service from 1988 onwards in a number of
applications. This represents the mid-range of pin
requirements which, it is envisaged, will vary between 2
and 12.

The design was to utilise the standard cable gland seal and
pressure-compensation system of the 'SE' and the existing
pin and socket inserts. Those items have been proven in
service and in qualification testing hence their use would
considerably reduce the number of new design features of the
connector.

The main feature to be designed specifically for this
application was the Push/Pull latching mechanism. Again,
existing technology was applied to this aspect of the
connector by utilising what is effectively the locking
mechanism applied to many hydraulic couplings, namely the
use of stellite ball-bearings and a sliding collar. This
allows the connector to be made-up by pushing on the main
body and released by gripping the collar and pulling. No
rotary motion is required. To facilitate handling, the
cable exit was made on a right angle.

The nose of the plug connector is machined in a spherical
form with a wall thickness of 5 mm. This allows for axial
mis-alignments of up to 30 degrees to be tolerated and
limits the effect of damage due to mis-handling or high
rates of approach.

The fixed receptacle has a V shaped entry to the alignment
keg-way to allow increased tolerance to radial
mis-alignments.

Both radial and axial mis-alignments are greatly reduced by
the use of docking mechanisms but the design is such that
mating is to be achieved both with and without docking.

Material selection is obviously vital for tolerance to
long-term immersion in sea water. At present, the main
components are manufactured from 316 L but changes to
such materials as ferallium are envisaged for production
connectors.

To provide protection prior to assembly of the connector
sub-sea, it is proposed that both Plug and Receptacle be
fitted with dummy connectors. The connection sequence would
thus be:

a.   Remove dummy from fixed receptacle and stow on ROV

b.   Remove Free Plug from its parking place (effectively a
      dummy receptacle) and mate with receptacle.

c   Mate dummy plug with dummy receptacle to retain their
      integrity for future use.

This sequence, whilst at first seeming complicated is
essential to maintain all components in first-class
condition by reducing contact with sea water and preventing
the entry of foreign bodies and debris, particularly to
the receptacle.

4.0 Progress to Date

Prototypes have been manufactured and subjected to the
tests detailed in Section 5. Initial results indicated that
a higher than expected mating force was required to obtain
satisfactory insulation resistance readings. It was not
fully realised that hand connection of the standard
connector using only the existing securing ring can exert a
load of up to 250 kg on the cup and cone seals. The
load required to achieve successful readings was initially
100 kg but this has since been reduced to 70 kg at which
level consistent IRs of better than 500 MΩ are produced.
It was found that at lower levels of mating force the IR,
although initially unacceptable, increased to higher levels
over a period of 12 hours. This was however, considered
intolerable for a production connector, hence the need for
the present level of 70 kg.

The basic components of the connector have remained
virtually unchanged throughout the programme, the only major
changes being the provision of a right-angled cable entry to
facilitate handling by a range of tools and manipulators and
modifications to sealing cone angles.

The present philosophy is, therefore, to accept the mating
force, given that it allows the use of standard technology
and will increase customer confidence.

5.0 Test Programme

5.1 Tests to Date

      Tests so far have been concentrated on establishing the
      minimum engagement load required to achieve
      consistently acceptable insulation resistance levels
      and to gain experience in the handling of the
      connector.

Modifications have been made to sealing cone angles
which resulted in greatly reduced mating forces. This
is due almost entirely to the elimination of a
hydraulic lock at each cup/cone interface, and is
demonstrated by the wet-mate/dry-mate load ratio being
reduced from 6 to 1 to less than 2 to 1.

Most recent tests have involved 100 matings at 1500 psi
and insulation resistances consistently in excess of
10 G ohms have been achieved.

5.2   Completion Of Test Programme

Having achieved and demonstrated consistent mating
performance to independent witnesses, the further
extension of the test programme will investigate the
mechanical performance of the connector rather than
electrical. This will take a number of forms:

1.   Multiple matings from extreme angle of entry.

2.   Tolerance to extreme rates of approach.

3.   Tolerance to foreign bodies and contamination.

4.   Handling trials with various types of ROV tooling.

5.   Long-term deployment off-shore in an installation
      that would allow periodic electrical checks to be
      carried out.

This programme is obviously not fixed, and may be varied at
any time following discussions with operators and end-users.

6.0   Future Developments

These can be divided into 3 sections thus:

(a) Further work on 'SE' Variant

(b) Development of 'CE' Variant

(c) Development of tooling

6.1 The work on the 'SE' variant is near to completion.
Following the conclusion of the tests outlined in 5.1 and
5.2 it is intended to deploy one of the prototypes sub-sea
for long term testing. The connector will be installed in
such a way that periodic monitoring is possible over an
extended period.

6.2   It has recently become policy amongst some of the major oil companies to demand redundancy of seals on sub-sea electrical connectors. Whilst this is provided on the cable to connector termination on the 'SE', the pin to socket interface has only one seal, ie the cup and cone arrangement. To provide redundancy at this point it is necessary to utilise the 'Controlled Environment' principle whereby a single oil-filled, pressure-balanced chamber surrounds all pins and individual chambers isolate each pin. Combining this feature with the compensated gland seal provides redundancy at both Cable/Connector and Pin/Socket interfaces. It is,therefore, our firm resolve to produce a 'CE' version of the ROV mateable connector which will have the additional benefit of reducing the mating force required.

The overall size will, of course, be slightly larger to accommodate the same number of electrical contacts and be rather more expensive. However, the connector will provide total redundancy of seals.

6.3   In parallel with the SE/CE developments, it is seen as a priority to develop a range of deployment and handling tools.

To provide the necessary mating force on the 'SE' version it is virtually essential to utilise a docked ROV. It would, therefore, be advantageous to provide a tool which would be attached to the connector to provide the necessary force without the necessity for a complete docking mechanism. This would have the additional advantage of allowing connection by non-dedicated ROVs and smaller vehicles.

7.0   Conclusion

The project as it stands today, has resulted in the development of a product which has, by design, brought together a number of well proven elements. Emphasis has been laid on minimum changes being made to existing components except where necessary to suit the peculiarities of the product.

In common with the other leading companies specialising in sub-sea connectors, it is our intention to continue development of this, and other products, in response to market requirements, and all comments from within the industry are most welcome.

Part 4
Safety

HYPERBARIC EVACUATION

V J HUMPHREY
Senior Diving Inspector
Health & Safety Executive
Offshore Safety Division
Greyfriars House
Gallowgate
Aberdeen
AB9 2ZU

ABSTRACT. A review of the hyperbaric evacuation methods and the responsibilities of persons appointed for the Health & Welfare of the offshore diving workforce: Post Cullen.

Train crashes, explosions, gas leaks and fire have all caused extensive loss of life and serious injury to large numbers of people world wide. The ability of rescue services to respond to disasters of this magnitude, depends on the availability of specific equipment, logistical support and the training of specialised rescue teams.

When incidents occur offshore, the problem is compounded by the fact that adverse weather conditions and the remoteness of the installation or vessel, can both initially delay the primary response from a dedicated safety vessel, if applicable and reduce the response time and effectiveness of co-ordinated rescue services.

For divers under pressure the rescue process is further complicated, because the evacuation of such personnel, must be conducted by transferring them under pressure with sufficient life support, to a place of safety away from the installation.

Lord Cullen, in his recommendations following the Piper Alpha disaster, stated that "operators should be required by Regulation to submit to the regulatory body for it's acceptance, an evacuation, escape and rescue analysis in respect of each of it's installations".

The Offshore Safety Division is presently of the opinion that in the interests of safety this recommendation should also be applied in respect to all diving support vessels (DSV), pipe laying and crane barges and other craft which may be employed to carry out air and saturation diving operations.

The analysis should specify the facilities and other arrangements which would be available for evacuation, escape and rescue of personnel in the event of an emergency which makes it necessary or advisable in the interests of safety for personnel to leave the installation.

In both the United Kingdom and Norway there is a legal requirement to provide a suitable means of hyperbaric evacuation. The legislation which specifies the requirement, in the UK, is The Diving Operations at Work Regulations 1981 as amended by The Diving Operations at Work (Amendment) Regulations 1990. Regulation 5

Volume 27: Subtech '91, 251–254.

paragraph (2)(b) states: "every diving contractor shall so far as is reasonably practicable ensure that:

emergency services are available including in particular in the cases of diving:-

(i) using saturation techniques or

(ii) at a depth exceeding 50 m,

Facilities for transferring the divers safely under a suitable pressure to a place where treatment can be given under pressure.

This instruction is further qualified by the following guidance on Regulations which is included in the HSE L6 document which replaces HS(R)(8).

Guidance Note 109 of the Regulations states: "special attention should be given in the rules to emergency procedures, including first-aid and medical assistance, and in particular to contingency planning for the evacuation of divers under pressure from an offshore installation or vessel".

Guidance Note 112 further states: "facilities should be available for the transfer of sick or injured persons under pressure where a compression chamber is required on site. Diving contractors should ensure also that their compression chamber is compatible with that of the transfer service and that arrangements have been made with a transfer service before the start of diving operations.

Air diving operations should be assessed individually to determine the potential risk to divers, especially when operations are carried out from production platforms or drilling rigs, to establish whether there is a requirement to provide a hyperbaric evacuation capability.

Guidance Note 113 states: "the contingency plans for evacuation of all divers under pressure must be capable of being implemented under all foreseeable circumstances. Whatever the chosen method of evacuation the diving contractor should provide adequate transportable chambers to transfer the total number of divers under pressure. All members of the diving team need to be fully conversant with the contingency plans and precise procedure for the evacuation of divers under pressure.

Operators should ensure that adequate training periods are allocated and diving contractors contingency plans must contain detailed instructions and procedures for the total evacuation, which includes the recovery of the hyperbaric craft and subsequent decompression of divers under pressure.

Guidance Note 114 states: "under some circumstances the facilities provided for the medical transfer of persons under pressure could provide additional back-up in the event of an extreme emergency, but the medical transfer system should never be the primary preplanned method of evacuation.

Because of the particular risks associated with production installations and drilling rigs, the owner/operator should properly evaluate, by "Hazard Risk Analysis", the necessity of sighting permanent or temporary saturation diving systems on board these types of facilities.

Where saturation diving systems are installed on oil and gas production platforms or drilling support vessels, there is always the inherent danger that the development of a major life threatening situation such as fire, collision or extremely adverse weather conditions, could have catastrophic effects on the safety of divers under pressure.

The owner of the installation has a responsibility to locate the diving system in an area which is as remote as is reasonably practicable from the production or drilling facility.   In situations where a sufficient degree of remoteness is impractical, consideration should be given to the installation of a physical barrier.

If a potentially dangerous situation arises which necessitates the abandonment of the production platform, drilling support vessel or DSV, then divers who are confined under pressure in saturation must be evacuated by the use of specialised equipment and written emergency procedures.   Additionally, the safety of those personnel responsible for both the recovery of the diving bell, if applicable, and the deployment of the hyperbaric rescue craft, have to be considered.   All appropriate installation communications should be fully integrated with dive control to allow the procedure to be implemented without delay.

In the case of drilling support vessels there should be an alternative means of evacuation other than the diving bell.

The ultimate responsibility for the safety of all personnel on any offshore installation lies with the owner, the concession owner or any other person who has a responsibility for the safety of personnel.   It then rests with the owners representative, **The Offshore Installation Manager (OIM)**.   Where diving support vessels (DSVs) and construction barges are engaged in offshore diving operations, **the responsibility for the evacuation of personnel lies with the Captain.**

In the fire situation, the installation's deluge system must have sufficient capacity to provide adequate protection, in all circumstances, for the whole of the diving complex, including the launch position, control and access ways.

Hyperbaric rescue craft should be equipped in accordance with the AODC/UKOOA design document and have some form of propulsion, which permits the rapid evacuation of all personnel under pressure, away from the potential hazard.   Deployment systems other than lifeboat gravity type davits should have both primary and secondary launching capabilities.

The Current Methods of Evacuation are:-

a)   Hyperbaric self propelled lifeboats.

b)   Towable hyperbaric evacuation craft.

c)   Standard diving bells.

d)   Through water transfer from one bell to another.

e)   Helicopter transfer by one man chamber.

Wildrake, Regalia and most recently the Semi 2, are examples of
where both fire and collision have threatened the safety of divers
under pressure.   There have also been instances where the fly-away
chamber has been brought to a state of readiness because of
extremely adverse weather conditions, to evacuate divers under
pressure if it became necessary.

Some of the evacuation methods are more acceptable than others.
The purpose built propelled rescue craft with it's external fire
fighting capability, medical lock and internal sanitary facilities
is far more acceptable to divers than the conventional non
propelled diving bell, which is both cramped and lacking in human
comforts.   The use of the fly-away one man chamber, for either
medical or emergency hyperbaric evacuation has limitations because
adverse weather conditions and an unstable situation offshore may
prevent it's use.

However, in a crisis situation, survival is the key issue.   The
risk of utilising any of the stated methods of evacuation far
outweighs the alternative of remaining on the installation.

RISK ASSESSMENT OF A HYPERBARIC EVACUATION SYSTEM

D J Burns, D M Deaves
WS Atkins Engineering Sciences Limited
Woodcote Grove, Ashley Road, Epsom, Surrey KT18 5BW

ABSTRACT. Risk Assessment is gaining increasing acceptance as a tool in addressing the safety of offshore operations. In addition to a complete assessment of an installation, which may form part of a Formal Safety Assessment, risk assessment techniques can be applied to specific safety related systems on a platform.

This paper presents the results of a study of the risks associated with the hyperbaric rescue of divers. It includes the identification of accident scenarios, and analysis of consequences of such events as ship collision, fire or explosion. A particular feature of this assessment is the identification of effects which would impede evacuation, such as buckling of steel members, availability of strategic personnel, etc. The study therefore included analysis of effects of incidents both on hardware, and also on personnel and procedures.

Incorporation of event frequencies then enabled calculations to be made of the risk to the divers during an emergency situation. In particular, comparison between the risks attached to different modes of evacuation enables decisions to be made, on safety grounds, with regard to the appropriate operational mode.

1. INTRODUCTION

This paper describes the application of risk assessment to a particular situation relating to diving operations. On the floating production platform under consideration, plans to increase diving activity had been met with regulatory concerns over the ability to evacuate a large number of divers in a platform emergency. The study which was undertaken enabled the system to be evaluated rationally in order to reach a decision on the feasibility of extending diving operations.

Although the presentation is based upon a specific case study, it is intended to demonstrate the more general use of risk assessment in an offshore context. This is discussed in the next section, followed by a description of the system which has been considered in the specific example included. The risk assessment has been described in the following three sections and the paper is concluded with a discussion both of the specific applications to this problem, and also the more general application of risk assessment to Formal Safety Assessments.

255

Volume 27: Subtech '91, 255–268.
© 1991 *Society for Underwater Technology*.

2.   THE USE OF RISK ASSESSMENT

Onshore regulations relating to the safety of 'major hazard' sites have now been in place for some years. These, the so-called 'CIMAH Regulations'[1], require the submission, by the operator, of a Safety Case in order to demonstrate that the hazards have been identified and properly controlled, and hence that the risks have been minimised. Although quantification is not mandatory, it is clearly considered to be an important ingredient of any Safety Case[2].

The recent publication of the Cullen Report[3] on the Inquiry into the Piper Alpha disaster has demonstrated the need to bring offshore legislation into line with that onshore, with a requirement for Formal Safety Assessment.   Again, whilst the level of risk quantification is not prescribed, it is clearly implied that some attempt at a quantified risk assessment (QRA) is required.   Whilst operators are only beginning to realise the full implications of this requirement, it is already evident that QRA can provide a valuable tool in a range of applications relating to offshore safety.

This subject has been discussed in a number of recent conferences, both in a general sense[4,5] and in specific examples, such as evacuation requirements [6]. This paper takes the general approach outlined in Burns et al[5] and applies it to the specific example of diver evacuation. The particular concern in this case was that existing evacuation facilities should be usable by a larger diving team than had hitherto been employed, without significantly increasing the risk.

3.   THE HYPERBARIC EVACUATION SYSTEM

The platform under consideration is assumed to be semi-submersible Floating Production Facility (FPF) with a diving spread consisting of the following main items of equipment:

Submersible Diving chamber (SDC)
Deck Decompression chamber (DDC)
Transfer chamber (TC)
Trunking system
Hyperbaric Rescue Craft (HRC)
Bell/Sat Control Cabin
Environmental Control and Hot Water Container
Main Umbilical Cable
Bell Handling System
Guide Wires

The Hyperbaric Rescue Craft (HRC) is generally a compression chamber housed inside a lifeboat. To be ready for a possible evacuation, the HRC's hyperbaric chamber is kept under pressure with heliox mixture and mated with the DDL via the trunking system. In an emergency evacuation, divers transfer under pressure to the HRC and the trunking clamp is release externally. The vessel is then launched by gravity into the sea. Some rescue vessels are provided with propulsion power, while others require towing. After removal to a safe area, they can be lifted from the water and mated with another decompression chamber. The HRC is normally kept at a lower pressure than the DDC in order to establish a seal against the door of the HRC. Two manways are fitted, one at the bottom and one at the top of the HRC.

The procedures under review here are those related to hyperbaric rescue, whereby the divers may be taken from within their normal diving system location to another system without any change in pressure, for medical or safety reasons[7]. Emergency response procedures will have been drawn up by the diving contractor, and are applied regardless of whether the cause of evacuation concerns the integrity of the structure or is a fire or explosion.

If the bell is in the water at the onset of an evacuation event, the shift supervisor instructs the divers to return to the bell, and issues instructions for recovering the bell as soon as possible, being careful of possible falling debris onto the bell, entanglement with subsea hoses and possible contact with structural members as a result of list. For the multi-diver situation, the divers left in the DDC are alerted by the Life Support Supervisor (LSS) to prepare to transfer to the HRC.

After recovering the divers from the bell, the LSS instructs the divers to transfer to the HRC. The LSS then supervises the sealing of the HRC trunking by overpressurising the DDC whilst the trunking door to the TC is held closed from within. After the last saturation diver has entered the HRC, he ensures that the HRC door is closed, the HRC pressurised and the trunking vented. The mating clamp between the HRC and TC is then opened by the diving crew on instruction from the LSS.

On final confirmation from the OIM and diving supervisor, the HRC, once unclamped, is lowered and moved clear of the platform.

4. SCENARIO IDENTIFICATION

In order to assess the probability of fatalities associated with the use of the HRC, the various scenarios leading to the need to evacuate divers have first been defined. The rescue of divers under pressure may arise from any of the following situations:

- FPF in danger of capsizing or sinking
- Presence of an unacceptable fire or explosion risk or an occurrence such as a blowout
- A fire or other disaster within the diving system
- A medical problem
- A lost bell

This paper is concerned only with the first two of these which particularly relate to the FPF under consideration. The remainder are considered to be risks which apply to all diving operations, irrespective of whether or not there is an emergency on the platform.

Figure 1 shows the qualitative relationship between causes of the need to evacuate divers and the possible ways in which the evacuation scenario could develop. Two broad categories of events leading up to a need for diver evacuation have been defined: hydrocarbon events and structural events. Examples of these have been discussed in the Norwegian guidelines on Conceptual Safety Assessments[8].

In Figure 1, the initiating hydrocarbon events are given as blowouts, riser failures, process pipework or vessel failures and non-process failures. Each of the initiating events

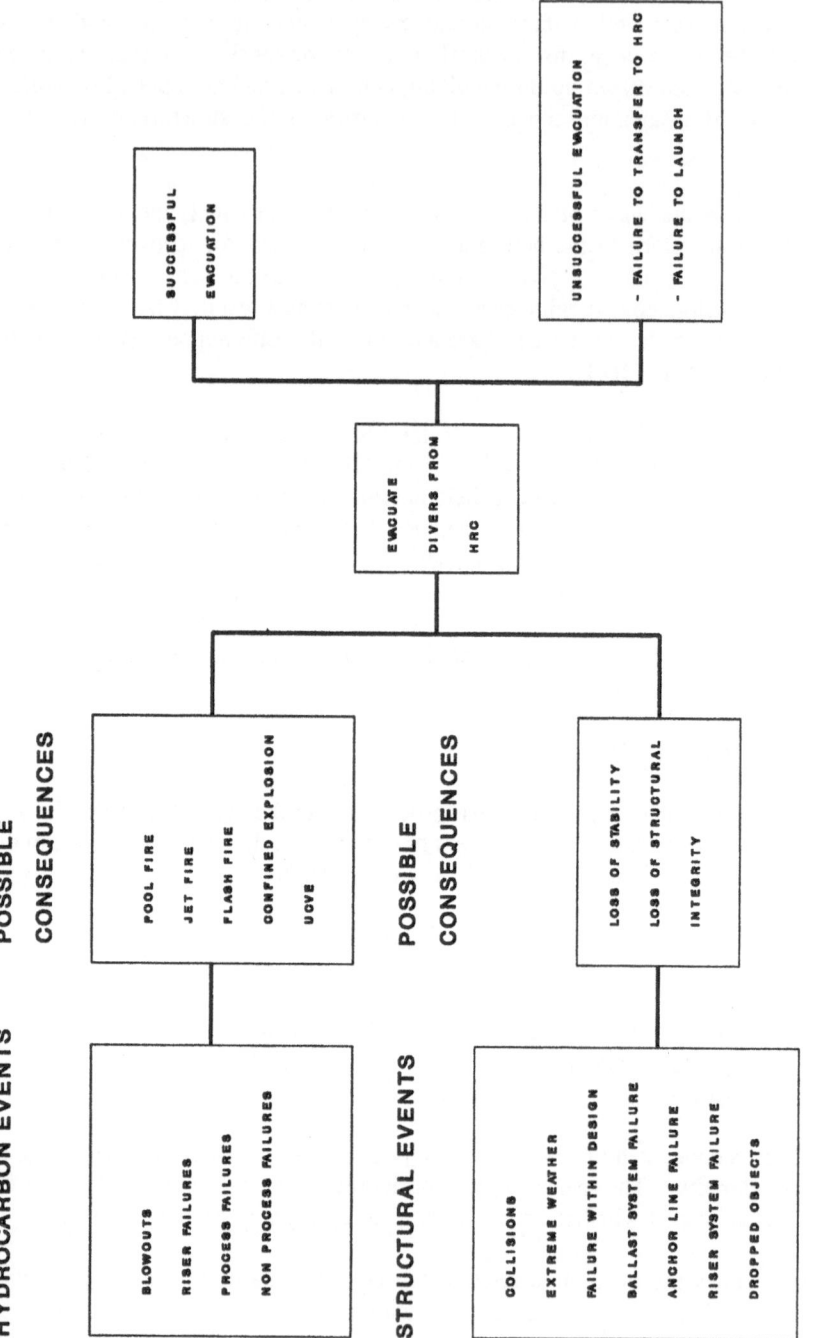

FIGURE 1   EVACUATION SCENARIOS

can give rise to one or more of the following consequences:

- Pool fire
- Jet fire
- Flash fire
- Confined vapour cloud explosion
- Unconfined vapour cloud explosion

The nature of the fire or explosion will depend on the type of release (eg. oil or gas), its pressure and temperature, the time at which it ignites and the surrounding structure.

4.1 Fire

A *pool fire* occurs when an accumulation of oil in a pool is ignited, resulting in a steady burning fire.

A *jet fire* occurs when gas is release from a rupture of a vessel or pipe into free air and rapidly ignited. The pressure of the release serves to generate a long flame which is stable under most conditions and is largely unaffected by the wind.

A *flash fire* occurs when a cloud of flammable gas in a mixture with air is ignited. The fire is usually of short duration as the flame travels rapidly through the cloud.

4.2 Explosion

Ignition of a gas cloud can result in damaging overpressures, which are considerably enhanced in confined regions. However, as there are no enclosed modules on the FPF, it is not considered that *confined explosions* should be included in any scenario.

A *vapour cloud explosion* occurs when the flame inside an ignited cloud of gas accelerates to a very high velocity, thus producing explosion effects. Once initiated, a VCE may cause considerable blast damage, (although less than for a confined explosion) and may also ignite a jet fire if the release is still continuing.

Overall structural failure of the FPF is assumed to occur if the explosion of a gas cloud occurs beneath the FPF. For a gas cloud located above or beside the FPF, only severe structural damage to the topside is expected.

In Figure 1 the initiating structural events are given as:

- collisions
- extreme weather conditions
- failure within design
- ballast system failure
- anchor line failure
- riser system failure
- dropped objects

Each of these initiating events can give rise to a loss of stability or loss of structural integrity necessitating an evacuation of the divers.

For the safe deployment of the HRC, the first four of these items have been considered to be significant. The effects on the operation are likely to be in one of two ways:

a)   Structural deformation causing fouling of the HRC on attempted release.
b)   Loss of stability causing inability to release the HRC.

For the first of these effects, the statistics of passing ships are considered, along with their potential effect in the event of a collision. Reference to a recent Dept of Energy study[9] indicates that most reported collisions had been of low impact energy ($<$ 1MJ), with a maximum of 11MJ. Since it is estimated that around 35-75MJ is required to cause column collapse[10], use of available statistics will result in conservative risk estimates.

Stability could be impaired during extreme weather, although it is concluded[10] that a semi-submersible designed to survive 100 year weather conditions with a considerable margin of intact stability would fail in such a manner as to impair HRC operations at a frequency of less than 10^{-4}/year. Ballast system failure could also result in a similar effect, although at a significant level this is only likely to be caused by human error.

5.   CONSEQUENCE ANALYSIS

The scenarios identified will have certain consequences as far as the HRC operation is concerned. They may or may not impair evacuation, and this is assessed using consequence analysis. The stages of such an analysis include calculation of release rates of hydrocarbon, calculation of damage effects on the system, and review of any mitigating features. This latter is done via event tree analysis, which enables all the conditions leading to non-evacuation to be identified.

5.1   Release Modelling

For an oil-producing facility, the presence of gas in the well-fluid means that any fracture of a well-fluid line will lead to both an oil spill and a gas emission. Above-and sub-platform releases are modelled. For the above platform releases, the ratio of solid flooring to grated flooring in the immediate vicinity of the equipment is considered in assessing the likely location of pool fires and BLEVES.

For each item of equipment, it is assumed that the areas likely to be breached are the pipework connection points. Release scenarios are thus based on the diameters of the main connecting pipelines.

For each rupture, the consequences depend upon:

a)   release size
b)   release location
c)   ignition characteristics
d)   explosion/pool/torch fire effects

When a pipeline rupture occurs, its contents will be discharged under pressure. Computer programs, such as EFFECTS[11], which include models for outflow, are used in this type of study. The calculation models included in EFFECTS are based on the report in Reference 11. Models are used independently and results can then be fed as input to the next appropriate model. For example, a liquid leak resulting in an evaporating pool, which then disperses before an explosion occurs would result in sequential use of the OUTFLOW, EVAPORATION, DISPERSION and EXPLOSION modules.

For gas releases above the sea, turbulent free jets will be formed. If the release is vertical, as is likely for a full bore rupture, the resulting plume will be deflected by the wind, and can be modelled by PLUME[12]. This includes the effects of atmospheric turbulence and predicts the extent of the flammable region of the issuing gas cloud, together with the mass and volume of gas, which is then fed into the EFFECTS program in order to predict damage ranges from fire or explosion.

Gas release ignition could result in any of the following effects:

-       torch (jet) flame emanating from point of release
-       fireball involving rapid deflagration of a large volume of gas
-       vapour cloud explosion involving a large volume of gas immediately surrounding the platform.

For releases below the sea, it is assumed that the jet momentum is rapidly destroyed and that the gas will rise to the surface under the buoyancy forces. When the gas reaches the surface, its downwind dispersion is calculated using PLUME.

Duration of a release will depend on the location and function or failure of the nearest isolation valves, and their associated shutdown signalling systems. An import riser failure would, in most cases, activate a subsea isolation valve at the bottom of the riser and a non-return valve at the top. As the inventory contained in each riser is relatively small, the hydrocarbon emission would be limited.

5.2   Hydrocarbon Damage Effects

Heat fluxes from oil pool fires, torch fires and flash fires are calculated using EFFECTS, as are explosion overpressures from vapour cloud explosions. Contours of heat flux levels and overpressures are plotted from the fire or explosion centre and overlaid on the installation layouts in order to estimate the threat to human life, and damage effects.

Times to failure in minutes for various structural items exposed to jet fires or pool fires are shown in Table 1[10].

Under explosion overpressures, the following assumptions are made:

-       Firewalls and steel walls are blow out at 2 psi (14kN/m²)
-       Decks are blown out at 9 psi (64kN/m²)
-       The columns of a floating platform are ruptured at 15 psi (107kN/m²)

	Jet Fire Flame	37.5kW/m² from Jet Fire	Pool Fire Flame	37.5kW/m² from Pool Fire
Unprotected structural steel beam (load-bearing)	10	20	10	30
Unprotected steel plate (non load-bearing	5	10	10	30
A60 firewall	10	20	30	60
A60(H) firewall	15	30	60	120
H120 firewall	60	120	120	240
Protected structural steel beam	15	30	60	120
Riser	10	20	10	30
FPF leg	15	60	30	120

TABLE 1   Times to failure in minutes (burn through or loss of load-bearing capacity)

	Radiation intensity (cal/cm²s)		
Explosion Yield	20kt	1Mt	20Mt
First degree burn	1.75	0.297	0.0886
Second degree burn	3.14	0.643	0.221
Slightly clothed (summer) few, if any, injuries	1.75	0.297	0.0886
Significant injury threshold	2.80	0.594	0.210
Lethality threshold	3.50	0.792	0.243
Near 50%	6.30	1.385	0.442
Near 100%	14.0	3.07	0.952
Effective time duration, s	1.43	10.01	45.2

TABLE 2   Fireball effects on personnel

Overall structural failure is assumed to occur as a result of the explosion of a gas cloud located beneath the deck. A flash fire is assumed to have no impact on the integrity of the process equipment.

Table 2[13] shows the estimated correlation between heat radiation intensity and burn injury. The threshold fatality limit for exposure to fire is generally taken to be $10kW/m^2$.

5.3   Structural Damage Effects

Structural events (see Figure 1) generally give more direct effects and their analysis is more deterministic than the hydrocarbon effects, which are more probabilistic.

Vessel collision, for example, will give an impact energy which can be calculated from its   size and velocity. Absorbed energies will differ from fixed jackets to floaters, but the magnitudes are relatively well-defined. Damage for a given impact energy is then estimated from design considerations. The main concern is loss of structural stability, although damage to process equipment or risers may otherwise occur, in which case the analysis proceeds according to Hydrocarbon Events in Section 3. Regarding earthquakes, the design will accommodate a given acceleration. An acceleration in excess of this design value is assumed to lead to structural instability and loss of the installation.

Similarly, the design will accommodate a storm, associated with a given return period. More severe storms, with a longer return period, are assumed to lead to a loss of the installation.

For failure within design, possible failure scenarios are, for a floating production unit:

- loss of one or more braces leading to loss of structural integrity
- loss of watertight integrity at any brace/column node or column/pontoon joint leading to rapid flooding.

Ballast system failure is analysed using Fault Tree Analysis in order to define the possible failure causes, leading to loss of stability. Anchor line failure in the mooring system for a FPF could lead to loss of stability, as could flexible riser system failure eg. risers clashing during operations or disconnect mechanisms not working when required, and excessive motion forces transferred to the FPF.

Dropped object studies are carried out from consideration of crane traverse areas, crane reliability and maximum impact loading. The identification of sensitive process equipment lacking dropped object protection is an important part of this study. Objects such as drill collars dropped overboard can lead to riser failure. The probability of dropped objects causing direct structural damage, however, is quite low.

6.   EVENT TREE AND RISK ANALYSIS

For each scenario under consideration, event trees are drawn to enable the ultimate consequences to be assessed. With a calculated frequency for the initiating event, and estimated probabilities for each conditions event, calculations can then be made of the risk of non-evacuation of diving personnel. Example event trees only are presented here; one for a typical hydrocarbon event, and one for a typical structural event; the complete analysis includes a comprehensive set of event trees.

6.1   Hydrocarbon events

The example event tree of Figure 2 relates to a typical gas leak. The conditions noted along the top of this diagram relate to the time of ignition and to whether the conditions are such as to lead to explosion. In view of the small distances involved, wind speed and direction are assumed not to be significant factors.

The event tree of Figure 2 relates only to the nature of the development of a hydrocarbon event once it has been initiated. Figure 3 shows the parameters which need to be considered subsequently before diver evacuation can be assured. Again, this is in the form of an event tree, the input to which would be taken from the output of Figure 2.

6.2   Structural events

In this case, the same events which were considered in Figure 3 will also be relevant. These have been combined with the parameters relating to platform stability to provide an integrated event tree, as shown in Figure 4.

6.3   Results

It is outside the scope of this paper to give complete results from the risk analysis which have been undertaken. However, the calculations have enabled estimates to be made of the additional risk to divers through possible non evacuation. This risk has been found to be of similar magnitude to the North Sea average.

7.   CONCLUSIONS

This paper has demonstrated the application of risk assessment techniques to the evacuation of divers via a hyperbaric rescue craft. In addition to the numerical value of risk which results, the analysis useful in assessing a number of contributory factors, for example:

a)   The importance of close co-ordination between members of the diving team during the operation of evacuation via the HRC.
b)   The optimal physical layout of the diving spread.
c)   The contribution to the risk to divers arising from attendant vessel collisions.

References

[1]   'Control of Industrial Major Accident Hazards Regulations', SI 1984/1902 HMSO, 1984.

[2]   HSR(21) - Guidelines on CIMAH regulations, HMSO, 1985.

[3]   Cullen, Lord. 'The Public Inquiry into the Piper Alpha Disaster', CM1310 HMSO (1990).

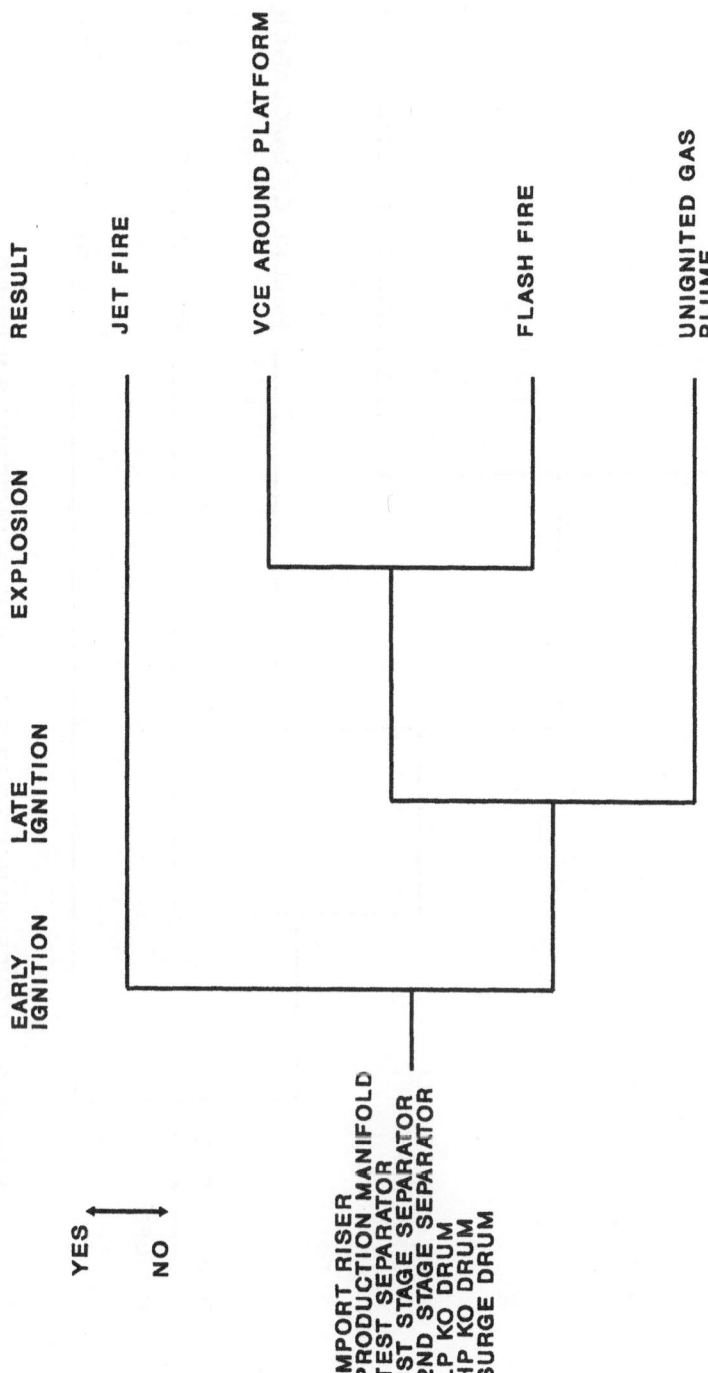

FIGURE 2   EVENT TREE SHOWING VARIOUS GAS LEAK SOURCES

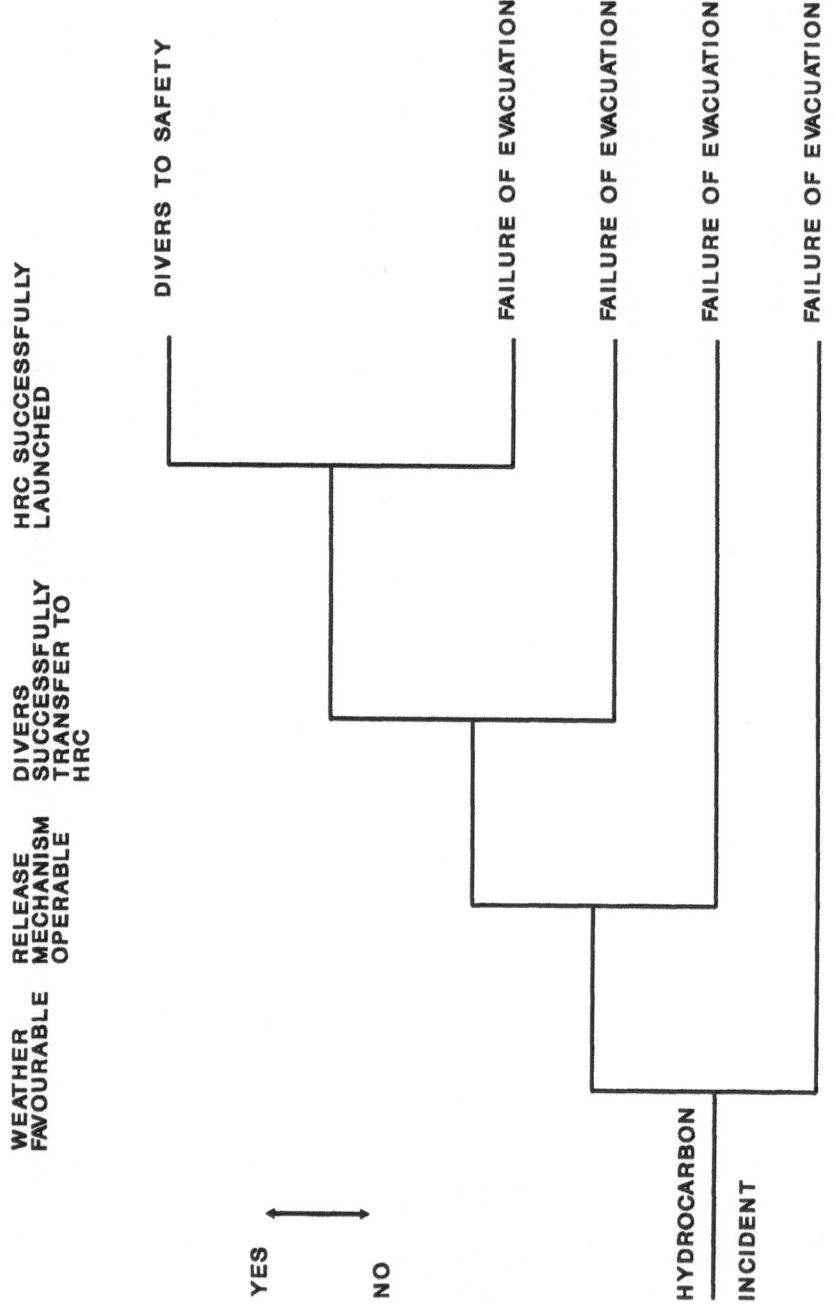

FIGURE 3   CONSEQUENCE TO DIVERS ONCE AN INCIDENT INVOLVING
THE RELEASE OF HYDROCARBONS HAS OCCURRED

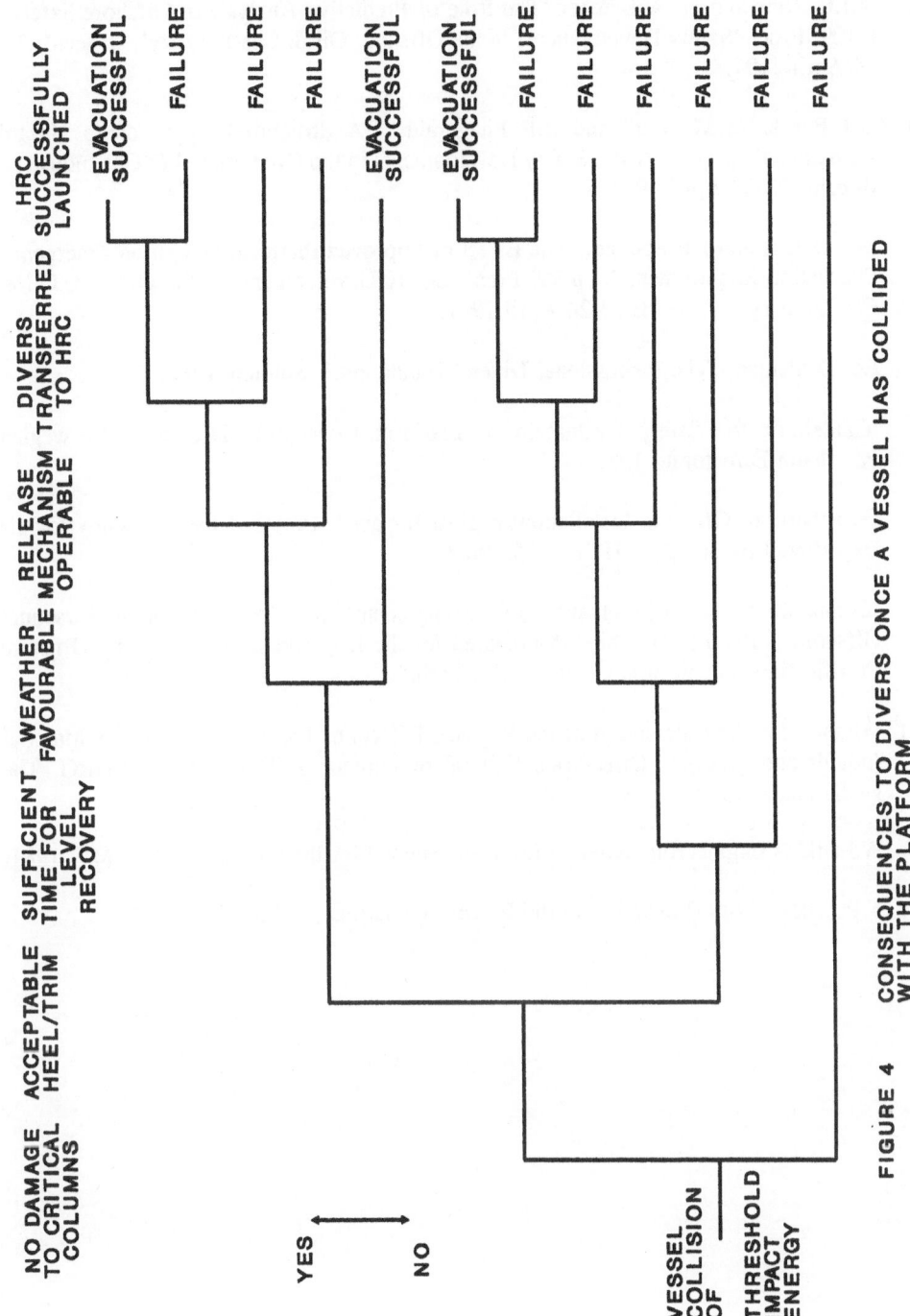

FIGURE 4   CONSEQUENCES TO DIVERS ONCE A VESSEL HAS COLLIDED WITH THE PLATFORM

[4]   N.I.C Rock and N. A Butcher 'The Role of Predictive Analysis in Offshore Safety.'
       p 109. Proc. 'Safety Developments in the Offshore Oil & Gas Industry'. IMechE 23-
       24 April 1991.

[5]   D.J Burns, M.M Grant and B.P Fitzgerald. 'A Structured Approach to Hazard
       Assessment'. p 115. Proc. 'Safety Developments in the Offshore Oil & Gas Industry.'
       IMechE 23-24 April 1991.

[6]   M.J Owens and J.R Spouge. 'The Effect of Improved Platform Design on Emergency
       Evacuation Requirements'. p 99. Proc. 'Safety Developments in the Offshore Oil &
       Gas Industry.' IMechE 23-24 April 1991.

[7]   Ed. D Sisman. 'The Professional Divers' Handbook.' Submex 1982.

[8]   'Guidelines for Safety Evaluation of Platform Conceptual Design.'   Norwegian
       Petroleum Directorate 1981.

[9]   'Protection of Offshore Installations against Impact.' Prepared by J.P Kenny for the
       Department of Energy, OTI 88-535, 1988.

[10]   'Comparative Safety Evaluation of Arrangements for Accommodating Personnel
       Offshore.' Report OTN-88-175 prepared for the Department of Energy by Offshore
       Certification Bureau in association with Technica.

[11]   'Method for the Calculation of the Physical Effects of Escape of Dangerous Material
       (liquids and gases).' Directorate General of Labour, 2273 HK VOORBURG, The
       Netherlands, 1979.

[12]   WS Atkins Engineering Sciences Limited. Manual for the programme PLUME, 1984.

[13]   F.P Lees. 'Loss Prevention in the Process Industries', 1980.

DIVING INCIDENT DATA - THE WAY AHEAD

R GILES
Health and Safety Executive
Offshore Safety Division
Diving Inspectorate

ABSTRACT

The Diving Inspectorate's aim, over many years, has been to collect data on incidents involving the diving operations for the offshore industry and, most recently, develop techniques to enable the assessment of safety. This paper examines, with examples, how this is being achieved and, furthermore, explains how feedback to the industry could be achieved in the future.

1. INTRODUCTION

1.1 Making an assessment of the safety of diving operations is an essential aspect of the monitoring and assurance of appropriate standards within the Offshore Diving Industry.

1.2 The foundations of such an approach are:

- Data available to describe the performance of current operations (historical data).

- Regulations, guidance notes, and operating manuals for current and new diving methods.

- Standard techniques available for carrying out assessments.

1.3 The Inspectorate's aim over many years has been to collect data and, most recently, develop techniques to enable the assessment of safety. This paper examines, with examples, how this has been achieved, and, furthermore, explains how feedback to the Industry could be achieved in the future.

Volume 27: Subtech '91, 269–273.
© 1991 *Society for Underwater Technology.*

2. REGULATIONS AND GUIDANCE

2.1 The Diving Inspectorate has been tackling the control of diving operations via the issue of regulations and diving safety memorandum and appropriate training. These have been generated since 1974 and are used as the basis of any operational safety assessment made by the Department.

2.2 In addition to this, the diving Industry has issued its own guidance documentation via the Association of Offshore Diving Contractors.

3. DATA

3.1 Since 1978 data has been collected on incidents involving the diving operations for the offshore industry. The data has been carefully sorted and assessed by the Diving Inspectorate's own Inspectors. In addition, the data has been coded and entered onto a computer controlled data base system so that up to 175 different items of information on any one incident can be retrieved. Thus the data has a high level of integrity and is quickly accessible to carry out day to day enquiries for reactive monitoring and control. It also enables more detailed searches to establish long term trends, and effectiveness of guidance and regulation, and perhaps, more importantly, to carry out formal safety assessments.

3.2 It is possible to use the data to provide a common basis for the measurement of risk associated with specific types of diving operations and for the comparison of alternative diving techniques. Whilst it is acknowledged that such data can never tell the whole story, it can be used very effectively in the comparative mode eg it can establish that one operation is 10 times more at risk than other operations.

4. TECHNIQUES

4.1 Many techniques have been developed over the years for systematic assessment of safety. Within the Diving Inspectorate, several of these have been investigated, ranging from the practical assessment of methods, materials and equipment, through to the more analytical methods associated with risk analysis.

4.2 In the case of the practical assessments, the feedback to field data and its subsequent analysis for trends has provided valuable confirmation that methods, materials, equipment and, where appropriate regulations are operating as expected.

4.3 In the case of the analytical assessments, the techniques of event and fault trees have been examined, together with more sequence orientated techniques such as, state probability diagrams.

4.4 Particular techniques may be suggested as relevant, once the formal safety assessment procedures have been finalized.

5.   EXAMPLES OF ASSESSMENT

5.1   An example of the application of the use of this data, to assess the risks of a specific part of the diving operation, has been to look in detail at the effectiveness of the standby diver.   Using the data base it was found possible to identify amongst the many individual details held:

- Incidents involving the standby diver. (Using specific data code)

- The circumstances surrounding the incident (Using the narrative stored with the incident details).

- The type of dive.

- The platform.

- The environmental conditions.

5.2 Initially, each narrative was examined to determine the exact circumstances of the incident. Then simple statistical techniques were used to examine the raw data.   Finally, confidence levels were applied to the probabilities derived in order that a degree of error was allowed for.

5.3 This analysis has enabled the establishment of specific probabilities associated with the effectiveness of the standby diver and can now be used in more complex operational studies.

5.4 In another case, risk assessments of two alternative methods of carrying out a specific diving operation have been made. As well as providing another example of the application of the diving data base, it also provided the opportunity to examine the techniques which would be applicable to the analysis of the risk. These techniques have been developed around the concepts of probability state diagrams, and have enabled a quantitative assessment of the relative risks to be produced. This assessment may in future be used directly to influence the generation of new guidance notes and operations regulations.

5.5 Thus we are now progressing significantly towards the
achievement of a common safety assessment tool, which, together
with a comprehensive data base means whereby formal safety
assessments may become a viable and practical way ahead. Some
aspects of data are still missing and to arrive at a completely
satisfactory set of statistics these would be required. In
particular it is considered important that more information be
collected on:-

- Number of dives and type of dive.

- Number of team members for each dive.

- Time taken for each dive.

6. FUTURE EXTENSIONS

6.1 Firstly it is hoped to establish a manual of diving risk data
which can be provided as a standard for formal safety assessments.
This will entail the derivation of incident rate associated with
each category of the data base. The exact format of presentation
and modeling is currently under review but in order to illustrate
a potential way head a simple example with hypothetical data is
given.

6.2 The example shows a particular type of dive for which
individual proportions for the occurrence of different category
codes have been derived. For example, under the category
'Environment', the code 01 represents 'Day time' and the code 02
represents 'Night time'. Using the base incident rate (also
derived from data base) it is possible to produce a simple
incident rate model which would relate the various categories
together.

6.3 For example, if we required to know the incident rate for a
dive where:

- The dive platform was a drilling rig
- The operation was routine maintenance
- The environment was any
- The location was the North North Sea
- The depth was 0-99 feet

6.4 Each could be determined from the table and used in a model
thus:

0.119 (platform) * 0.077 (purpose) * 1 (environment) *
0.794 (location) * 0.187 (depth) * 0.223 (basic rate)
        = 0.000303 incidents/dive

6.5 Secondly it is intended to provide guidance on methods for carrying out the assessments. It is conceivable that worked examples together with computer software could eventually be provided.

6.6 Thirdly it is intended to provide common standards of acceptance for risk levels for different diving operations. The Department may provide assessments of 'standard' dive types as part of the assessment package.

7. BENEFITS TO THE DIVING INSPECTORATE

7.1 Given the application of the data collected and the techniques developed the Inspectorate sees the following benefits will accrue:

- Consistent and comparable safety assessments.

- Monitoring of safety performance for existing facilities.

- Prediction capability for new operations.

- Monitoring capability for new regulations.

8. BENEFITS TO THE DIVER

- Safer diving with less potential for accidents and fatalities.

9. CONCLUSIONS

System safety assessments for the Diving Inspectorate are not seen as a revolutionary new concept, but rather as a natural progression from a cycle of data collection, analysis, regulation and guidance which has been effectively applied over the last 12 years.

AN END TO DECOMPRESSION SICKNESS?
A NEW APPROACH TO THE DECOMPRESSION DISORDERS

T.J.R. FRANCIS, D.J. SMITH and J.J.W. SYKES
Institute of Naval Medicine
Alverstoke
Gosport
Hampshire PO12 2DL
England

ABSTRACT.   The current classification of the decompression disorders, from which our diagnostic "labels" are derived, has shortcomings.   In particular, it requires the diagnostician to make difficult decisions with respect to the location of presumed neurological lesions and the mechanism of the disease process.   It is now recognised that in the great majority of clinical settings, these decisions are virtually impossible to make with certainty.   In addition, it is considered that the dichotomy between 'Type I' decompression sickness (DCS) and 'Type II' DCS is artificial.   Each group contains a variety of conditions with no known commonality of pathophysiology, it is widely recognised that symptoms from the two categories may coexist and that 'Type I' may progress to 'Type II'.   The consequences of these shortcomings are that treatment algorithms are applied inconsistently, the epidemiology of these conditions remains obscure and occasionally, cases are managed inappropriately.   A solution is to abandon the current terminology in favour of a descriptive system.   In this presentation such a system is proposed.   It is considered that the adoption of such a system will lead, through better understanding, to improved management of the decompression disorders.

1. Introduction

The current means by which we classify decompression sickness dates back only 30 years to the experience gained during the construction of the Dartford Tunnel in London. Golding *et al* [1] proposed a system for decompression sickness based upon perceived severity of the cases which arose in the caisson workers employed in the construction of the tunnel.   Only symptoms considered 'sufficiently severe to bring the man back for treatment' were considered to be decompression sickness. They divided the cases into two types: Type 1, or simple 'bends' and Type II, which were more serious or complicated cases which displayed vertigo (a spinning sensation); shock; paralysis; epigastric pain and shortness of breath. Their system is still in use today essentially unchanged. Traditionally, decompression sickness has been distinguished from the barotraumata and a summary of the current classification of the decompression disorders is

275

presented in Table 1.

TABLE 1. The present classification of the decompression disorders
(after Elliott & Kindwall [2] and Farmer [3])

DECOMPRESSION SICKNESS

    <u>TYPE I</u>

    MUSCULOSKELETAL (including 'niggles')
    CUTANEOUS   Transient pruritus
                                  Circulatory manifestations (Cutis Marmorata)

    LYMPHATIC
    MALAISE/ANOREXIA/FATIGUE

    <u>TYPE II</u>

    PULMONARY
    NEUROLOGICAL

                                 Spinal cord
                                 Cerebral
                                 Cranial nerve
                                 Labyrinthine disturbance (inner ear DCS)
                                 Peripheral nerve
                                 Migraine-like symptoms
                                 Girdle pain

    HAEMOCONCENTREATION AND HYPOVOLAEMIC SHOCK

BAROTRAUMA

    OTOLOGICAL

                                 External Ear Canal
                                 Middle Ear
                                 Inner Ear

    PARANASAL SINUS
    DENTAL
    PULMONARY

                                 Mediastinal emphysema
                                 Pneumothorax
                                 Arterial gas embolism

---

2. Problems with the Current Classification

The deficiencies and consequences of this classification have been reviewed previously [4]. In summary, it requires the diagnostician to make difficult decisions, particularly where the nervous system is involved. These include determining the location of the lesion (e.g. "cerebral" or "spinal cord") and the mechanism of injury (e.g. "decompression sickness" or "arterial gas embolism"). At a recent Undersea and Hyperbaric Medical Society Workshop [5] it was recognised that in the great majority of clinical settings, such decisions are virtually impossible to make with certainty. The consequence of this is that the existing diagnostic "labels" can not be applied rigorously. As a result, treatment algorithms are inconsistently applied and communication between divers, physicians and medical researchers is compromised.

The existing dichotomy between 'Type I' or mild decompression sickness (DCS) and 'Type II'or serious DCS is spurious. Each group contains a variety of conditions with no known commonality of pathophysiology. It is widely recognised that symptoms from the two categories may coexist and that 'Type I' may progress to 'Type II'. Consequently, it is considered that the use of these terms is not just confusing, but potentially dangerous if divers are lulled into delaying or failing to report a symptom as a result of using terms such as 'pain-only', 'mild', 'non-serious', or 'Type I'. These terms also lack any inherent meaning, they have to be learned. Consequently, communication with medical personnel who have not been indoctrinated into their use is made difficult. The decompression disorders are potentially highly dynamic conditions, yet the terminology currently used takes no account of this. It can be argued that it is this dynamic quality (e.g. progressive or relapsing) that is a better index of the urgency of a case than whether it is 'Type I' or 'Type II'.

The main reason why the existing classification has been retained is that its use has been made seductive. Treatment tables have been applied, more or less as a reflex, depending upon whether the "diagnosis" is 'Type I', 'Type II' or 'AGE'. Fitness to return to employment as a diver (or as an aviator) has been based upon the same, or manipulated, diagnostic categories. This has resulted in patients being shoe-horned, occasionally with much difficulty, into these artificial and very limited groups more for administrative convenience than to truly reflect what is wrong with the patient.

At this workshop it was concluded that the present system of classification based on medical cause should be abandoned and that a descriptive definition of the decompression disorders be adopted. The workshop proposed that the current terms: Decompression sickness 'Type I'; 'Type II' and Arterial Gas Embolism be abandoned in favour of the term 'Decompression Illness' which, for terminology purposes, is modulated by terms to describe the evolution and manifestations of the disease.

3. A Proposed Protocol

The protocol consists of a matrix which provides a formalised *aide memoir* for data collection and from which a terminology has been derived which is capable of being used to describe a wide variety of decompression conditions. It was considered that the following key information is required to describe a case of decompression illness adequately: The evolution of the case; the clinical manifestation(s); the time to onset of each manifestation; the gas burden and whether or not there is evidence of barotrauma. Additional important information includes: The response to recompression and the results of any investigations. A summary of this system is in Table 2 and definitions are provided below.

TABLE 2. A descriptive definition of the decompression disorders

---

ACUTE DECOMPRESSION ILLNESS

    1. EVOLUTION

                                      Progressive
                                      Static
                                      Spontaneously Improving
                                      Relapsing

    2. MANIFESTATION(S)

                                        Pain
                                                  Limb pain
                                                  Girdle pain
                                      Cutaneous
                                      Neurological
                                                  Audiovestibular
                                      Pulmonary
                                      Lymphatic
                                      Constitutional

    3. TIME OF ONSET

    4. GAS BURDEN

    5. EVIDENCE OF BAROTRAUMA

BAROTRAUMA

    PULMONARY

                                        Mediastinal emphysema
                                      Subcutaneous emphysema
                                      Pneumothorax
    OTOLOGICAL

                                        External ear
                                      Middle ear
                                      Inner ear
    SINUS
    GASTROINTESTINAL

3.1. EVOLUTION

The evolution of a case refers to the development of the condition PRIOR TO RECOMPRESSION. This information is best recorded as the case evolves. Because decompression illness is frequently dynamic, the evolution may change from one observation to the next. Thus, a condition which initially presents as being "progressive", as the patient becomes increasingly aware that something is wrong, may stabilise so that it can then be described as "static". The patient may subsequently undergo a substantial improvement, occasionally to the extent of a complete resolution of the symptoms. This can be described as "spontaneously improving". Occasionally, the symptoms return or new symptoms appear, in which case the condition would be described as "relapsing".

3.1.1. *Progressive.* A condition may be described as progressive if the number or severity of symptoms or signs is increasing. Examples include limb pain which is becoming increasingly severe, or which involves more sites, or a neurological presentation in which the loss of function is becoming more profound or extensive. The development of a new manifestation, such as a neurological symptom or sign in addition to limb pain also represents progression of the condition. Additional description may also be useful such as whether the progression is rapid or slow.

3.1.2. *Static.* This is self explanatory. The condition is not changing substantially.

3.1.3. *Spontaneously Improving.* It is common for a number of presentations of decompression illness to improve without recompression. Sometimes this may be to the point of apparent recovery, although this may only be transient. Because the intensity of symptoms of decompression illness can fluctuate, substantial improvement must occur for this term to be applied. As with other terms to describe the evolution of a case, this should only be used to describe events prior to recompression.

3.1.4. *Relapsing.* Occasionally, cases which have improved spontaneously undergo a secondary deterioration, particularly with some neurological manifestations. This term is used to describe such cases. When a condition gets worse in the absence of any spontaneous improvement, it should be described as "progressive".

3.2. MANIFESTATIONS

The manifestations describe the organ systems or parts of the body which are affected. These terms are used to describe the disease complex at the time of the report.

3.2.1. *Pain.* Pain is probably the most frequent manifestation of decompression illness. It is used to describe the deep aching pain in or around one or more joints which may begin during decompression or after completion of a dive. Unlike the pain of musculoskeletal injury, limb pain decompression illness is generally not exacerbated by movement of the affected joint. The pain may range from mild, barely detectable discomfort to a steady, boring, nearly unbearable pain. Limb pain should be distinguished from "Girdle Pain". This is a poorly localised, aching or 'constricting' sensation which is generally in the abdomen, pelvis or, occasionally, in the chest. Girdle pain in the context of decompression illness is generally considered ominous since it is frequently a harbinger of neurological deterioration.

3.2.2. *Neurological.* Involvement of the nervous system may be subtle, multifocal and consequently very difficult to localize. There are also a number of different possible mechanisms which may be involved in the development of the pathology of these conditions. Consequently, it is important that the terminology used presumes neither a location or a mechanism. Involvement of the nervous system can be broken down into the loss of certain functions: Higher functions, which would include aberration of thought processes or affect, loss of memory, difficulty talking etc.; Alteration to the level of consciousness including seizures; loss of coordination; loss of strength or sensation; dysfunction of special senses; loss of control of bladder or anal function. All of these can be further described, however, this properly forms part of record keeping rather than the terminology of decompression illness.

A distinct syndrome within the neurological category is involvement of the Audio-Vestibular system. This consists of: vertigo (a sense of rotation or spinning), tinnitus (ringing in the ears), nystagmus (jerking movements of the eyes) and loss of hearing after a dive. Nausea and vomiting may accompany these symptoms but, of themselves, are not sufficient to imply audiovestibular involvement in decompression illness. Again, more than one mechanism may be responsible for this manifestation and it may be very difficult, without elaborate invrestigation, to determine the site of injury. However, if a cause can be established, such as round window fistula, then the more specific diagnosis should be made.

3.2.3. *Pulmonary.* Involvement of the lungs in decompression illness may be because of two quite distinct processes: lung rupture due to barotrauma and the cardiopulmonary consequences of massive venous gas embolism. Although the mechanisms involved are distinctly different, it may be difficult for lay personnel to distinguish between them initially because many of the symptoms and signs are shared. However, modern diving practices result in pulmonary decompression illness due to the latter mechanism very rarely indeed. The symptoms or signs which imply pulmonary involvement in decompression illness are: chest pain, cough, haemoptysis (coughing up blood or blood-stained sputum), shortness of breath, cyanosis (blueness or duskiness of the skin, lips or mucuous membranes), pneumothorax (gas trapped in the pleural space in the chest), subcutaneous emphysema ('crackling' under the skin) of the neck and, occasionally, voice change. When describing pulmonary decompression illness, it is important to note whether there is clinical or radiological evidence of a pneumothorax or mediastinal emphysema, since this is known to be a consequence of lung rupture.

3.2.4. *Cutaneous.* The skin may be affected by diving in a number of ways. Two very common manifestations of decompression, but which are not generally regarded as illnesses, are suit "squeeze" and itching in the absence of a rash. The term "cutaneous" decompression illness should be used to describe the condition which generally presents with severe itching around the shoulders or over the trunk which, after a time, develops into an erythematous rash and which may progress to blueish mottling or marbling of the skin. When further describing the condition, it is desirable to describe the location of the disorder.

3.2.5. *Lymphatic.* This term may be used to describe cases in which there is painful swelling of individual or discrete groups of lymph nodes or rare cases where there is extensive oedema of one or more limbs.

3.2.6. *Constitutional.* There are a number of non-specific symptoms which occur after diving which, if severe or if accompanied by other manifestations, may be considered part of the decompression illness syndrome. These include headache, fatigue, malaise (which may include nausea and, possibly, vomiting) and lack of appetite.

Very rarely, there may be other manifestations of decompression illness may occur. Such conditions should be described using appropriate medical terminology.

3.3. TIME OF ONSET.

The time of onset can provide a great deal of information to medical personnel regarding mechanisms of disease and, possibly, the outcome of some cases. Following hyperbaric exposures, this should be the time from reaching the surface to the onset of the manifestation.

If the manifestation occurs during ascent then it should be recorded as such.

3.4. GAS BURDEN.

The gas burden is an estimate of the residual inert gas load present on surfacing. At present, recording the dive profile is the most useful index available.

3.5. EVIDENCE OF BAROTRAUMA

Clinical or radiographic evidence of barotrauma should be documented, particularly where there are pulmonary or audiovestibular manifestations. Where there is such evidence, the barotraumata are diagnosed as before.

To complete an accident record, it is important to record the outcome of recompression and the results of additional clinical investigations.

4. The Terminology

Lengthy descriptions are unwieldy for communication purposes, an abbreviated label is needed until the natural syndromes are identified. The general form of the proposed terminology is as follows:

Acute [Evolution Term], [Manifestation Term(s)], **Decompression Illness**

The term "acute" is used to distinguish these conditions from the possible chronic consequences of decompression such as osteonecrosis. The phrase decompression illness (DCI) incorporates the familiar terms decompression sickness and arterial gas embolism. The evolution term is used exactly as defined above. The number of manifestation terms which are used will depend both on the condition and the context in which the terminology is employed. In a condition where there are only one or two manifestations, it is appropriate to use those which apply e.g. Acute static, cutaneous and neurological decompression illness. In complex cases it may be appropriate to use the term 'multisystem'.

The amount of detail in the description will depend on the purpose for which the terminology is being used. As a diagnostic label, the above terminology should suffice. There are situations where more information is frequently needed, such as during the transmission of information over the telephone or radio during a consultation. In this situation, the three additional key pieces of information: the time of onset, the gas burden, and any evidence of barotrauma can be added. These are likely to be valuable in discriminating between the various syndromes. An example of such a brief report is:

Acute progressive limb pain and neurological decompression illness presenting 20 minutes after surfacing with a moderate gas burden and no evidence of barotrauma.

For treatment reports and database purposes a more detailed report is likely to be necessary.

5. Advantages of this System.

The main advantages of this system is that it contains no guesswork with respect to either the mechanism or the anatomical location of the disease process. Furthermore, it employs terms which are readily understood within the medical community. It does not require the first person to attend the patient to have a great deal of experience or expertise to use it properly. Communication between divers and non-diving medical personnel will be facilitated by its use because all the terms have inherent meaning. However, basic instruction is critical for the system to be used properly. Such instruction is made relatively simple because there is no need to brain wash people to accept untenable rules and assumptions. This system permits the description of a dynamic, changing condition without difficulty.

We have been unable to classify decompression illness reliably to date which has limited our progress towards understanding these intriguing conditions. It is hoped that by using a readily understood descriptive terminology, consistent and accurate 'diagnoses' should now be possible which will improve the management of cases. More importantly, if these data are then collected and collated, the natural syndromes associated with decompression will become readily apparent. If we learn what the natural syndromes are, we will be able to direct our efforts more effectively towards preventing decompression illness more effectively in the future.

6. References:

1. Golding, F.C., Griffiths, P., Hempleman, H.V., Paton, W.D.M. and Walder, D.N. (1960) Decompression sickness during the construction of the Dartford Tunnel. *Br J Industr Med* 17, 167-180.

2. Elliott, D.H. and Kindwall, E.P. (1982) Manifestations of the decompression disorders, in: Bennett, P.B. and Elliott, D.H. (eds.) *The physiology and medicine of diving*, 3rd. edition. Ballière Tindall, London, pp. 461-472.

3. Farmer, J.C. (1990) Ear and sinus problems in diving, in: Bove, A.A. and Davis, J.C. (eds.) *Diving Medicine*, 2nd edition. W.B. Saunders, London, pp. 200-222.

4. Gorman, D.F. (1991) A proposed classification of the decompression illnesses, in: Francis, T.J.R. and Smith, D.J. (eds.) *Describing decompression illness*, 42nd Workshop of the Undersea and Hyperbaric Medical Society, Bethesda, MD, pp. 6-9.

5. Francis, T.J.R. and Smith, D.J. (eds.) (1991) *Describing decompression illness*, 42nd Workshop of the Undersea and Hyperbaric Medical Society, Bethesda, MD.

A NEW APPROACH TO AIR DIVING

P.B. JAMES, C.F. LAFFERTY, A.O. BRUBAKK

Wolfson Institute, Ninewells Medical School, University of Dundee, British Gas PLC, London, Dept. Biomedical Engineering, University of Trondheim.

ABSTRACT. A new air decompression table has been developed for a dive to 40 m.s.w. for one hour, which employs a switch to a helium and oxygen mixture and oxygen during decompression. The safety of the table has been validated using two types of ultrasonic monitoring.

1. Introduction

The problem of neurological decompression sickness in the air range has now been acknowledged in the U.K. and action taken to limit exposures to both compressed-air and nitrox mixtures.[1] However the restrictions applied to in-water and surface-decompression air dives severely limit the useful work that can be done. They also force more dives to be undertaken in order to complete the same amount of work. Increasing the number of dives increases the risks from hazards other than decompression sickness.[2] About half of all the diving fatalities in the oil and gas industry have been related to surface-orientated diving. Fortunately this has led to the successful development of transfer-under-pressure techniques in the air diving range.

Unfortunately, diving on air well within the current limits cannot guarantee that subtle neurological damage will not occur and, from an examination of the data there is no safety margin. The recommended limits for in-water decompression only slightly exceed the no-stop decompression curve for air dives. However even adherence to the no-stop decompression limits does not provide protection against neurological decompression sickness as the case of a diver who developed symptoms after a dive of 30 minutes at 30 f.s.w. has demonstrated.[3]

A second problem is that the therapy of decompression illness, especially in the air range, is unsatisfactory. In 1964 it was found that the air therapy tables recorded a failure rate of about 46% in the treatment of serious decompression sickness.[4] The minimal-recompression oxygen breathing tables U.S.N. 5 and 6 were introduced to solve this problem in 1965. It was soon recognised that oxygen at an equivalent presssure to 18 m.s.w. (60 f.s.w.) may paradoxically cause deterioration of a diver suffering from decompression sickness and this was recorded in the U.S. Navy Diving Manual released in 1970. Both human and experimental studies indicate that there may be an irreversible component of severe dysbaric illness. Currently, therefore, the Industry has both unsatisfactory working procedures and inadequate schedules for therapy.

Compressed-air as a diving gas has been limited to 50 m.s.w. (165 f.s.w.) by U.K. legislation, not because of decompression sickness, but because at greater depths narcosis becomes significant. Clearly, because it is inexpensive and the diving equipment needed is simple, many would still prefer to retain compressed-air in shallow-water diving. However, it can only be retained if the problem of decompression sickness can be solved.

283

Volume 27: Subtech '91, 283–287.

2. Decompression Sickness and Decompression Tables

The risk of decompression sickness was recognised by Bert in 1875 [5] to be due to nitrogen being released from solution that is, gas phase, due to supersaturation. He observed bubbles in the blood from a severed jugular vein in a dog following a decompression. Haldane, [6] after defining the no–stop decompression exposure, originally stated that a decompression on a 2:1 ratio was associated with bubbles but not symptoms. However in a subsequent publication he stated that no gas was formed on this decompression ratio.[7]

It would be possible to calculate decompression tables reliably using Haldanian methods if gas actually remained in solution for a given supersaturation ratio. Unfortunately, both the occurrence of decompression sickness and the use of ultrasound have shown that gas does form when supersaturation is present. When gas remains in solution it is available for transport in the blood, ready to be safely converted into gas phase after crossing the alveolar wall and exhaled. However, if gas comes out of solution in the tissues in a 'pocket', it can only be transported to the lungs, either when it redissolves, or when it is released into the circulation and transported by the blood as bubbles. The transport of gas to the lung as bubbles is actually more efficient, in terms of molar concentration of gas carried per unit volume of blood, than its transport in solution, but carries the risk of bubble arterialisation through the lungs. There is clear evidence of the separation of gas from solution in both the tissues and blood in standard air decompressions. The elimination of the nitrogen on normal decompression procedures therefore involves both dissolved and separated gas.

The testing of most tables has relied on the development of Type 1 decompression sickness as the end–point. Unfortunately, as the survey of decompression sickness in the North Sea has shown,[3] in deep air diving Type 2 symptoms are often the initial presentation. As Type 2 symptoms are most likely to be due to intravascular gas, ultrasonic monitoring has been adopted to ensure safety.

3. The Scientific Basis for Decompression Table Development

3.1 INHERENT UNSATURATION

The one parameter that can be calculated and even measured for steady–state exposures in a constantly perfused tissue is the inherent unsaturation. This concept, first recognised as important in decompression in 1938 by Momsen and Behnke [8] in the development of the U.S. Navy helium and oxygen diving tables, was objectively demonstrated at tissue level by Hills and LeMessurier.[9] Arterial blood carries oxygen in solution at a gas tension close to the partial pressure of the oxygen in the respired gas. The oxygen is metabolised after transfer to the tissues, but the gas tension is not fully replaced by the carbon dioxide produced, and this gas is also much more soluble than oxygen. The sum of all the gas tensions present in the tissues therefore never equals the sum of the partial pressures of the gas in the lung, hence the term unsaturation. At normal atmospheric pressure (1 ATA) the unsaturation is 55 m.m. Hg.

Fortunately the tissues protected by the greatest metabolic demand and which have the most constant blood flow are in the nervous system. The level of unsaturation is proportional to the partial pressure of oxygen up to an as yet undetermined point. On any decompression beyond the limit of the inherent unsaturation for a given tissue, the quantity of gas remaining in solution or separating as gas phase cannot be determined. Also, when gas phase is present, the relative amounts of gas that are transported in solution, or as bubbles to the lung cannot be determined. These variables account for much of the unpredictability of decompression and it is clear that it is impossible to separate the formulation of decompression tables from the pathophysiology of decompression sickness.

3.2 GAS SWITCHING ON DECOMPRESSION

Supersaturation on decompression provides   the driving   force   for the elimination of the inert gas from the tissues but carries the unpredictable risk of gas separation. Clearly if a proportion of the tissue inert gas could be removed without a decompression then, with an appropriate adjustment of the reduction in the absolute pressure, the risk of significant gas formation can be reduced. This can be undertaken by changing the diluent inert gas respired at constant pressure. Switching to pure oxygen is already used in air diving, but this is only possible at 15 m.s.w. (50 f.s.w.) or less, because of oxygen toxicity. This means that for dives deeper than this an initial decompression has to be undertaken with the risk of gas formation, due to supersaturation. A change of breathing gas to a mixture of another inert gas and oxygen can however be undertaken at depths in the air range.

When a gas switch is made, washout of the first gas will be accompanied by uptake of the second. Ideally, the two gases should have the same physical characteristics, so that the wash-in and wash-out rates, assuming constant blood flow in the tissue, are identical. Unfortunately the inert gas closest to nitrogen is neon which is prohibitively expensive. Helium is transferred more quickly by the lung and may be taken up faster by aqueous tissues than nitrogen, but, fortunately, is less soluble than nitrogen in blood, so that there is some limitation introduced by its transport. However the data available indicates that the rates for the two gases are not greatly different and suggests that a switch could be made safely at pressures relevant to air diving. If the gas switch is not covered by the oxygen window then an abrupt switch will be accompanied by the summation of the two gas tensions in the tissues and this supersaturation may lead to gas formation.

A gas change from nitrogen and oxygen mixtures to heliox, has been made safely in a research programme for the deep submergence rescue vehicle in the U.S.Navy. It has been recognised that a stricken submarine may become pressurised and that the oxygen content of the trapped air may fall to normoxic values. Accordingly, tests have been undertaken of abrupt switches from a 17 hour nitrox saturation to heliox, isobarically, at 40 m.s.w. (132 f.s.w.) and at 60 m.s.w. (198 f.s.w.) in goats.[10] The oxygen partial pressure used was only 0.3 ATA. Occasional venous bubbles were produced at 40 m.s.w. (132 f.s.w.) after a delay of 20–60 minutes. Bubbles were always detected after the switch at 60 m.s.w. (198 f.s.w.), but no signs of decompression sickness were produced in either case. For these experiments the air depth equivalent for the 40 m.s.w. exposure at the partial pressure of oxygen (0.3 ATA) used was 48.75 m.s.w. (158 f.s.w.) and the duration of exposure was 17 hours. Adding the partial pressures of the nitrogen and helium at the point of the switch gives a total gas tension of 9.4 ATA (pN_2 4.7 + pHe 4.7), at an absolute pressure of 5 ATA – a ratio of 1.88:1. Human data on isobaric gas switching is also available from the D.S.R.V. development programme. [11]Abrupt switches, which generated itching, were made at 30 m.s.w. using nearly normoxic mixtures. Ideally the switch from air to heliox should be made slowly to minimise the countertransport supersaturation, but the most important factor is to use a much higher partial pressure of oxygen to maximise the oxygen window.

3.3 THE FORMAT OF THE DECOMPRESSION TABLE USING HELIOX AND OXYGEN

A table has been developed for a bottom depth of 40 metres and a duration of one hour. The first upward excursion which allows for a bell to be lifted and a seal obtained before transfer into the deck chamber is dictated by the level of the unsaturation, particularly in the nervous system. When diving to 40 metres (132 f.s.w.) with the gas tension values involved expressed in m.m. Hg, the unsaturation can be calculated to be 683 m.m.Hg. and, provided the tissues metabolise oxygen fully, the initial ascent can be gas free. This value is equivalent to an upward excursion of 8.9 m.s.w. (29 f.s.w.) However, as steady-state conditions are not reached in many tissues with a bottom time of one hour it was decided to allow the first stoppage to be at 30 m.s.w. (100 f.s.w). The gas switch was then made, after a ten minute period on air at 30 metres, to a 50/50 helium and oxygen mixture, which has both a high partial pressure of oxygen (pO_2 2.0 ATA) and a correspondingly lower partial pressure of helium, (pHe 2.0 ATA). At the depth of the gas switch the combined gas tensions for the nitrogen and helium are 5.2 ATA (pN_2 3.2 + pHe 2.0), at an absolute pressure of 4 ATA. This is a ratio of 1.3:1. It is recognised that gas may form transiently in some tissues when the gases are summated if the blood flow is intermittent and evidence was obtained of this effect in the preliminary study. However the use of an elevated partial pressure of oxygen is an effective treatment for this separated gas. The next stoppage was at 20 metres and after a similar period at this new pressure, another upward excursion was made to allow oxygen breathing at 10 m.s.w. The maximum partial pressure of oxygen used therefore was 2 ATA with most of the period of oxygen breathing being at 1.5 ATA.

4. Subjects and Methods

Twenty commercial divers who were in possession of a current certificate of fitness took part in the study. Four divers were used for each pressurisation. Two divers undertook the in–water dive, each monitored by a tender in the chamber above the wet pot. The tenders were initially at 31 metres of sea water equivalent pressure, but, as the divers emerged, the complex was pressurised to 40 m.s.w. to complete one hour at this pressure for the divers.

After completion of the period at 40 m.s.w., the initial upward excursion was undertaken to 30 metres still breathing air. The divers then transferred into the main chamber to breathe 50/50 heliox on the built–in–breathing system (B.I.B.S.) using AGA masks. This heliox breathing period at 30 metres was only 10 minutes and the gas was continued during the decompression to 20 metres over ten minutes and for the second stop of 50 minutes at this pressure. The decompression to ten metres was also on heliox and again took ten minutes. The divers then breathed oxygen for fifty minutes before decompressing to the surface.

All the divers were monitored sequentially during the decompression using Doppler ultrasound. This was continued for two hours after surfacing. A second ultrasound technique which employed scanning was also used after surfacing. This system allows the visualisation of the heart and can detect single bubbles in transit through into the pulmonary artery. All the data was recorded on magnetic tape for subsequent analysis.

5. Results

There were no cases of decompression sickness and no bubbles were detected in any diver during the chamber decompression. Only in one of the twenty divers were a few bubbles detected. They were first seen on the scanning ultrasound detector about twenty minutes after the diver had finished the chamber decompression. At about one hour they could also be heard on the Doppler ultrasonic equipment and they just achieved a Spencer Grade 1 score. They disappeared soon after this time.

6. Conclusions

1. It has been shown that it is safe to switch abruptly to a helium and oxygen breathing mixture during decompression from an air dive at depths that preclude the use of pure oxygen.

2. The switch to helium and oxygen does not produce symptoms or bubbles in the pulmonary artery.

3. A decompression can be undertaken using heliox and oxygen for a one hour dive using compressed–air to 40 metres with very few bubbles in one of twenty divers.

References

1. Department of Energy. (1990) London Safety Memorandum No.2.

2. James, P.B. (1984) Some hazards of shallow diving. Proc 6th Underwater Engineering Symposium

A.O.D.C. Aberdeen 23,1.

3. Giles R, Shields T.G., Duff P.M., Wilcox S.E. (1989) Decompression sickness from commercial diving air – diving operations on the U.K. continental shelf during 1982 to 1988. Robert Gordons Institute of Technology, Aberdeen.

4. Goodman M.W and Workman R.D. (1965) Minimal recompression oxygen breathing approach to the treatment of decompression sickness in divers and aviators. Research Report 5–65 U.S. Experimental Diving Unit, Washington.

5. Bert P. La Pression Barometrique. (1875) Translated by: Hitchcock MA, Hitchcock FA. Undersea Medical Society, Bethesda Maryland 1978.

6. Haldane J.S. (1907) Report to the Admiralty on Deep Water Diving. HMSO London.

7. Haldane J.S.(1922) Effects of high atmospheric pressures. Chapter XII, Respiration, Yale University Press.

8. Momsen C.B. and Behnke A.R.(1938) Report on helium–oxygen mixtures for diving. E.D.U. Report, NY5/S94(132) Washington.

9. Hills B.A. and LeMessurier D.H. (1968) Unsaturation in living tissue relative to the pressure and composition of inhaled gas and its significance in decompression theory. Clin Sci,36,185–195.

10. D'Aoust B.G., Smith K.H., Swanson H.T. et al. (1979) Prolonged bubble production by transient isobaric counter–equilibration of helium against nitrogen.Undersea Biomed Res,6,109–125.

11. Hamilton R.W., Adams GM, Harvey C.A. and Knight D.R.(1982) A composite study of shallow saturation diving. Naval Submarine Medical Research Laboratory, Groton.

HADES, Highest Accumulated Decompression Score.

A. O. Brubakk and J.E. Jacobsen.
Environmental Physiology Group, Department of Biomedical Engineering, University of Trondheim, 7006 Trondheim, Norway and Stolt Nielsen Seaway a/s, PO Box 370, 5501 Haugesund, Norway.

ABSTRACT. The HADES project was initiated in order to develop methods for evaluating decompression tables and in particular to study the effect of diving activity and environmental factors upon decompression outcome. The database contains detailed information about over 2000 saturation dives performed on two diving vessels over a time period of six years by a stable population of divers. The experimental model system enables the detailed study of the effect of decompression procedures with changing temperatures, muscular activity and gas composition. In this model, vascular gas bubbles are used as an indication of decompression stress. Initial evaluation of the diving data indicate that a subgroup of divers performing a larger proportion of the dives, in particular dives with stopped decompressions, have a much larger incidence of decompression sickness. The experimental system is able to maintain precise control over environmental factors, initial results indicate excellent correlation with results from human dives.

1. INTRODUCTION.

No diving activity is possible without decompression. This self-evident fact is pointed out to underline the importance of this part of the diving procedure, a part that can not be eliminated by any equipment. Over the years, considerable time and effort has been invested in trying to develop procedures that are safe and effective. The decompression schedules used in diving today have a high level of safety and will to a large degree prevent serious injury. However, recent data indicate that changes can occur both in the lungs [1] and in the brain [2] can be detected in individuals who never have experienced clinical decompression sickness.

The HADES project was initiated following a series of decompression incidents in one of Seaways diving ships. Without any apparent reason or change in diving procedure, a profile that had been used successfully for many years suddenly started to give an unacceptable high incidence of DCS as is seen from TABLE 1.

The tables used were linear with a decompression speed of 30 msw/24 hours to 15 msw, ppO² was .5 bar. From 15 msw to surface the speed was 13.5 msw/24 hours. Night stops were used.

The high incidence in 1988 led us to reduce the saturation decompression speed from 30 to 27 msw/24 hours, this was followed by an increase in the number of DCS-incidence in 1989. A reduction in the allowable excursion distance dramatically reduced the amount of decompression problems.

TABLE 1.   INCIDENCE OF DECOMPRESSION SICKNESS
            HeOx SATURATION 30-70 MSW

	DCS I	DCS II
1985	0	0
1986	1	0
1987	3	0
1988	2	1
1989	1	5
1990	0	0

These episodes led us to perform an evaluation of our diving procedures. As no apparent reason could be found for the sudden increase in the incidence of DCS, we had to rethink some of the basic assumptions we had used when establishing our tables. We were forced to admit that some of the empirical adjustments to the tables that we and others had performed over the years had a very weak base. In fact, although decompression procedures were developed to reduce or eliminate gas bubbles, they are evaluated upon the basis of producing few acute clinical symptoms. Even quite small pressure changes can produce gas bubbles. From the work of Daniels [3] and Eckenhoff [4] we knew that pressure differences between 40 and 70 kPa (0.4 - 0.7 bar) are sufficient to produce gas bubbles on air, to our knowledge similar data do not exist for heliox.

Based upon our evaluation of the literature, we concluded that:

- the conditions under which a dive is performed can considerably influence decompression outcome, in fact this may be at least equally important than gasmix and depth.

- the absence of acute signs of decompression sickness is not an adequate protection against long term effects. We therefore suggest that the term GRD (Gas Related Disease) is introduced, where DCS is one manifestation.

- clinical signs of decompression sickness is not an adequate indicator of decompression stress

- vascular gas bubbles is the only measurable indicator of decompression stress.

- the lungs ability to eliminate gas is a major determinant of decompression outcome and this ability can be influenced by gas embolization.

- gas bubbles, once they have formed, have a very long life-time and their presence can influence gas elimination and decompression outcome. Thus, repeated decompressions can lead to an accumulation of gas and each individual has a highest acceptable decompression score (HADES) at any particular moment in time.

Based on the above, the HADES project was initiated.

The main aim of the HADES project was to

- Develop a method to record and evaluate diving activity.

- Develop an experimental model to evaluate the effect of environmental factors upon decompression.

- Develop a set of diving procedures that takes environmental factors as well as the diving history of the diver into account.

2. METHODS

2.1. The HADES Dive Data Base (HADIDAB).

The aim of HADIDAB is to store operational diving data that allow a detailed analysis of each individual diver's exposure. The database shall contain detailed information about the pressure profile as well as about environmental factors like oxygen, carbondioxide and inert gas partial pressures, humidity and temperatures.

The basis for the information in the database are the diving logs. Each dive is given a unique identification number and the identity of each diver participating in that dive is recorded. The following logs are used.

> Saturation summary log
> Saturation decompression log
> Saturation control room log
> Bell lockout log
> Bell control room log
> Air dive log

This application has been developed using the Paradox 3^R database program. In addition to having a robust data base manager, it offers a full-featured applications language (The Paradox Applications Language) that enables the writing of complex programs to tie together data. The data is presented in a simple row and column format, similar to what you get from a good spreadsheet program. The table-querying system, query by example, makes the process of asking questions of the data relatively simple.

All data has been entered from the logs. Following that, all data has been checked for accuracy by individuals with extensive knowledge about the diving activity.

The hardware used is a 386 IBM compatible PC with 4 Mb Ram and a 100 Mb Harddisk.

2.2. The experimental model.

Based upon our evaluation of the literature, we have concluded that vascular gas bubbles is of significant importance. This concept is described in detail in two recent reports [5,6].

2.2.1. Choice of animal model

As already pointed out by Boycott et al [7], it is necessary to use a large animal for decompression studies. The reason for this is that the respiratory exchange and hence the uptake and elimination of gases through the lung is roughly proportional to the ratio of the body surface to weight. The smaller the animal, the larger this ratio will be and hence the greater the respiratory exchange and consequently the circulatory rate. This can be the reason for the fact that small animals like rats and mice are much more resistant to decompression sickness than man [8]

Many different animals species have previously been used for decompression research (goat, sheep, dog). We decided to use the pig as an experimental animal based on the fact that this animal has a circulatory and ventilatory system very similar to that of man [9]. This experimental animal has previously been used by Fife et al [10]. They have demonstrated that decompression sickness occurs on roughly the same decompression profiles in man and in the pig. They further point out that the results in their animals were very reproducible and that changes in body weight did not seem to influence the results.

We have developed an animal model based on a lightly anesthetized, spontaneously breathing pig, where we are able to monitor all relevant ventilatory and circulatory variables.

2.2.2. The chamber system

A 300 l pressure chamber capable of performing experiments in the pressure range of 20 - 10.000 kPa (0.2 - 100 bar) has been constructed. This chamber has a sufficient size to accommodate pigs up to approximately 25 kg weight.

2.2.3. The control system

In decompression experiments, the precise control of environmental parameters are of importance. Furthermore is necessary to have a system for logging all recorded variables in a way that enables us to relate them to a specific part of the dive. Such a system has been developed. It is programmed in Labview and runs on a Macintosh FX computer with 8 Mb Ram. This system automatically controls temperature and pressure in the chamber according to predetermined values and performs recording and storing of data. The data can be transmitted to a spreadsheet for further analysis.

Many decompression procedures call for changes in breathing gas during particular parts of the dive. An automated system that enables this to be controlled from outside the chamber has been constructed. The animal is breathing from a bag system inside the chamber, all gas switching is computer controlled.

2.2.4. The physiological measurement system.

Blood pressures are measured in the pulmonary artery and the arterial system. Further pressure measurements, including the measurement of spinal fluid pressure is being introduced.

Continuous measurement of mixed and end-expiratory oxygen and carbondioxide is being measured as well as expired nitrogen and helium content.

Ventilation is measured with a pneumotachograph, more accurate instruments for ventilatory measurement is being constructed.

Blood gases are continuously monitored, a chromatographic method for analysis of inert gas content of blood is being developed.

2.2.5. The ultrasonic detection system.

Gas bubbles are detected in the heart and the pulmonary circulation using a 7.5 MHz ultrasonic transducer mounted on a gastroscope. This transducer can record images as well as Doppler shifts. The transducer is connected to a CFM750 ultrasound scanner (Vingmed Sound a/s, Horten, Norway). A computerized system that automatically counts gas bubbles in the images has been constructed.

3. RESULTS.

The database now contains the pressure profiles of all dives performed on two diving ship over a period of six years. In this time period both vessels were on long term contracts with a stable work force of divers. The vessels cover two scenarios. One was doing heliox saturation dives to 30-60 msw combined with air diving, while the other performed diving in the range between 70-180 msw with little air diving. 140-150 divers on each vessel have performed about 2000 sat dives. The data give a nearly complete exposure picture of all divers over this time period.

The majority of the data for one of the vessels have been verified and are ready for analysis. This information has been used for comparing of the diving activity of those individuals who had DCS (Hit divers) with those that participated in a dive leading to DCS (Hit-mate divers) and with those that never had DCS (Clean divers).

In this time period 142 divers performed 1026 dives. The mean age of the divers varied from 33 to 36 years of age, there was no difference in age between the three groups. The dives was performed at storage depths between 40 and 60 msw,

Only initial results can be presented here, as the analysis of the data is still in progress. As can be seen from Fig 1, the duration of the dives seemed to show a reduction over the observation period. There was an increase in the number of sat dives performed by the observed divers, from an average of 2.1 dives in 1987 to 3.7 in 1989. The group with DCS problems did, however, have a much higher dive frequency than the clean group. The Hit and Hit-mate group performed an average of 11.9 and 13.0 dives/diver over the timeperiod, while the Clean group only performed 6.1 dives/diver.

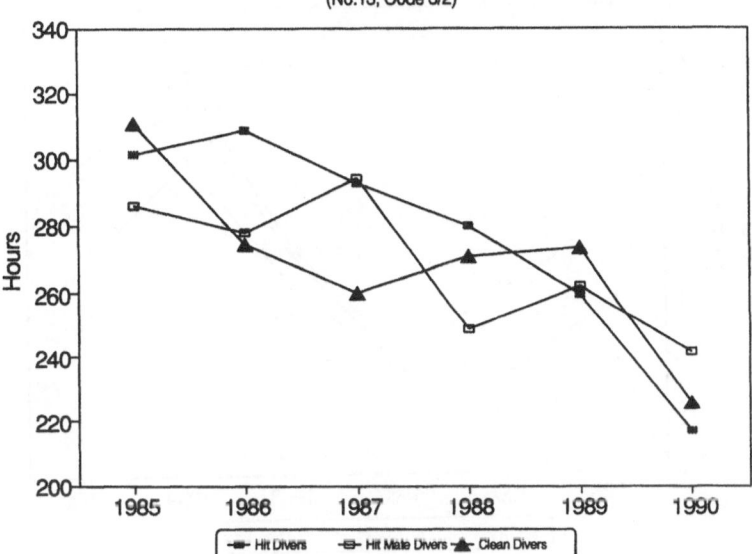

Average Duration of Dives
(No.13, Code 3/2)

FIG 1. Average duration of saturation dives from 1985-1990.

As can be seen from Fig 2, the Hit and Hit-mate divers accordingly spent more time decompressing than the Clean divers.Over the whole observation period, the Hit and Hit-mate group spent twice as much time decompressing than the Clean divers.

Another interesting development could also be seen from these data. The number of decompressions pr dive did not show any discernable pattern. However, the percentage of interrupted decompressions (non-contiguous decompressions) rose sharply in 1989, in particular for the Hit divers, as can be seen from FIG 3.

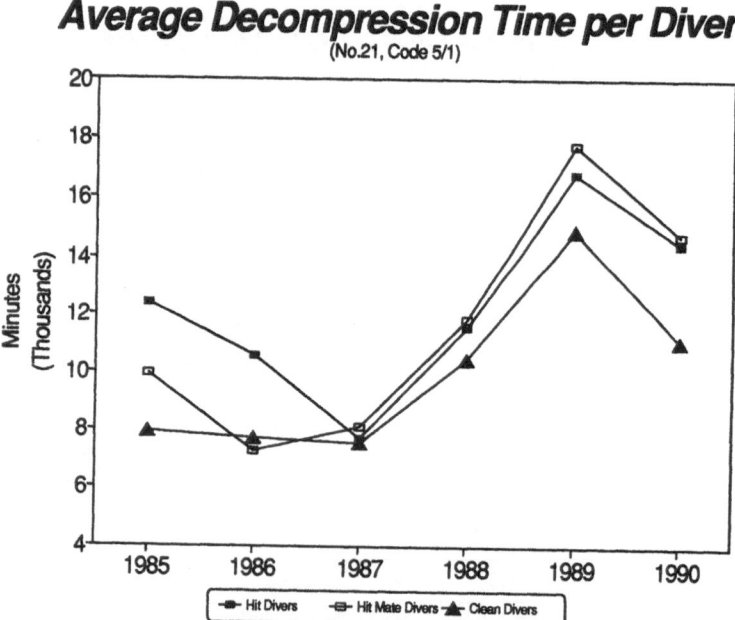

Fig. 2. Average decompression time over the six year period.

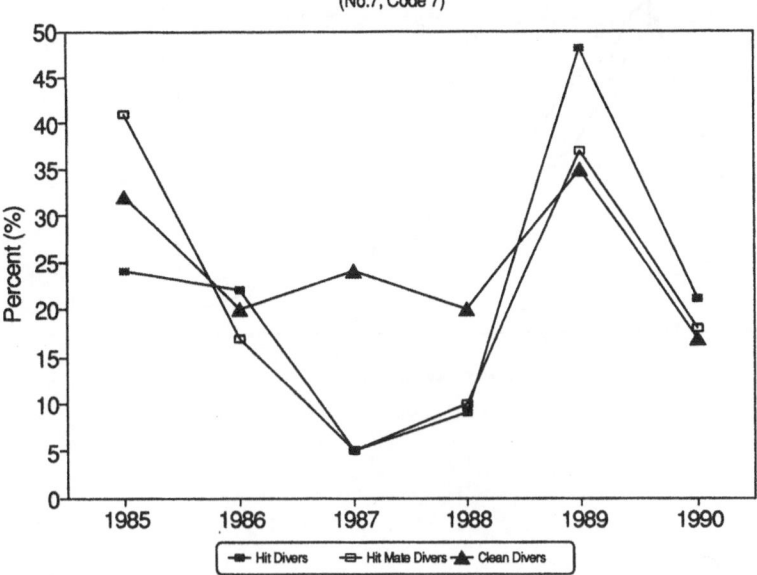

Fig 3. Average amount of non-contiguous decompressions (see text) in the six year period.

The experimental set-up has been tested in a series of air dives. The control system is able to control pressure within +/- 0.1 bar and the animals body temperature to +/-0.2 °C. The automatic breathing system is able to accurately change breathing gas over a few breaths, demonstrated by the rapid change in He and N_2 content in the expired gas. Following dives to 5 Bar for 30 minutes, a considerable number of bubbles are produced. Fig 4 shows a comparison between the changes in pulmonary artery pressure and bubble count, demonstrating a close relationship. Changes in ventilation and gas exchange could also be documented. Initial experiments where gas bubble formation has been compared to that observed in experimental dives on humans, indicate that both the incidence, amount and time course of gas bubbles in the pulmonary artery is similar.

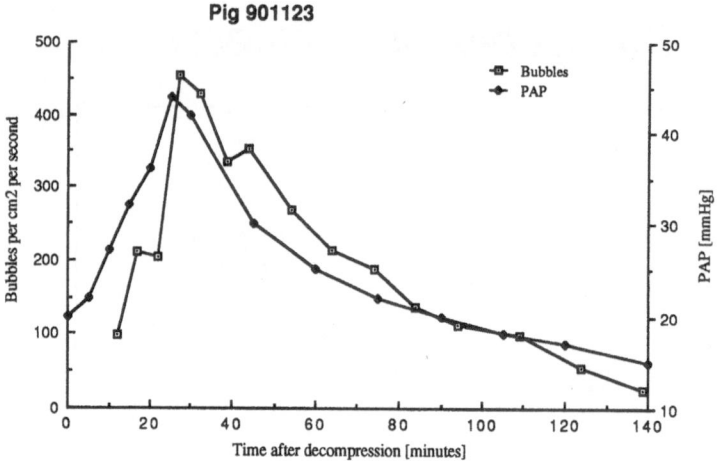

Fig 4. Changes in Pulmonary Artery pressure (PAP) and number of gas bubbles in the Pulmonary Artery following a no-decompression dive to 40 msw for 30 minutes.

4. DISCUSSION.

There has till now been relatively little effort invested in documenting the relationship between environmental factors and the outcome of decompression. The approach described in this paper is a combination of two different methods. First, we wanted to esablish a database for analyzing commercial diving profiles in great detail. Second, we wanted to establish an experimental model, where the effect of environmental factors can be studied.

The main problem in establishing the database was the inaccuracies of the logs. Some errors were found in most of them. The correction and verification of these logs required detailed analysis by individuals with extensive knowledge of the operations and the procedures used. Our experience has led us to have considerable distrust in any evaluations made from operational data unless the verification process has been both rigorous and extensive. This work has clearly demonstrated that the logging procedure has to be made automatic, where all environmental data are recorded in real time. The system developed for our experimental chambers fulfill these criteria.

The preliminary analysis of the database, showed that those individuals that had an extensive diving activity also were the ones with decompression problems. Although this conclusion is not too surprising, the result gives us confidence in the method of analysis. What is perhaps surprising, is that there seems to be a clear difference between the groups, indicating that the tables used may produce a high incidence of DCS in a subgroup of individuals who perform more diving. Many uncertainties are however involved in this analysis. The diagnosis of DCS is based upon the clinical symptoms at the time of the dive and on the fact that the symptoms were considered severe enough to warrant treatment. The diagnosis of DCS II is thus based on the fact that these individuals had symptoms from the central nervous system. Another problem is the large difference in numbers of the different groups. DCS will be a rare event in most saturation diving operations, and only a much larger database can solve this problem.

One of the main problems in all epidemiological research is the lack of accurate exposure data. This is of particular importance in diving due to the allegations that diving can lead to permanent disabilities. Only careful correlation of exposure data with the observed changes can give us an indication of the validity of this claim.

In order to demonstrate the effect of the total diving environment upon the organism, experimental work must be performed. By measuring the amount of bubbles formed under different conditions and the biological effect of these bubbles, an evaluation of the safety of different procedures can be made.

Our experimental model is based upon 18-23 kg pigs. The body surface area of pigs, with a weight of 20 kg and a length of 80 cm is approximately 0.6 m^2. A man of 65 kg and a height of about 175 cm gives a body surface area of approximately 1.8. This gives a pig area/weight ratio of 0.03 and an area/weight ratio in man of 0.028, in other words approximately the same. Oxygen consumption and cardiac output in man and pig is also approximately the same in absolute values.

These data would indicate that the rate of uptake and elimination of pig and man is approximately the same. Due to the fact that the weight of the experimental animal is smaller than that of a grown man, the saturation time is probably shorter. The decompression requirements seem, however, to be similar as documented by Fife [10]. Our own observations support this. If these observations can be further verified, this experimental model can serve as an excellent and reproducible way to test decompression procedures, thus improving safety and reducing cost of new table developments.

Decompression sickness is used as the endpoint of nearly all decompression procedures. It is well documented that decompression sickness is subject to extreme variability, both among individuals and in one individual at different times. It can thus be that what is regarded as variations in susceptibility to decompression sickness in reality is a variation in the ability to tolerate air embolism.

Using ultrasound, the amount of bubbles produced can be quantified. Gas bubble production can be used as an indicator of decompression stress. The computer based analysis system used in this study can give an accurate evaluation of the amount of gas present in the circulatory system. The gasbubble production and their effects can be documented experimentally and compared to our measurements in divers.

The HADES database will be further developed and more detailed information about dives will be recorded. The information will be used to calculate an exposure dose for decompression.

In the experimental work, the effect of gas bubble production upon subsequent decompressions and gas elimination will be evaluated. Effects of temperature, muscular contractions and different gas mixes will be studied.

All data will be used to develop a model for decompression that relate environmental and previous exposure to the final decompression to surface.

5. ACKNOWLEDGEMENTS.

The HADES project is being executed by Seaway Stolt Nielsen a/s and SINTEF-UNIMED, Section for Extreme Work Environments. It is sponsored by Phillips Petroleum Norway. Without the support and enthusiasm of Dr. Tor Nome this project would probably never have been established.

Many institutions and individuals have participated in this work and their contribution is gladly acknowledged.

The experimental decompression system has been developed with the support of SINTEF-UNIMED, The University of Trondheim, STATOIL (the FUDT project), Norwegian Space Center and Phillips Petroleum Norway. Members of the experimental team are Anne Vik, Randi Reinertsen, Anne-Lise Ustad, Astrid Hjelde, Bjørn Jenssen, Atle Kleven and Arnfinn Sira. Dr. Tom Hennessy, City University, London, has participated in the development of the theoretical models and Dr. Valerie Flook, University of Aberdeen, has participated in the building and testing of the gas analysis system.

At SEAWAY a number of individuals have participated in the development of the database and in the registration and verification of the data. We want to acknowledge the contribution of James McLellan, Ben Knutsen, Jan E. Nakkestad, Andy Morgan, Brian Bishop, Arne Mjelde, Torgrim Rustad, Magne Vågslid and Marit Saltvedt.

6. REFERENCES.

1. Thorsen, E., Segadal, K., Kambestad, B., and Gulsvik, A. (1990) 'Diver's lung function: small airways disease?', Brit. J. Industr. Med 47, 519-23.

2. Todnem, K., Nyland, H., Kambestad, B.K, and Aarli, J.A. (1990) 'Influence of occupational diving upon the nervous system: an epidemiological study', Brit. J. Industr. Med 47, 708-714.

3. Daniels, S. (1986) 'Bubble formation in animals during decompression', in A. O. Brubakk, J. Kanwisher and G. Sundnes (eds.), Diving in animals and man, Tapir Publishers: Trondheim, pp. 229-264.

4. Eckenhoff, R.G., Olstad, C.S., and Carrod, G. (1990) 'Human dose-response relationship for decompression and endogenous bubble formation', J. Appl. Physiol. 69, 914-918.

5. Hennessy, T. (1991) 'HADES: Model of decompression' SINTEF Report STF23 F91008, Trondheim.

6. Brubakk, A.O. (1991) 'HADES (Highest Accumulated Decompression Score). Background and main theories' SINTEF Report STF23 F91009, Trondheim.

7. Boycott, A.E., Damant, G.C.C., and Haldane, J.S. (1908) 'The prevention of compressed air illness' J. Hygiene 8, 342-443.

8. Berghage, T.E., David, T.D., and Dyson, C.V. (1979) 'Species differences in decompression' Undersea Biomed. Res. 6, 1-13.

9. Douglas, W.R. (1972) 'Of pigs and men and research: a review of applications and analogies of the pig, Sus Scrota, in human medical research' Space Life Sci. 3, 226-234.

10. Fife, W.P, Mezzino, M.J., and Naylor, R. (1978) 'Development and operational validation of accelerated decompression tables' in C. W. Schilling and M. W. Beckett (eds.), Underwater Physiology VI, FASEB, Bethesda, Maryland, pp. 359-366.

INSHORE DIVER SAFETY - A REVIEW OF RECENT DEVELOPMENTS

D R Lamont
HM Principal Specialist Inspector
(Construction Engineering)
Technology Division
Health and Safety Executive
Bootle

ABSTRACT. The paper reviews reported accident and decompression sickness figures, legal proceedings and other enforcement matters relating to the UK inshore during industry since 1989. It also outlines changes in legislation and the statutory diver training scheme during that period.

1.   INTRODUCTION

Most of the published work on commercial diving safety in the UK arises from the offshore diving industry. The inshore sector has a less glamorous image and consequently has attracted less attention in the literature in recent times. The UK amateur diving organisations are active in producing data on the diving activities of their members however they only deal with the amateur sector.

This paper is a review of activity by the Health and Safety Executive affecting divers in the inshore/onshore commercial diving sector. This sector covers all non oil and gas related commercial diving in territorial waters and on land.

The inshore/onshore sector embraces a very wide range of diving operations including fish farm diving around the north and west of Scotland, civil engineering diving on a variety of outfalls, structures, bridges and reservoirs etc both around the coast and on inland sites, the construction and repair of docks, harbours and inland waterways, ship repair and maintenance, operational diving by police and military units, underwater scientific research, commercial diving schools, the commercial instruction of amateur divers and various diving activities in the leisure industry.

Volume 27: Subtech '91, 303–310.
© 1991 *Society for Underwater Technology.*

2.   LEGISLATION

The Diving Operations at Work Regulations SI 1981:399 (DOWR) apply equally to the inshore/onshore sector as well as to work offshore in the North Sea. Likewise The Diving Operations at Work (Amendment) Regulations 1990 apply.

3.   ENFORCEMENT RESPONSIBILITY

Enforcement of DOWR is by the Field Operations Division of the Health and Safety Executive. Its Inspectors are probably best known as "HM Inspectors of Factories". In each of HSEs 20 Area offices, there is at least one Inspector nominated to deal with diving related matters in addition to his other duties. Recently HSE has increased the resources allocated to the inspection of inshore diving operations and this should lead to higher levels of compliance with the requirements of DOWR by the Industry.

4.   ACCIDENTS

The reporting of industrial accidents and occupational ill-health throughout all UK industry is governed by the Reporting of Injuries, Diseases and Dangerous Occurrences Regulations 1985 (RIDDOR).

Whilst all fatal accidents involving commercial divers are thought to be known by HSE it is believed that many injury-only accidents go unreported. As there are no statutory reportable dangerous occurrences involving divers, there is a lack of useful information on diving accidents.

4.1 FATAL ACCIDENTS

Since January 1989 the number of fatal accidents resulting directly from commercial diving operations has been 6 in 1989, 1 in 1990 and 4 in 1991 to date. **The corresponding figures for the offshore sector are 0 in 1989, 0 in 1990 and 0 in 1991 to date.** The difference between the fatal accident figures for the two sectors is not always appreciated.

The circumstances surrounding these accidents are summarised below:

1989

A young uncertificated diver recovering the equipment from the sea bed below a jetty under repair.
Police diver trainee undertaking extended aptitude training.
Two police divers undertaking the recovery of a sunken barge in a gravel pit as a training exercise.
Clam diver collapsing immediately following a number of deep and long dives to harvest clams.
Fish farm diver trapped in nets which he was placing around a cage on fish farm.

1990

Diver trainee whilst undergoing Part IV training at commercial diver training school.

1991

Civil engineering diver trapped in sluice on large culvert.
Amateur diver trainee whilst undergoing "commercial instruction of amateurs".
Amateur diver trainee whilst undergoing "commercial instruction of amateurs".
Amateur diver trainee whilst undergoing "commercial instruction of amateurs".

With the exception of the fatality involving the civil engineering diver earlier this year all the fatalities outlined above have involved the use of self contained underwater breathing apparatus. Another common factor in the 1989 and 1990 fatal accidents was that there was no voice communication between the working diver and the diving supervisor (if present). In 3 of the incidents there was no standby diver immediately available.

So far this year there have been 3 fatal accidents involving the "commercial instruction of amateur divers". This activity is regulated by Certificate of Exemption DOW/4/81(General) issued by the Health And Safety Executive. In addition HSE awards a restricted Part IV certificate on the basis of specific amateur diving instructional qualification. Enforcement of health and safety legislation on the premises of amateur diving schools however is frequently the responsibility of the local authority as these premises are non-industrial. In late 1990 internal discussions within HSE led to the decision that HSE should consider carrying out an overview of safety in the commercial instruction of amateurs during the Summer of 1991. Investigations into these accidents have already led to enforcement proceedings being instigated and further action is planned. Consequently the matter is sub judice and cannot be discussed further in this paper.

4.2 NON-FATAL ACCIDENTS

In the period since 1989 only a handful of non-fatal accidents have been reported to HSE. These included a diver being trapped when a concrete block fell on his umbilical, a diver suffering lung damage when a blast of air entered his helmet possibly as a result of ice formation in the panel regulator and an incident involving the failure of an air supply to a chamber when surface decompression was being carried out. The pattern of accident data available to HSE is clear - the diver is either killed or if he survives the accident goes unreported.

5.   DIVER HEALTH

Barotrauma and decompression sickness are reportable diseases under RIDDOR (Regulation 5, Schedule 2). The level of reporting is believed to be low. One reason for this is that before a disease is reportable the condition must be diagnosed by a doctor. A doctor is not always involved in the treatment of decompression sickness on a diving site. In addition ignorance of the statutory requirement to report decompression sickness is common. No reliable figures exist within HSE for decompression sickness in the inshore/inland sector. The unofficial information "grapevine" in diving advises HSE of between 5 and 10 cases of decompression sickness per annum.

Amongst the cases of decompression sickness notified to HSE has been a handful of cases affecting divers in the fish farm industry. These divers have appeared to have had Type II decompression sickness symptoms after having carried out a large number of short shallow dives in a short period of time. HSE is having research carried out into this apparent problem. It is hoped that guidance to the industry will result. As the fish farm industry is a relatively new industry little was known about the diving techniques used until a survey of diving practices on fish farms was carried out by HSE in 1989 (Lamont (1991)).

6.   ENFORCEMENT OF DOWR

During the period under review a number of HSE inspectors have been particularly active in enforcing DOWR. The industry has in the past been critical of HSE's ability to deal with "cowboy" diving contractors. Recent publicity given to a number of successful prosecutions has started to reverse this view.

6.1 PROSECUTION

A number of prosecutions under DOWR have been taken. These have been well publicised in the diving press with the result that a greater awareness of the requirements of the regulations appears to have spread through the industry.

Some of the prosecutions have followed the investigation of fatal accidents, others the result of preventative inspection. At least one prosecution resulted from inspectors visiting a site for totally unconnected purposes and coming across diving operations in the course of their visit.

Successful prosecutions taken following accident investigations include:

A diving contractor from south west Scotland being fined £350 for failure to provide a standby diver.
A diver fined £50 for not having a certificate of Medical Fitness to Dive.

A fish farm operator (diving contractor) fined a total of £1,000 for 7 breaches of DOWR including failure to provide a standby diver and diving supervisor.
A commercial diver training school fined £500 having been found guilty of failing to provide a sufficiently experienced standby diver and sufficient emergency air cylinders and demand valves (appeal pending).

In the above examples the level of penalty may appear low but is in line with that often imposed by the Courts for breaches of Health and Safety legislation.

Not all commercial divers go to training school to obtain a Diver Training Certificate. Some obtain work on the strength of forged certificates or photocopies of certificates. An inspector visiting a number of construction sites where diving was known to be taking place compared the certificates of divers on the sites with records held centrally by HSE. As a result a number of divers were successfully prosecuted for diving without having valid Training Certificates and certificates of Medical Fitness to Dive and in some cases for being in possession of forged certificates. Despite fines of around £150 per offence being imposed on the divers one diver was caught diving in Scotland on false documentation within weeks of having being fined in a South Wales court for similar offences.

In addition a total of three diving contractors who had employed these divers were successfully prosecuted. Penalties including costs of up to £2,800 were imposed.

As a result of these cases diving contractors became more aware that they should always see the original of a divers certificate to verify their authenticity before employing the diver.

In north west England 2 Inspectors visiting a site by appointment for a meeting noticed a diving operation taking place adjacent to the site access. Unfortunately the Inspectors did not notice a standby diver kitted up and in immediate readiness to dive. As a result the firm was successfully prosecuted and fined £500 for failing to provide a standby diver.

6.2 PROHIBITION NOTICES

Not all enforcement action ends up in the courts. Recently an Inspector in the south of England served Prohibition Notices on uncertificated divers running an "amateur" diving school for profit and an enterprising diver alleged to be cleaning the hulls of pleasure boats but without the benefit of a diving supervisor or standby diver. In both these cases the illegal activity has been stopped despite there being insufficient evidence to mount a prosecution.

6.3 INDUSTRY-BASED INITIATIVES

HSE through its Field Operations Division is currently increasing the number of inspectors in the Factory Inspectorate and the (Agricultural Inspectorate) involved in the inspection of diving operations. In addition HSE is supporting industry-based initiatives aimed at improving safety standards in diving in the fish farming industry and in docks and harbours.

For the past year it has been HSE policy that divers in the fish farm industry diving in scuba are required to use full face masks and have voice communication to the diving supervisor on the surface. The next stage may well be the introduction of surface supplied diving equipment for this industry.

With the dock and harbour industry the initiative is aimed at ensuring divers have the appropriate level of training for the work they are doing. In the guidance booklet "Diving Operations at Work - Guidance on Regulations"[2] HSE sets out what it considers to be appropriate standards of training for different types of underwater work. For example Part 1 is considered to be an appropriate standard for divers engaged in heavy manual work, the use of power tools and civil engineering work. However it is not uncommon for such work to be carried out by Part 3 or Part 4 divers. Recently HSE has agreed with employers representatives that their divers will be brought to the appropriate standard by early 1993.

7.   LEGISLATION

7.1 FIRST AID

The one significant change in legislation has been the coming into force in April 1991 of the Diving Operations at Work (Amendment) Regulations 1990. These Regulations make provision for diving contractors to provide at the dive site first aid equipment and where appropriate trained personnel (diver medics). In addition all divers must have (by 1993) a Certificate of Diving First Aid which is renewable on a 3 yearly basis. Guidance on the requirements for first aid equipment is contained in the booklet "Diving Operations at Work Guidance on Regulations"[2]. Apart from the commercial diving schools a number of first aid and offshore survival training establishments have been approved by HSE to provide the necessary training.

At time of writing it is too early to comment on the effectiveness of these regulations.

7.2 REGISTRATION AND NOTIFICATION

The registration and notification of diving operations was the subject of a consultative document published in 1990 by HSE . The document set out proposals for the registration of all the diving contractors and the notification by the diving contractor to HSE of certain specified diving operations. At present HSE is considering the comments received as a result of the consultation exercise and is considering its response.

8. DIVING TRAINING

1991 saw the introduction of a competence based approach to diver training. This scheme follows the principles set out by the National Council for Vocational Qualifications which is active in industrial training throughout UK industry. In consultation with both sides of the diving industry HSE is publishing four "Standards for Assessing Diver Competence" to correspondence with the respective parts of Schedule 4 of DOWR. The competence standards are based on the 1986 Diver Training Standards[3]. No alterations to the topics covered by the respective standards have been made but the new documents set out in more detail the competencies to be achieved. The change has resulted in additional work for the commercial diver training schools (and inspectors) but should lead to more consistency in the skill levels of the trainees. The 1986 Diver Training Standard lacked detail on the practical aspects of diving training to be covered. As diving is a practical activity, this deficiency has been rectified.

9. EXPERIENCE ROUTE COMPETENCE ASSESSMENTS

With the introduction of the competence based training regime the route by which experienced divers (mostly from abroad or from the armed services) could apply for certification by experience assessment has been changed. In the past the assessment consisted of an appraisal against set criteria of a divers logged experience. From Summer 1991 onwards a diver will have to undergo a theoretical and practical "Experience Route Competent Assessment" (ERCA) following a satisfactorily assessment of his logged experience. During the ERCA a diver will have to demonstrate that he has (at least) the same level of competence in all topics as a trainee satisfactorily completing a diver training course at the same standard.

At the time of writing the paper no divers had been certificated as a result of the experience route.

10. GENERAL CERTIFICATES OF EXEMPTION

The general Certificates of Exemption covering scientific diving, underwater archeology, underwater journalism and the commercial instruction of amateurs are shortly to be revised to reflect problems found by HSE in their interpretation and enforcement, changes in the commercial climate surrounding the exempted activities, and recent and proposed changes in legislation. The results of this revision should be published in Autumn 1991.

At present commercial divers working under the general exemptions are excluded from the Experience Route Competence Assessment scheme however the situation may well change in the near future.

11. REFERENCES

1.   Lamont D R (1991) "Diving Practices in the Fish Farming Industry" Progress in Underwater Science Vol 16.

2.   "Diving Operations at Work - Guidance on Regulations" HMSO 1991, ISBN 0-11-885599-9.

3.   "Diver Training Standard" PTS I-IV HSE, 1986.

Part 5
Tomorrow's World

OPTIMISATION OF UNDERWATER INSPECTION PROGRAMMES

KEN WOOLLEY
Quest Consulting Limited
Willowbank House
Willowbank Road
Aberdeen AB1 2TT

ABSTRACT. Underwater inspection of offshore oil and gas structures in the North Sea has evolved over the last 20 years. The initial techniques were fairly primitive by comparison with current methods.

This paper examines the development of underwater inspection in the North Sea from the late 1960's to the present day and considers how programmes can be rationalised. It looks at both techniques used and training given. Current inspection methods are reviewed with particular emphasis on steel structures.

There is generally greater opportunity to optimise inspection costs on steel structures, be they floating or fixed. Concrete gravity platforms are discussed but by their very nature the inspection requirements for concrete are less onerous.

The key to optimisation of underwater inspection programmes begins at the conceptual or design phase of a new platform. Input by qualified engineers can significantly reduce inspection and maintenance costs downstream.

Substantial financial savings can also be made by reviewing and rationalising inspection programmes during the service life of offshore structures. Quest Consulting Ltd. has extensive experience in this area and its staff have worked in this capacity for a number of North Sea Operators over the past ten years.

It is important that with any inspection programme, the reason for inspection is clearly defined. This should be to ensure the continuing 'fitness for purpose' of the structure by most the cost effective methods. To achieve this, the engineer must have sympathy and an understanding for the loads and possible degradation of the structure.

This paper concludes with some of the methods available and details the data required to fully optimise an underwater inspection programme.

Volume 27: Subtech '91, 313–317.

1. Introduction

Quest Consulting Limited is a professional engineering consultancy based in Aberdeen. The optimisation of underwater inspection programmes is an area that we are particularly skilled in. Individually we have been involved in underwater activities since the late 60's though the majority of our work has been conducted over the last ten years.

We have been involved in the rationalisation and streamlining of underwater inspection programmes for a number of North Sea clients. However, for the purpose of this presentation we will reflect on the following areas:-

Brief History of North Sea Inspection
Current Inspection Techniques
Requirement for Inspection
Optimisation by Design
Optimisation during Service Life
Conclusions
For reasons of client confidentiality any examples given will be non-specific.

2. Brief History of North Sea Inspection

Underwater Inspection has come a long way over the past 20 years. In the late 1960's and early 1970's underwater inspection often comprised the deployment of divers using scuba and limited, if any, contact with the surface. (There were no ROV'S as we know them today, although manned submersibles were being used). Typically the diver would be briefed prior to diving , take whatever equipment that was necessary to execute the task and then plunge overboard. Early diving practices comprised a general swim round, weld cleaning by needle gun followed by a general visual inspection, cathode potential readings by a very early model of a bathycorrometer (with moving a moving coil voltmeter) and still photography.

From the mid 1970's as development got underway in the Central and Northern North Sea, diving and inspection techniques expanded and improved. Non-destructive testing (NDT) practices employed in other industries, for example, magnetic particle inspection (MPI) were introduced into North Sea diving. Professional diving schools were set up and divers were trained in inspection. Inspection certification was initially carried out by the certifying authorities, e.g. Lloyd's, DnV and then under the CSWIP scheme (phase 7) starting with 3.1D.

Perhaps the greatest improvement in underwater inspection control was the advent of hat mounted TV's which enabled the surface team to guide a diver through the inspection. Underwater television also contributed to diver safety.

Methods and systems improved during the 1980's and now in the 1990's we have a major industry with a whole suite of activities available to provide detailed, quality information about the status of offshore steel and concrete, fixed or floating North Sea structures.

3. Current Inspection Techniques

The present range of possible techniques are exhaustive. Some of these are briefly described below:

General Visual Inspection - a swim round by diver or ROV using video and still photography, noting general condition without removing any obscuring mud or marine growth. It will often include a debris survey, marine growth survey , a seabed survey and cathode potential (CP) survey.

Close Visual Inspection - a detailed survey at key locations following removal of marine growth and any obscuring mud or debris.

Magnetic Particle Inspection - a popular NDT method typically used to detect surface breaking cracks in nodal welds.

Eddy Current Inspection - an alternative technique to detect surface breaking fatigue cracks in steel structures.

Flooded Member Detection - various methods exist to determine whether a steel bracing contains water. The theory being that a through-wall penetration may exist which may be detrimental to the structural integrity.

Wall Thickness - an ultrasonic technique to detect wall thickness, particularly if wall thinning is suspected.

A Scan Ultrasonics - a sophisticated surface technique used in a similar manner underwater to size steel wall defects.

Radiography - another surface technique which may be applied underwater to size steel wall defects.

Acoustic Emission - remote, continuous monitoring technique which may detect crack propagation in steel structures.

Vibration Monitoring - measuring changes in the natural frequency of structural components by remote methods.

Other techniques include photogrammetry, remote monitoring, current measurements, etc.

The above techniques are mainly for steel structures. Concrete gravity structures are less inspection sensitive generally requiring only a visual check for scour and damage with additional information gathered on the various appurtenances.

Having described the techniques and general wealth of knowledge available based on the application of these techniques we will now turn our attention to the requirement for inspection.

4. Requirement for Inspection.

Why inspect?

Inspection of offshore structures is carried out to ensure the continuous safety of personnel on board and comply with legislation to provide a certificate of fitness every five years.   In addition, the Operator seeks to protect its investment by ensuring the structure is capable of maintaining production and that the structure remains fit for purpose.

This is a very important statement.

The prime objective of an underwater inspection programme is to clearly identify and accurately monitor potential problem areas or known problem areas.

5. Optimisation by Design.

An underwater inspection programme is best optimised by qualified input during the design phase of an offshore structure.   Careful consideration to detail can significantly reduce inspection costs offshore.

An example of 'significant' is a potential saving of millions of dollars during the in-service life of a structure.

The 'input' should be provided by experienced underwater engineers consulting with the project design team at the conceptual phase.

Any recommendations are platform specific depending on water depth, location, type of construction, materials used, type of corrosion protection, location of services, component referencing,etc.

Subsequent access to key components for inspection is important and consideration should be given to the relative costs of air and saturation diving, the use of ROV's and the future use of robotic intervention.

6. Optimisation during Service Life.

Significant savings can be made for new-build structures. However, there are of the order of 200 to 300 existing oil and gas structures in the North Sea.

Are we over-inspecting existing structures? An interesting question.

It is very important that any underwater inspection is properly focussed.   The inspection should be carried out for engineering reasons and not for inspection's sake.

When a structure is installed a fairly detailed programme of inspection should be carried out according to a well designed plan to ensure that the structure is behaving as designed and to establish baseline information for subsequent analysis. Once there are three to four year's of service inspection data, then the programme should be reviewed and evaluated and possible rationalisation carried out. This might increase the level or method of inspection in certain areas of concern and reduce it elsewhere.

Optimisation of inspection is a matter of engineering judgement based on design data, fabrication records, service history, redundancy/consequence analysis, inspection techniques and costs. The product of optimisation is often reduced costs with significant savings by the Operator.

Furthermore, the certifying authority is usually keen to see a well organised programme which is varied according to findings and engineering assessments to ensure the continuing 'fitness for purpose' of the structure.

7. Conclusions.

Underwater inspection has come a long way in the last 20 years. The techniques and programmes of the late 1960's and early 1970's now appear primitive.

A vast number of techniques and solutions are available for underwater inspection.

However, it is important that inspection is carried out for engineering reasons and not inspection for inspection's sake.

REAL TIME PHOTOGRAMMETRY - A TECHNIQUE FOR TODAY OR TOMORROW?

JOHN TURNER, DAVID J YULE, JOE ZANRE
Camera Alive Limited
Campus 3
Aberdeen Science and Technology Park
Balgownie Drive
Bridge of Don
Aberdeen AB22 8GW

ABSTRACT. Photogrammetry has been in regular use since 1980 to produce three dimensional measurements underwater, for a variety of tasks. The paper will briefly summarise the history of the technique, from its inspection orientated origins to the latest engineering related, deep water, ROV surveys, in order to analyse the strengths and weaknesses of the technique. This analysis will introduce the technological improvements, leading to the satisfaction of market demands, that can be made possible with advancing technology. The development of Real Time Photogrammetry, from the feasibility study to offshore trials, will then be introduced, emphasising the enormous possibilities of the technique and its component technologies: electronic still imaging; digital image data transmission and storage; real time and still stereo viewing; and near instant three dimensional measurements of complex objects. The by-products of the development will also be introduced, which include total electronic reporting systems and the database controlled replay of visual data within a networked computer system, with suggestions as to how these could influence structural and pipeline inspection, as well as engineering surveys in today's, as well as tomorrow's, world.

1. INTRODUCTION

1.1 What is Photogrammetry?

Photogrammetry is a visual imaging technique for the acquisition of accurate three-dimensional measurements without contact with the subject.

Photogrammetry is not a new technique, and has been in use for more than 100 years for the compilation of maps from aerial photography, for topographical surveying and for industrial uses, such as the quality control of large castings.

Volume 27: Subtech '91, 319–331.

1.2 Why is it used Underwater?

Underwater Photogrammetry provides a fast and cost effective method of obtaining precise three-dimensional measurements, for inspection or engineering purposes. It is necessary only to photograph a subject with a pair of photogrammetric cameras, or with a single photogrammetric camera from two or more camera positions, in such a way as to achieve stereo coverage, to allow accurate measurements to be taken from the photographs.

- It is a non-contact technique, and so there is little or no contact or interference with the subject.

- It allows measurements of subjects that are difficult to measure by conventional techniques, because of their shape or position.

- It uses separate data acquisition and data processing phases and so it keeps underwater operational time to a minimum.

- Its use is not complicated by a more complex shapes, for example if many measurements are required to accurately define the shape of impact damage which included multiple impacts, tears etc.

- Total effort does not increase proportionally with the number of points to be measured.

- The analysis process is selective, and so can be tailored to the exact requirement of the application, even though the information density stored on photographic film is enormous.

- The information can be stored archivally on the photographic film, to be extracted, or redefined, at any time.

- The information may be presented in many forms: as discrete measurements; sections; contours; plan and elevation drawings; isometrics; contours etc, to best suit the exact requirements of the end user.

Underwater photogrammetry is a tool that can be applied with a diver or an ROV. In many cases the technique allows an ROV to carry out a task that has previously had to be carried out with the use of a diver.

1.3 For what is it used?

Photogrammetry has been offered as an underwater tool since 1980 and has been used by practically every operating company in the North Sea for a wide variety of applications. It has also been used in practically every region of the world in which oil is produced offshore.

Underwater photogrammetry was primarily used as an inspection tool, for the evaluation of corrosion, weld defects, small areas of damage and similar tasks. In this mode the subject could be photographed using a single stereo pair of photographs and the results presented in a tabular or graphical form. Measurements within areas approximately 350mm x 500mm could be taken to a reproducible accuracy of 0.1mm ±0.3mm.

The technique was then developed to allow larger areas to be photographed such that dents or larger areas of damage could be measured. In this mode the subject was still photographed in a single stereo pair of photographs. Measurements within areas approximately 1.4 x 1.7m could be taken to a reproducible accuracy of 1mm ±2mm.

The technique has now been developed to provide sophisticated structural surveys, made up from a number of stereo pairs of photographs. This has allowed the measurement of node geometry, structural surveys to allow clamps to be fabricated with a metal to metal fit, and as-installed surveys of items as large as complete wellheads, using ROVs to take the photographs. The area of coverage is only limited by cost and measurement tolerances of ±5mm are common, with ±2mm possible.

1.4 Strengths and Weaknesses.

Underwater Photogrammetry has particular strengths in obtaining three-dimensional measurements operationally. These can be summarised, in addition to the points introduced in section 1.2, as:

- Accuracy;

- Objectivity;

- Cost Effectiveness;

- Ease of presentation of data;

- Sophistication;

- Use in deep water.

It also has particular weaknesses, which have prevented an even wider utilisation of the technique. These can be summarised as follows:

- Delays between data acquisition and presentation of result;

- Necessity for highly skilled operators and expensive equipment lead to high costs and bottlenecks.

2.0 REAL TIME PHOTOGRAMMETRY - FEASIBILITY STUDY

2.1 The Reasoning

Within a conventional photogrammetric process the cameras were deployed, the photographs taken, the film recovered and chemically processed all before the quality of the images and their suitability

Photogrammetry

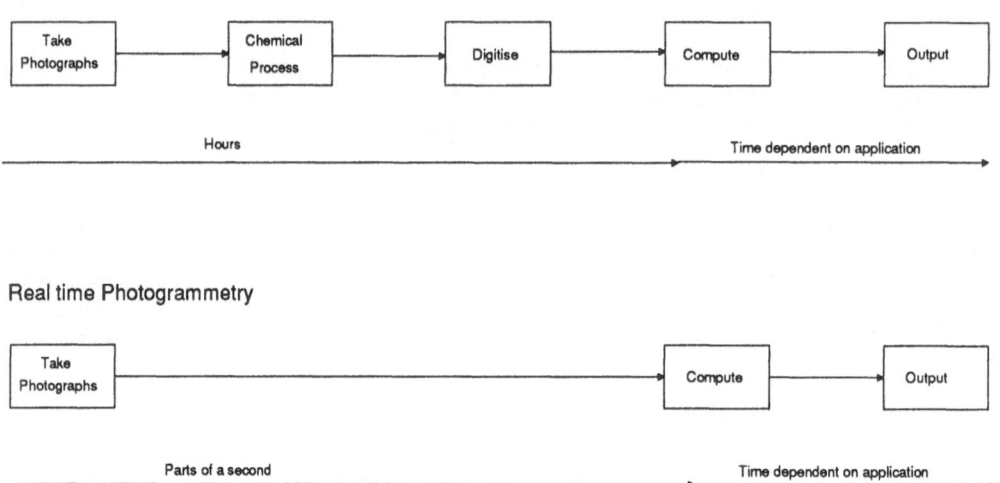

Real time Photogrammetry

Figure 1. Schematic to illustrate how the application of Real Time Photogrammetry speeds up
the measurement process.

for analysis could be evaluated. These stereo photographs had then to be selectively digitised, usually onshore even though offshore systems with limited data processing capability were available, before dimensional data could be provided and before the format of the results could be decided. Technology does not stand still, and the offshore industry is always keen to adopt technology that will lead to improved data or cost savings. The analysis of the major processes of an underwater photogrammetric system identified some that could be eliminated or shortened with the introduction of new technology. This has been illustrated in figure 1.

It should be emphasised that it is only recently that the technology has come into existence at a price to enable Real Time Photogrammetry to become a reality. It is a marriage of differing technologies from a variety of disciplines, some of which would not, at first sight, appear to be compatible.

2.2 The Camera

As a first stage a feasibility study indicated that the photographic cameras could be replaced with electronic still cameras. These differ completely from video cameras, or from the still video cameras now being introduced to the domestic and news-gathering market, for the following reasons:

- A video camera is an analogue device. A still video camera is an analogue device which only records a field or frame of video in an analogue form. An electronic still camera is a digital device where the only analogue process is the build up and retention of the initial charge wells in the Charge Coupled Device (CCD) when imaging light falls on it.

- A video camera relies on line synchronisation pulses and timings from analogue components in order to extract the image from the imaging device. An electronic still camera extracts the data from each discrete imaging element (pixel) individually under full handshake control.

- A video camera has a relatively low resolution imaging device. In the case of a typical Charge Coupled Device (CCD) now used in most video cameras this will have a format commonly of 500 x 400 pixels. An electronic still camera will have a CCD density in excess of 1000 x 1000 pixels.

- A video camera can resolve 64 levels of grey. An electronic still camera can resolve at least 256 levels of grey.

- A video camera is subject to noise at all stages in its process, while an electronic still camera is only subject to noise within two of its processes, both of which are easily controlled so that the effect of noise can be minimised.

The process of Real Time Photogrammetry dictated that it was essential to use a digital camera, to enable the camera to be calibrated, and also to ensure that this calibrated image data file was not corrupted within transmission and storage. Additional benefits of the electronic still camera are that it would give all the advantages of the use of video, in comparison with the use of photography, particularly real time quality control in both the quality of the image and also the subject imaged. It would also give most of the advantages of the use of photography in that the image resolution would be high, although would be monochrome rather than colour; the image would not be degraded within the transmission system or by storage and that it could easily be incorporated into a reporting system, particularly an electronic reporting system.

2.3 The System Specification

The specification of an electronic still camera had the additional advantage that the image was fully digitised as a cause of the process. It did not have to be selectively digitised under operator control as for conventional photogrammetric measurements. The whole image, with its enormous density of imaging points, could be used to directly present three-dimensional coordinate systems to graphical software as soon as a match could be achieved between the left and the right hand images that make up the stereo pair.

The feasibility study also indicated that it would be possible to replace the expensive analytical plotter, an optical/mechanical instrument manufactured to high tolerances and not transportable or suitable for use offshore, with an image processing system.

The feasibility study defined the likely system components as in figure 2. Underwater would be two electronic photogrammetric cameras. These would either use existing video lighting, or a photographic flash for illumination. A laser would optionally project a pattern onto the subject to allow faster image matching. The digital images would be digitally transmitted to a personal computer based image processing system, which would be linked to a high resolution screen fitted with a liquid crystal shutter to allow stereo viewing. Image storage would be on optical disk. The images would be processed using standard photogrammetric algorithms with the software written so that the data

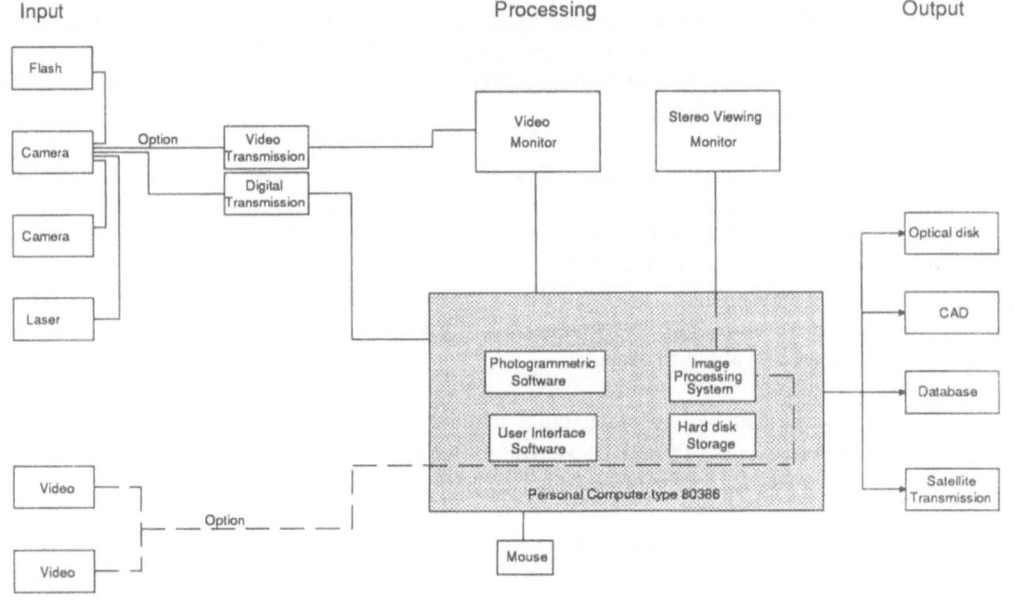

Figure 2. Schematic of a Typical Underwater System Configuration.

could be processed in real time. A user interface system would allow engineers totally unfamiliar with the theory of photogrammetry to extract the three dimensional information that they required.

As an option the system would allow real time images in colour from a pair of conventional video cameras to be viewed in stereo.

The system, as defined in the feasibility study, allowed the retention of all the strengths of the photogrammetric process, as well as overcoming its two major weaknesses.

3.0 REAL TIME PHOTOGRAMMETRY - THE DEVELOPMENT

3.1 Introduction

The development of the real time photogrammetric system progressed along the lines indicated in the feasibility study. The main areas of work were:

- The camera;

- The transmission system;

- The image processing system;

- The stereo viewing hardware;

- The photogrammetric software;

- The user interface software.

Each component of the system will be considered in turn, to indicate the problems to be faced with the introduction of the system and the advantages that each part of the system will give to offshore operations. The advantages of the overall use of the system will then be considered.

3.2 The Camera

There is only one manufacturer of high resolution area CCDs within the UK; only one other in Europe; and only one other with units readily available (for non-military uses) in the USA. The CCD chosen for the camera was a 1242 x 1152 device provided by EEV Limited. This was chosen because of the preference to work with a UK manufacturer; the desire of the manufacturer to provide specialist development facilities to allow the use of the CCD to be optimised; and the planned availability of a 2000 x 2000 pixel device that could be controlled with the same electronics.

The device chosen could be equated to the performance of a 35mm photographic film, particularly as it was a similar physical size to the film, with a format of 27.9mm x 26mm against the 35mm film format of 36mm x 24mm. Its physical size also meant that it would be suitable for use with conventional underwater optical systems for 35mm photographic cameras.

The CCD has a faceplate sensitivity of 6 lux which is similar to a conventional underwater colour video camera. It can hence be used with conventional video lighting, although flash would still be used to freeze the motion of rapidly moving subjects and to increase depth of field.

A particular aspect of photogrammetry is that the camera has to be calibrated to take into account both physical and optical imperfections. In the case of the electronic camera there were additional imperfections encountered. A correction has to be applied to the dark current, the background noise within the CCD which affects each pixel differently, and also the difference in sensitivity of each pixel. While the conventional photogrammetric calibration technique was able to take account of the former imperfections, the latter had to be corrected by measuring them for each CCD individually, under a variety of different temperature and lighting conditions, and using software to correct the individual output from each pixel.

An important part of the calibration process is the ability to be able to identify where each discrete piece of intensity information, generated by a discrete pixel, has come from as a function of the overall geometry of the CCD. This is termed pixel to pixel correlation between the camera and the image processing system.

The analogue charge registered in each pixel is proportional to the intensity of light falling on that pixel. The total picture is made up of the complete pattern of these intensity values, and hence the ability to correlate between the light intensity and the analogue charge held by each pixel is critical for image quality.

The camera as designed is currently unique in that it is the only truly operational underwater photogrammetric camera that samples the analogue signal generated by light falling onto a pixel under full handshake control. The analogue to digital (A/D) conversion is also carried out under full handshake control so that, within the 8 bit binary image file that is the result of the process, it is possible to identify the coordinate of each pixel, and so to calibrate the system as precisely as a photographic camera. This is illustrated in figure 3.

The A/D conversion is carried out as a 10 bit conversion. This means that the A/D convertor can recognise 1024 grey levels (10 bit signal) and so can precisely allocate a digital value to the analogue system. As it is sufficient for the real time photogrammetric system to work with 256 grey levels (8 bit), the two least significant bits are used only to ensure the complete accuracy of the 8 bit digital file, and so adds to the overall quality of the system.

The image is held in a buffer memory which mirrors the CCD. However, instead of holding it in an analogue form, as it would be in the CCD, the memory holds it in a digital form.

The last part of the camera, the microcontroller, handles all communication functions as well as the control of other camera functions such as focus and exposure; and storing the calibration information.

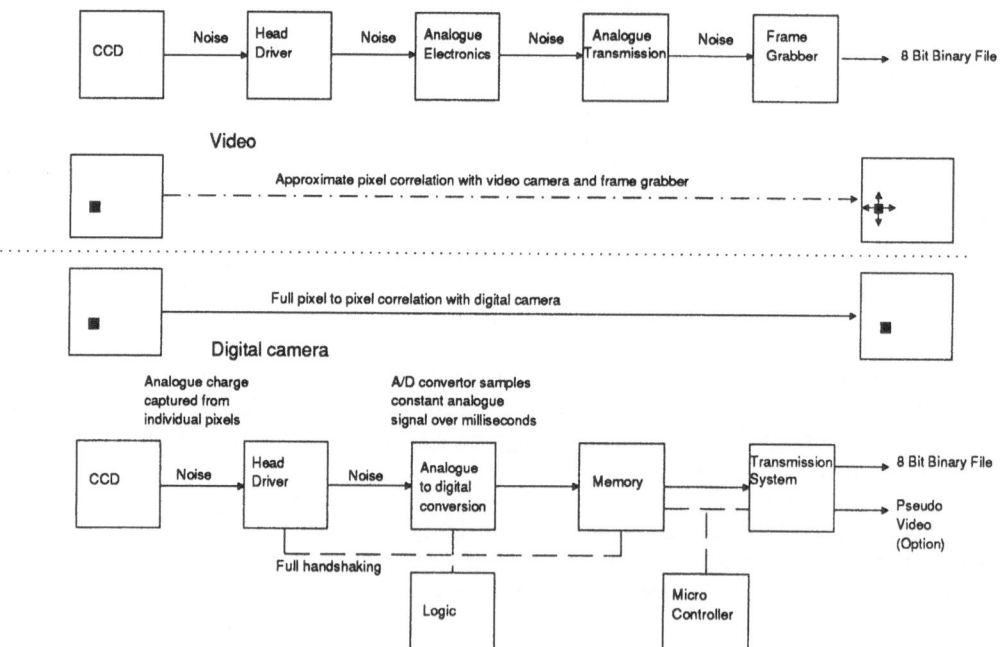

Figure 3. Principle of the Electronic Digital Still Camera contrasted with Video.

3.3 The Transmission System

It was found that conventional twisted pair transmission systems would not be adequate for fast image transmission. These systems commonly utilise the RS422 standard for digital transmission and would take of the order of two minutes to transmit one image from the camera.

A dedicated transmission system has been designed for the camera for this type of cable. It is almost double the speed of a conventional RS422 system and uses an automatic error correction algorithm. The use of this algorithm almost doubles the transmission time, but ensures that, on a noisy line, signals do not have to be echoed repeatedly until the correct image has been received.

It was found that fibre optic systems were more than capable of handling the full data output of the camera at the maximum ten frames a second.

To ensure that a "viewfinder" system was available for continuous monitoring, even with the older transmission systems, the microprocessor was programmed to sample pixels selectively in order to simulate a video signal of lower resolution. This is shown in figure 3.

3.4 The Image Processing System.

The feasibility study had shown that the conventional IBM compatible personal computer could not handle the data produced from the camera at a speed sufficient for real time processing. It was taking approximately 15 seconds to write one image to the screen with the software addressed the VGA display chip through BIOS. Although workstations and mini-computers were considered, the problem was still seen to be that nothing was suitable for handling large image files quickly. It was hence necessary to install an image processing system, based on the Texas Instruments TMS 34020 graphics chip, into the bus of the personal computer. This lead to an increase in speed of approximately 2000 times, as well as giving the potential of applying standard image processing enhancements in real time.

The image processing is really a powerful computer in itself, although it has been designed just to handle images rather than all the normal tasks commonly associated with computer use.

The topside system components, which comprise of a personal computer, the installed image processing system and the stereo viewing screen are illustrated in figure 4.

3.5 The Stereo Viewing Hardware.

Conventional stereo viewing systems utilise two monitors with mirrors to allow the user to gain qualitative three-dimensional information. They suffer from a number of problems including being difficult to use; being tiring to use for long periods; and being restricted to one user at a time. They were not considered suitable for this application. Other stereo viewing systems use liquid crystal glasses synchronised to a single display in such a manner that sequential images are viewed by left and right eye sequentially, so giving a stereo effect. However, they normally suffer in that they are of a lower resolution than, or are updated more slowly than, a conventional monitor. This is because a conventional monitor cannot be driven fast enough to update the full screen and so an interlace

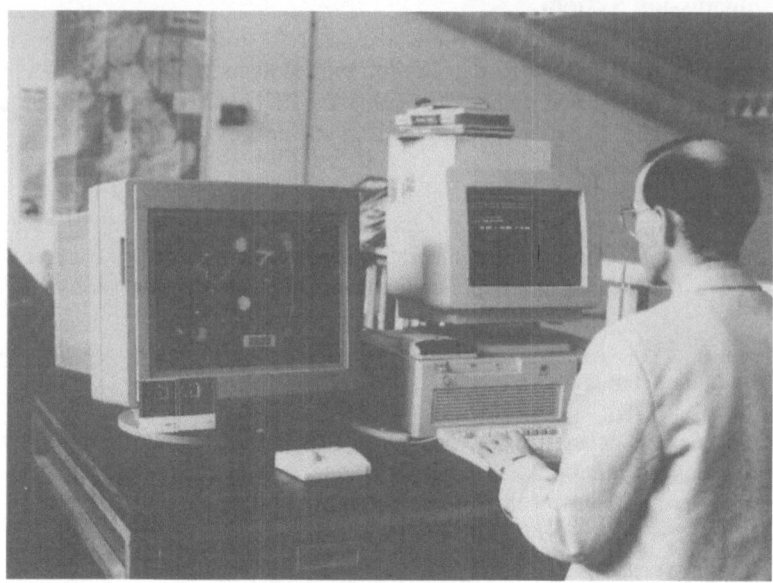

Figure 4. The topside system which comprises, from left to right, the stereo viewing screen and the personal computer, in which is installed the image processing system.

system, which halves the potential resolution of the monitor, or a slower refresh speed, is used. The result is either a low resolution display or a jerky image movement.

The system selected retains the full image quality of a high quality graphics monitor with a resolution of 1280 x 1024 pixels and 256 grey levels. The monitor is fitted with a liquid crystal screen which can be cross polarised in synchronisation to the rate at which the image processing system displays alternate left and right hand images on the screen. The user has only to wearing cross polarised glasses to view high resolution images.

Using a slightly reduced window of 1024 x 512 pixels, which suits the screen aspect ratio and the image processing system memory configuration, these images have been driven successfully at speeds up to 120 Hz, so eliminating all jerkiness of movement. The system has several additional advantages. It can be viewed by several people in stereo at one time, hence allowing three-dimensional information to be discussed. It can be used in conjunction with a real time frame grabber installed within the image processing system to display the output from two conventional video cameras in stereo, and it will allow 24 bit colour images to be used.

3.6 The Photogrammetric Software.

The photogrammetric software is used to calculate the three- dimensional coordinate system from the two-dimensional coordinate systems provided by each camera. To do this it has to perform the following functions:

- It has to be able to interrogate each camera to receive the calibration information unique to each camera. This is done when the system is powered up and goes through its diagnostic checks.

- It has to be able to match both images to ensure that homologous points on each of the left and right hand image are used for the calculation. This matching process is aided by the use of a projected laser pattern in some circumstances.

- It has to be able to manipulate the images to correct for any tilts, rotations and other misalignment of the two images.

- It has to be able to apply all camera calibrations.

The real time system must carry out the calculation for each time the image is moved on the screen, potentially 120 times a second. This has led to the adoption of advanced programming techniques which use the standard algorithms in innovative manners, to ensure that the speed criteria can be maintained.

3.7 The user interface software.

One of the most important weaknesses preventing the widespread use of conventional photogrammetric systems was that they required skilled personnel to extract the data from the photographs. This increased the cost of the utilisation of the systems, as well as creating bottlenecks. A major objective in the development of the real time photogrammetric system was to ensure that it could be operated by the engineer who required the dimensional information. The design of the user interface system was hence critical to the success of the system.

After the evaluation of several systems, it was decided to standardise on the Windows 3 user interface and to hence write the software so that it could become a Windows 3 application. This interface was chosen as it is now becoming one of the industry standard interfaces, and so users would find that the software would have the same "feel" as other packages that they had used. It also gave the advantage that many of the input and output routines for driving the program by a "mouse" and for outputting to a range of printers or other programs were already in place. The only disadvantage was the difficulty in becoming familiar with the concept of programming under Windows.

The interface has been written so that the operator required only to point using a "mouse" to use the program. The only typing that is required is for storage of file names.

4.0 THE POTENTIAL

4.1 Meeting the Initial Demand

The system meets the initial demand, which was for a means of being able to obtain accurate three-dimensional measurements in real time without the involvement of specialist personnel. It will be able to replace a conventional photogrammetric system for all but the most accurate or most complex of tasks, and will also allow the use of photogrammetry for general inspection tasks and for spoolpiece measurements where the conventional technique was too expensive, or provided the information with too long a lead time respectively.

4.2 Automation and Supervisory Control.

The ability to analyse relative three-dimensional coordinates will allow many ROV related tasks to be automated, either partially or completely. This will have many consequences to speeding up and de-skilling ROV manipulative tasks. For example, having defined the path along a weld which is required to be cleaned, and the distance to either side of the weld which has to be cleaned, the ROV would be able to monitor and control the cleaning process so that, even though poor local visibility prevented the operator from viewing the surface to be cleaned, single pass cleaning could be performed. It could even be performed simultaneously with two systems, so increasing the efficiency in the utilisation of the basic ROV work platform.

There are many other tasks that the ROV could perform automatically as long as some form of supervisory control was available. It would be possible to ensure that many subsea completion systems were designed in such a manner that this type of intervention became the norm.

4.3 The Technologies.

Many discrete technologies have been developed as a result of the development of Real Time Photogrammetry. Some of these will be able to be applied independent of the system as a whole, and will provide the opportunity to improve current methods of working.

The electronic still camera, although available in monochrome only at present, has sufficient resolution to replace the use of stills photography. The advantage of the camera, because the image is stored and transmitted digitally, is that original quality is not lost in transmission or storage as it is with video. A image could be transmitted over satellite lines to provide a high quality illustration of a problem and to allow immediate feedback into the offshore process.

Digital storage of images, on optical or magnetic disk, could also parallel the use of computers to plan and report on inspection contracts. As databases are currently used to extract data on features that have to be inspected, and to correlate the data to look for trends and other events, so could the images be controlled in exactly the same manner. An engineer working on a single monitor would be able to view the data and the images simultaneously, and engineers working from the same databases would be able to view the information simultaneously, no matter where in the world they were.

The stereo viewing system is also a great improvement on historical systems. The ability to view the subject in stereo makes interpretation easier and more reliable, and the ability to monitor manipulative tasks in stereo ensures that they are carried out efficiently.

5.0 CONCLUSION

Real Time Photogrammetry, with its composite and attendant technologies is now available to be applied by the underwater industries. It is a technology for Today, but it will probably be Tomorrow before the attendant ROV control and manipulative systems are developed to the extent where they are able to take full benefit of the data provided by this intelligent visual imaging system. The tools are now available for others to start to construct.

ALTERNATIVE APPROACHES TO PIPELINE SURVEY: THE PIPELINE ENGINEER'S VIEW

J.H.A. BAKER
Technical Developments Manager
J P Kenny Caledonia Limited
ABERDEEN
AB1 2DB

ABSTRACT: Annual ROV pipeline survey specifications tend to place considerable emphasis on visual records and the commentary made by the survey crew. However, the latter is based largely on the former which are, in turn, restricted by the field of view and the uncertainty of orientation of the boom cameras. Thus a slow and laborious process yields only limited and subjective results. In addition, re-visiting old records is not only extremely time-consuming, tedious and costly, but does not facilitate any comparison from year to year. This paper, which is based upon three recent studies, looks at the limitations of current survey techniques and explores, from the point of view of the pipeline engineer, some alternative and innovative approaches aimed at improving pipeline survey. By taking advantage of recently-developed techniques, it will be possible to obtain and present more pertinent data. The way these data can be used to advance the understanding of the dynamic interactions between pipelines and the seabed are discussed.

1. Introduction

Every year, hundreds of kilometres of subsea pipeline are surveyed to ensure, on behalf of the operators and, ultimately, the relevant regulatory authorities, that the pipelines are neither damaged nor in a perilous state. There are two principal survey methods used:

o   Side-scan sonar survey, and

o   Video survey by ROV.

In general, the former is used to conduct a preliminary survey of the whole length, usually annually, while the latter is used only to carry out more detailed investigations of areas of particular interest or worry. However, full video surveys are generally carried out every fifth year at least.

Once a pipeline has been installed, the specialist pipeline engineer tends to be called in only when a particular problem is identified. By

333

Volume 27: Subtech '91, 333–345.

this time, the information available from the various annual and major
surveys may not present the most useful picture, and the resulting
analysis may have to be quite subjective.

This paper looks at the capabilities and limitations of each survey
method, from the point of view of the pipeline engineer, and then
considers how alternative techniques could enhance the presentation of
survey data. The longer term aim would be to utilise survey results for
prevention rather than cure.

2. Current Survey Techniques

Annual pipeline surveys are required to identify, as far as possible
[1]:

o   Pipeline movement;
o   Freespans;
o   Exposure of previously buried lengths;
o   Loss of weight coat;
o   Damage to the pipeline, or fittings;
o   Local debris;
o   Excessive marine growth, and
o   The state of the cathodic protection (c.p.) system.

These regulatory requirements tend to result in surveys being conducted
in the sense of: "is it still alright?".

Annual surveys are usually conducted by side-scan sonar. These are
performed by towing the fish parallel to the pipeline at a set altitude
and offset. Shadow techniques are then applied to identify any exposure
or significant lateral movement. Local debris will be highlighted,
while the existence of freespans or localised damage may also show up.

Side-scan sonar surveys are relatively simple, fast and inexpensive,
but do have disadvantages. For a start, even to an experienced analyst,
degree of exposure must be a subjective assessment. No two analysts can
be expected to assess the same condition in the same way, particularly
if the records are from successive years and made with different sonars.
Thus comparisons from year to year could result in misleading
conclusions and, for this reason, some operators (for example, BP) no
longer record degree of exposure. This introduces its own ambiguity
since "pipeline exposed" can mean anything from the crown of the
pipeline just visible below a thin veneer of sediment to the pipeline
lying fully exposed upon the seabed.

A second disadvantage is that, even when the pipeline is fully
exposed, a freespan will only be evident if the gap below it is
significant, and will be difficult to detect if the pipeline is lying in
a trench. Equally, side-scan sonar traces give little indication of the
transverse profile of any exposure. These surveys normally make no
attempt to measure the cathodic protection.

Conversely, side-scan sonar is good at highlighting debris, and some
forms of damage. Significant lateral movement can also show up
clearly.

Should any feature from the side-scan sonar survey give cause for concern, an ROV will be mobilised to carry out a visual survey. In addition to the forward-looking camera, an ROV equipped for pipeline survey is fitted with a pair of boom-mounted cameras on pan-and-tilt units, one each side, an ultrasonic trench profiler, and a c.p. probe. An ROV survey will generally be performed for the full length of a pipeline (major survey) every five years.

In comparison to a side-scan sonar survey, an ROV survey is slow and expensive. Emphasis is placed on the video records, and the accompanying log and commentary, but the latter are based on the former which, for reasons explained below, may not be very reliable.

The range and field of view of a colour closed circuit television (CCTV) tube is limited. Thus, as the boom cameras are panned and tilted, the only visual reference is usually the pipeline itself. As a result, any assessment of degree of exposure can only be made with respect to the very item which is being assessed. Lack of orientation and concept of size therefore results, once again, in a very subjective assessment. Also, the limited field of view makes any attempt to estimate the profile of any exposure extremely unreliable.

During an ROV survey, however, the survey crew are assisted by the ultrasonic profiler which presents a regularly, and frequently, updated transverse profile. These profiles are transferred to the video records only at infrequent and somewhat random intervals, and seldom held for long enough for their import to be comprehended. Hard copy reports, likewise, contain profiles only at large intervals.

Freespans are identified by the light from one boom being visible from the opposite boom camera when its lights are dimmed (generally termed "light creep"). This method is not foolproof when gaps are small but, in general, is a useful technique for identifying lack of support. However, as will be seen below, lack of support does not necessarily mean a freespan in the accepted sense.

Video techniques are useful for identifying marine growth, debris (within the field and range of view), damage to, and loss of, weight coat. The c.p. probe facilitates continuous profiling, and detailed investigations of areas of damage. However, except in extreme cases, video will not identify lateral movement of the pipeline.

A common problem is that the records from different surveys may not be aligned. This could easily be rectified by fitting some form of physical Kilometre Post (KP) markers, such as electro-magnetic reflectors or even miniature transponders, to the pipe.

It can thus be seen that both techniques have advantages (sometimes not fully realised) and disadvantages, but neither is ideal. However, before going on to address improvements in techniques, it is necessary to think briefly about the sense in which these surveys are conducted. As previously mentioned, all the features of current survey techniques are concerned with the state of the pipeline as it is (at the time of the survey). This approach provides a means for identifying cures to problems, but not for preventing problems. To do the latter, it is necessary to plot the history of the pipeline, generally by comparison from survey to survey. In this regard, video records are of limited value since the human brain is not capable of useful comparison between video sequences viewed consecutively. To put this into context, it is necessary to consider the physical processes involved.

3. Seabed Sedimentary Processes

Once a pipeline has been laid and trenched (if necessary), then, provided it is stable and does not suffer mechanical interference, any subsequent change in its state will be the result of erosion or deposition due to sedimentary processes. A pipeline, or an open trench, both constitute an irregularity on the seabed which interferes with the flow. In each case, whether this irregularity leads to erosion or deposition is a function of the mass transport balance. If the net removal of sediment is less than the transport, deposition will occur: advantage is taken of this mechanism in the natural backfill of an open trench. If the net removal is greater, erosion will occur: in the case of untrenched pipelines, this is generally due to scour action.

Under steady flow conditions, the presence of a pipeline on the seabed causes the flow to deflect, and to form an eddy low on the upstream side of the pipe. The deflected flow then separates from the pipe, and the wake creates a second eddy in the lee of the pipe. These two eddies scour the sediment from both sides of the pipe (see Figure 1(a)).

The energy in these eddies is able to lift the sediment to the boundary layer flow which may then transport it, leading to a net erosion (see Figure 1(b)). If this process continues, the face and lee erosion may well meet below the pipe to create a tunnel (see Figure 1(c)). The tunnel then causes a jet flow which will rapidly remove further sediment. This process is known as Tunnel Erosion.

The sediment removed by the tunnel flow may be deposited a short distance downstream to form a bank (as in Figure 1(c)). On the other hand, if the flow velocity is large enough, wake turbulence may continue the scour action some distance downstream before deposition occurs. This is known as Lee Erosion (see Figure 1(d)).

If the predominant flow is in one direction, the scour profile will be asymmetric (as in Figures 1(c) and 1(d)). A weak reverse flow will tend to be deflected by the spoil bank and the profile will still be asymmetric (particularly if only Tunnel Erosion has occurred). If the flow reverses with similar velocity (as in tidal flow) however, a symmetrical profile is likely to emerge with a small spoil bank at each side (see Figure 1(e)).

These scour mechanisms are largely two-dimensional. The break-through into a tunnel tends to occur at localised points where the sediment is weaker, or perhaps the pipe lies across small ripples. However, once the tunnel has been created, the erosion rapidly becomes three-dimensional, removing sediment laterally. Thus, the tunnel will increase in width below the pipe. Initially, the vertical and lateral rates of scour will be similar, but the vertical rate will tend to reduce once a certain depth below the pipe has been achieved, and the majority of the erosion will then be lateral (see Figures 2(a), (b) and (c)). This depth depends on the type of flow: in steady flow, it is of the order of $0.6D$ (where D = pipe diameter) [2], but wave action will tend to reduce this (possibly to as little as $0.2D$ in pure oscillatory flow) [3], whereas vortex-induced oscillations may increase it to the

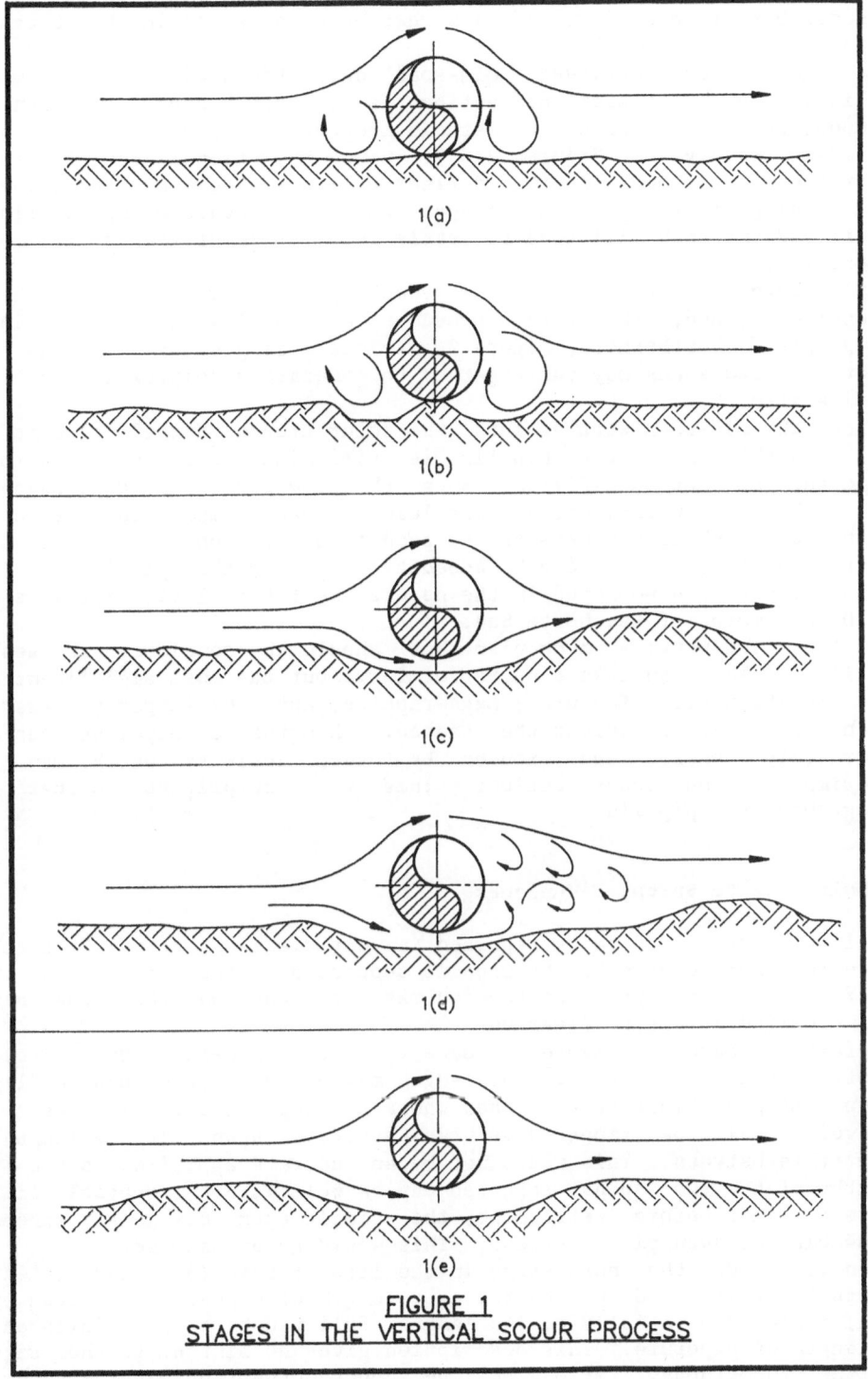

1(a)

1(b)

1(c)

1(d)

1(e)

FIGURE 1
STAGES IN THE VERTICAL SCOUR PROCESS

order of, typically, 1.05D [4]. Local factors can modify these values considerably in practice, and 3.0D has been observed in the Southern North Sea.

As the process continues, mid-span deflections will initially be negligible but, because deflection is a fourth-order function of unsupported length, beyond a certain length, mid-span deflection will increase rapidly (see Figure 2(d)). Since the flow is now forced to negotiate a spoil bank and then deflect well below ambient seabed level, the closing of the gap at mid-span will not induce much additional scour, and the span will tend to settle into the scour trench [5] (see Figure 2(e)).

It should be noted that the scour process is extremely three-dimensional and, therefore, produces a complex three-dimensional topography. Furthermore, Figure 2 depicts a single span whereas, in practice, two spans may run together to generate a complex longitudinal profile also.

Once the mid-span settles into the scour trench, tunnel flow is cut off. In this area, the pipeline is similar to one that has been trenched, but not back-filled. When the flow crosses a depression in the seabed, flow retardation may lead to deposition. In this case, whether a trench is man-made or the product of erosion, back-fill will occur, possibly to a depth adequate to bury the pipeline. This mechanism has been employed in the natural self-burial of pipelines in the Dutch sector of the North Sea.

A final sedimentary mechanism that should be considered is seabed mobility. This can take a number of forms but the most significant are the flow-transverse features, mega-ripples, and the larger sand-waves, which can migrate across the seabed. Lengths of pipeline can be alternately buried and exposed by such activity which may be superimposed on scour action. This will complicate further the topography of a pipeline.

4. Relevance to Survey Techniques

Pipeline surveys tend to be reported in very clear-cut terms: the pipe is either buried or not, the pipe is supported or there is a span. The descriptions in the previous section make it clear that the actual state of a pipeline is not so clear-cut.

Referring back to Figure 2, case (d) would generally be correctly reported as a span, although the length may be inaccurate because "light creep" is indistinct at the ends where the gaps are small. Case (e) however, would be reported as two separate spans with a length of support in between. This would not be an accurate appraisal since the lengths of the two "spans" reported may be well within acceptable limits whereas, just before settlement, the single span may have exceeded allowable, or even yield, stress. This would go unremarked.

Consider now the next stage in the life of case (e), as depicted in Figure 3. This would be reported as a length of exposure, followed by a span, followed by a fully buried length, followed by a span, followed by a length of exposure. This description gives no hint as to the origins of the topography. It may well be supposed that the pipe is actually becoming unburied and scoured to a state of spanning whereas, in fact,

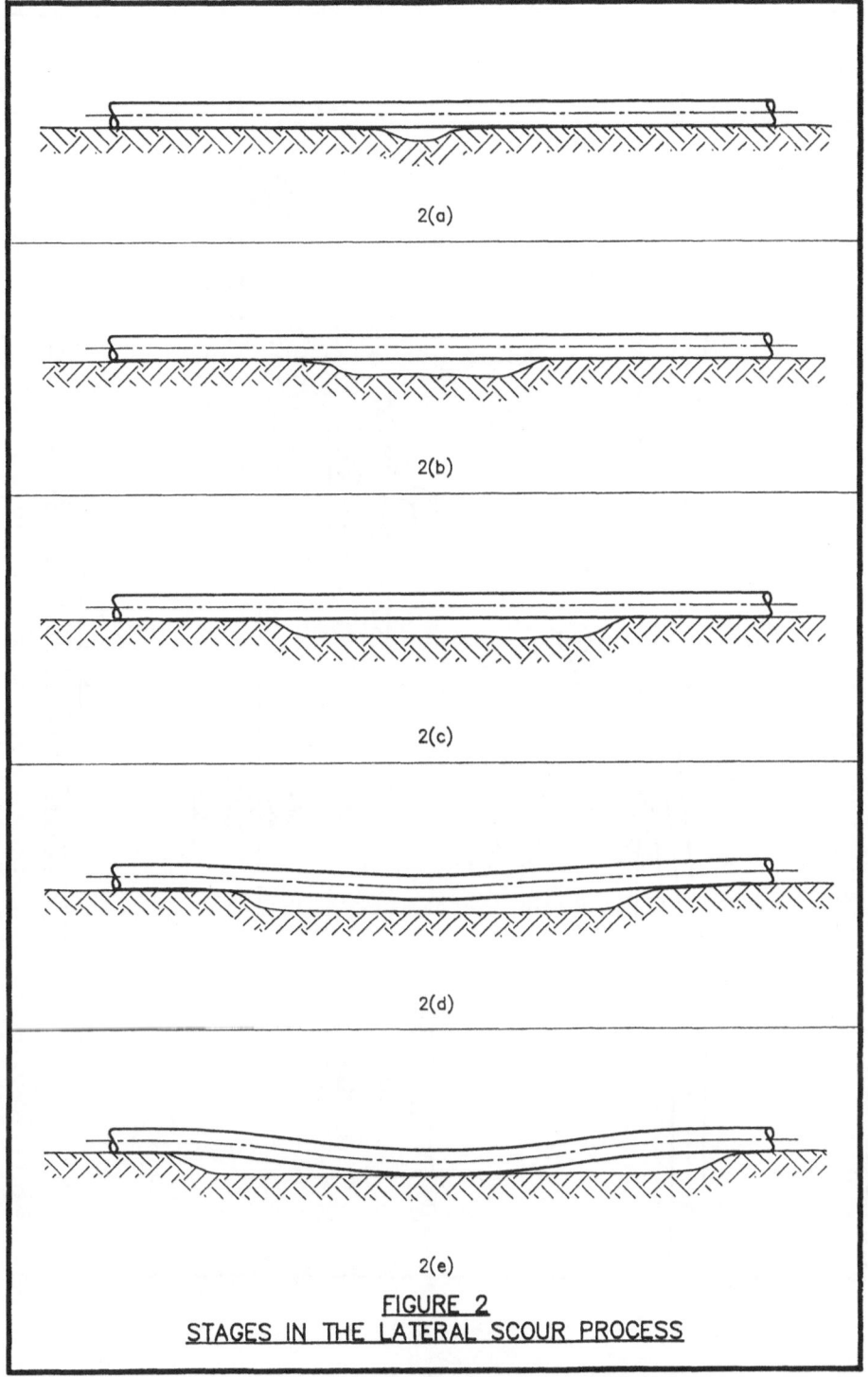

2(a)

2(b)

2(c)

2(d)

2(e)

FIGURE 2
STAGES IN THE LATERAL SCOUR PROCESS

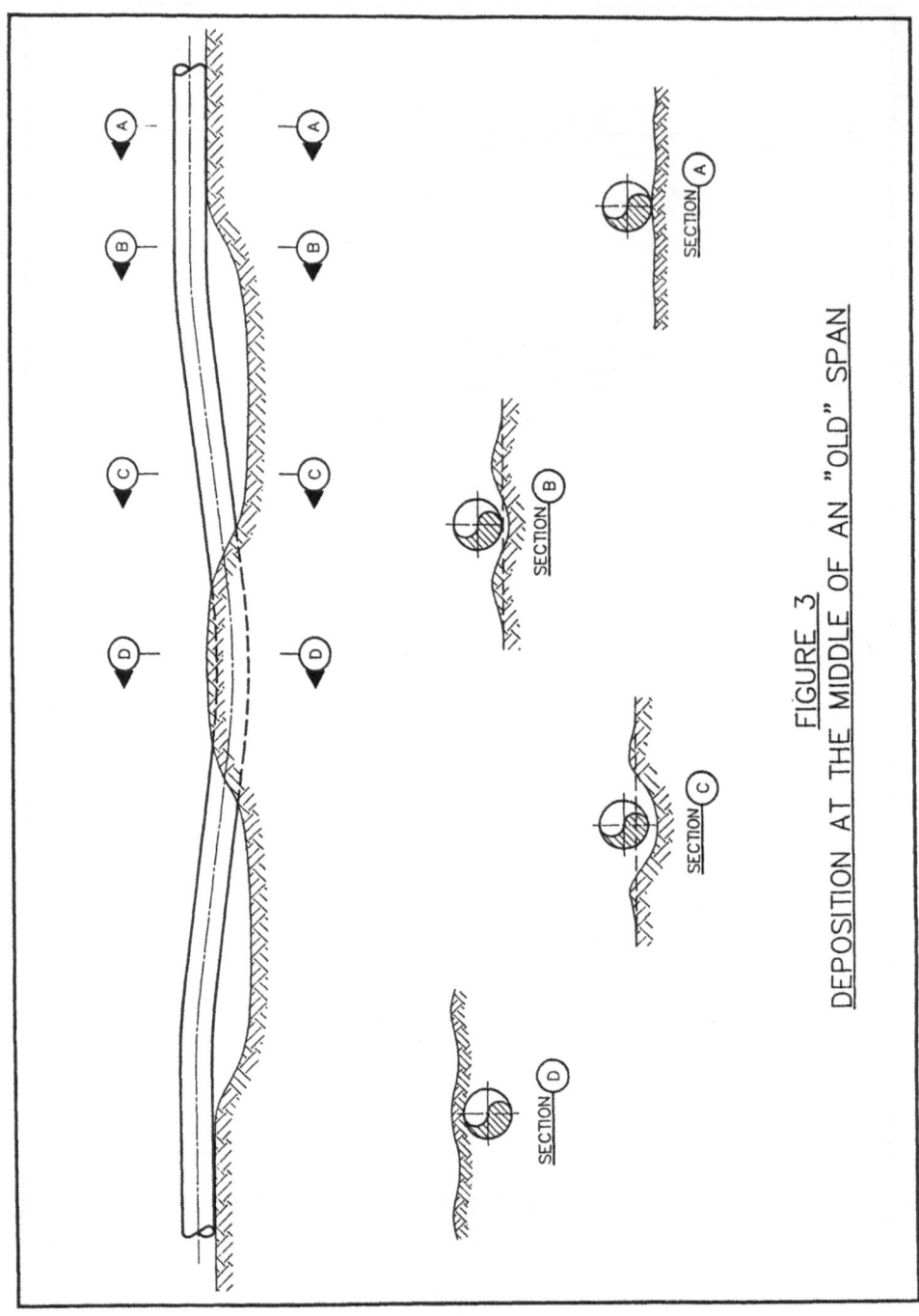

FIGURE 3

DEPOSITION AT THE MIDDLE OF AN "OLD" SPAN

the opposite is true. As has been noted previously, the range and field of view of ROV-mounted CCTV cameras is limited and, as has been apparent during two detailed pipeline stability review studies recently carried out for BP, this can mean that, for instance, the difference in trench configuration between points B and C may well go undetected.

This shows that a description of the current state of a pipeline, based on a commentary and survey log which are, in turn, based on video observations, may be very misleading. If the pipeline is to be analysed, it is necessary to have a better knowledge of the topography, and to be able to compare this to previous survey results.

5. Alternative Survey Techniques

By far the most useful data generated by an ROV survey are the transverse profiles. It is unfortunate that these are not, at present, used to their full advantage. During a survey, the profile is updated on the monitor every thirty seconds or so, but this information is only channelled to the video recorders sporadically. Its absence is very obvious if the video records have to be revisited. It is recommended, therefore, that split-screen techniques are utilised to ensure that the profiles are continuously displayed on the video records.

Although these profiles are recorded digitally, they tend not to be provided in hard copy with the survey report: in general, one profile per hundred metres or so is given as an example on the charts. Recommendations for improved techniques, developed below, will supercede this suggestion, but current techniques would be enhanced if all profiles were produced at a reasonable scale in hard copy.

However, as has been described in the previous section, the topography of a pipeline subject to seabed sedimentary processes is highly three-dimensional. A more significant recommendation, therefore, is that the digitised profiles be combined to create a three-dimensional surface plot. A wire diagram presentation, whose orientation can be altered in x, y and z on the screen, would suffice (see Figure 4) although better presentations could be developed. All the techniques required to achieve this exist: they simply need to be harnessed.

The digitised profiles could all be aligned on the basis of a flat seabed. However, a further enhancement would be to base the alignment on the actual longitudinal profile. In theory, this would produce an absolute model of the pipeline topography.

There is a further problem however. In order to produce a profile of the pipe and its immediate surroundings with reasonable resolution, the acoustic profiler heads need to be mounted fairly close to the seabed. In this case their transverse range is limited and will not extend to the ambient seabed, especially if there is any lee erosion. However, to ensure the full width of profile (at least twelve diameters to each side), the profiler heads must be mounted high up on the ROV and resolution will be impaired.

One solution to this is to run the ROV along the pipeline to obtain the high resolution profiles, and then to fly it above the pipe to obtain broader profiles. This has been done in at least one case.

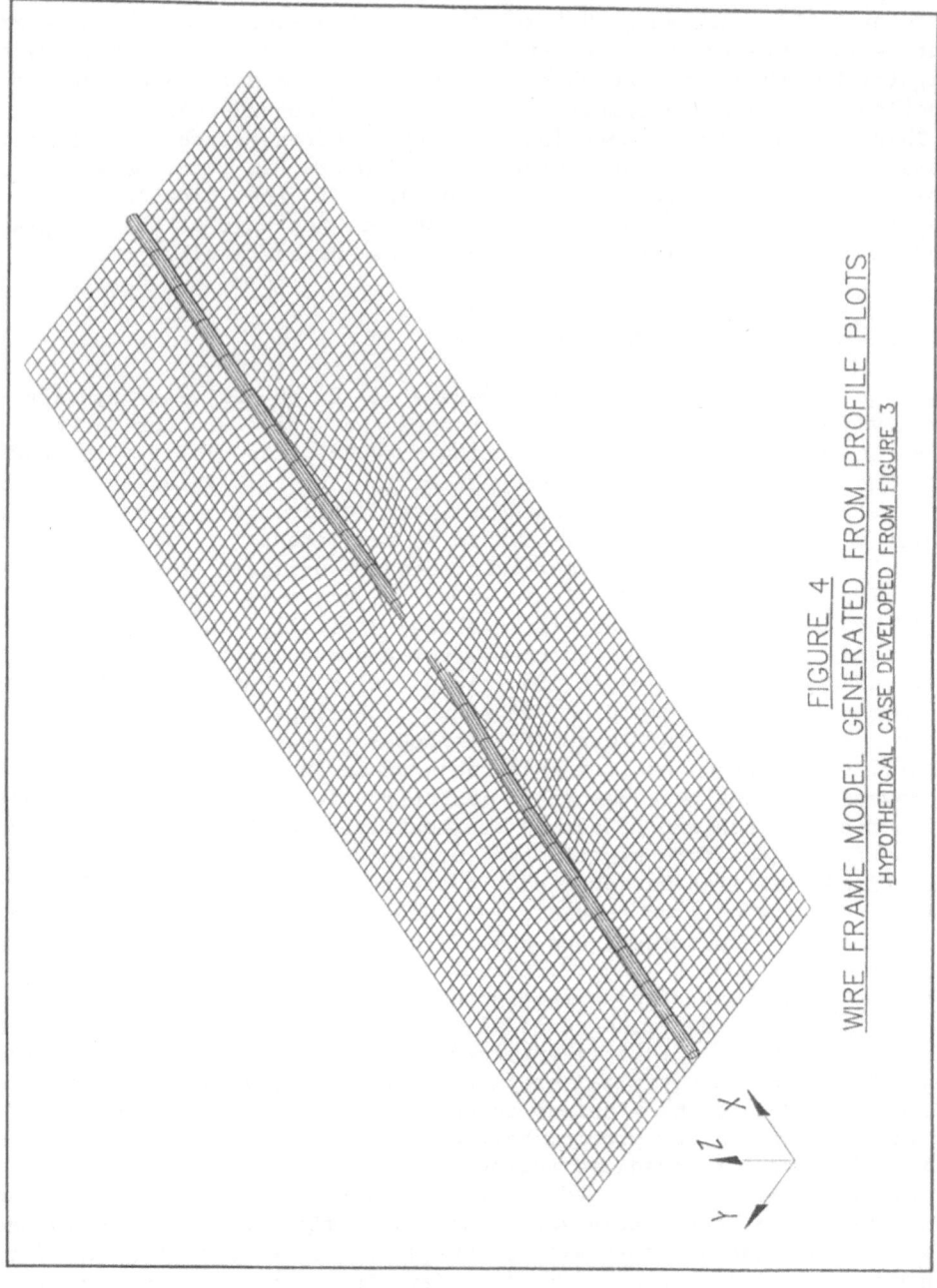

FIGURE 4
WIRE FRAME MODEL GENERATED FROM PROFILE PLOTS
HYPOTHETICAL CASE DEVELOPED FROM FIGURE 3

Alternatively, two sets of acoustic profiler heads could be fitted, one low down at the front of the ROV, and one high up aft. Transmission and data processing would have to be synchronised one hundred and eighty degrees out of phase to avoid interference. However, this might stretch the number of telemetry channels on the ROV to the limit.

Another technique which has recently been introduced is laser profiling. This is said to provide much higher resolution at much higher speeds. If initial appraisals prove successful, this may well become the industry standard.

Given one or two sets of profiles processed to create a 3-D surface plot of the topography, the next question is: would video still be necessary? Clearly, the ROV pilot needs to see where he is going, but boom cameras and video records could become redundant. A still camera and strobe could be triggered at intervals to sample seabed soil type, marine growth and the state of the weight coat. If particular problem areas were identified, they could be revisited with video.

This suggestion may not be relevant if a tethered ROV is used for the survey. But it leads to the ultimate in this series of recommendations, which is that the industry should work towards the use of Autonomous Underwater Vehicles (AUVs) to perform pipeline surveys. Such a vehicle could be launched at one end of a pipeline: it would fly along the pipeline and return at a different altitude to be recovered at the launch point (where it could wait out bad weather if necessary).

Using a pulse induction pipetracker, the AUV would not need to know whether the pipe was buried or not, but could return depth-of-burial records if it were. A still camera could sample at set intervals (as above) while the laser profiler would record the topography at a much higher rate than an acoustic profiler, enabling the AUV to transit at higher speed. This exercise will be more efficient if electronic KP markers are fitted to the pipe to facilitate comparison from year to year.

Once the data have been (rapidly) processed into a 3-D surface plot, the current state of the pipeline can be assessed, and compared to previous datasets to determine the history. An ROV equipped with CCTV cameras would then be mobilised if any specific aspects required visual investigation. It is considered that an AUV equipped with a laser profiler, backed up by ROV-mounted video if necessary, would satisfy all regulatory requirements with regard to pipeline survey. It should also be noted that an AUV could fly a pipeline at, say, monthly intervals at very low cost if launched from an adjacent platform, or the shore. This could even be done in adverse weather conditions.

The main benefit of adopting the various recommendations presented herein is that they will lead to a much better appreciation of the state of a pipeline. Comparison with previous surveys would easily show, for instance, whether a pipeline was unburying, or a span was growing, enabling timely remedial action. In this way, potential problems will be prevented, rather than actual problems cured.

A further benefit is that such information will increase the general understanding of the seabed sedimentary processes and pipeline/seabed interactions discussed in Section 3.

6. Conclusions

This paper can be summarised by the following conclusions:

o   Current pipeline survey techniques have a number of limitations that impair the comprehension of the state of a pipeline. Limited ability to obtain pipeline history by comparison of surveys leads to the cure of problems rather than prevention.

o   Transverse profiles are the most useful data obtained from pipeline surveys. These should be made more available in current survey techniques.

o   Digitised transverse profiles should be combined to generate three-dimensional surface plots of pipelines and their topography. These would facilitate a better appreciation of the current state of a pipeline, and comparison with previous surveys would improve the understanding of the underlying mechanisms by providing a history of the pipeline.

o   Laser profiling may provide greater resolution, and enable higher survey speeds. Multi-altitude surveying will further improve the dataset.

o   The survey industry should move towards the use of Antonomous Underwater Vehicles for pipeline survey. These will be able to carry out laser profiling and photographic sampling relatively cheaply, and several times per year, if necessary.

o   As well as providing better information to pipeline operators, adoption of these recommendations will also enhance the general understanding of seabed sedimentary processes and pipeline/seabed interactions.

7. Acknowledgements

The author would like to thank Dr. A J Grass of University College London, and Mr G Pritchard of BP Exploration for their helpful comments. Thanks are also due to the Directors of J P Kenny Caledonia Limited for their comments, and for permission to publish this paper.

8. References

[1] "Submarine Pipelines Guidance Notes" (1984) The Pipeline Inspectorate, Petroleum Engineering Division, Department of Energy.

[2] Kjeldsen S.P. et al (1974): "Experiments with Local Scour around submarine pipelines in a uniform current". Report No:

STF 60 A 73085. Norwegian Hydrodynamic Laboratories, Trondheim, Norway.

[3] Bijker E.W. Leeuwestein W. (1983) "Interaction between Pipelines and the Seabed under the Influence of Waves and Currents": Symposium Proceedings: 'Seabed Mechanics' IUTAM/ IUGG. Section 7. pp 235-242.

[4] Kristiansen O.(1988): "Current induced vibrations and scour of pipelines on a sand bottom". Thesis. Institute of Structural Engineering, University of Trondheim.

[5] Bijker E.W. (1976) "Wave-Seabed-Structure Interaction." Conference Proceedings: "Behaviour of Offshore Structures", Norwegian Institute of technology pp: 830-845.

PIPELINE FREESPAN MONITORING

B.G.MURRAY
Development Engineering International
Wellheads Road
Farburn Industrial Estate
Dyce, Aberdeen, AB2 0HG

ABSTRACT. A technique has been developed for acquiring and analysing the vibration signature of pipeline freespans. The information derived can provide a cost effective means of avoiding unnecessary free span rectification work.

When assessing pipeline freespans, two aspects require to be considered separately. First, the stresses occurring in the span as a result of its own weight and current drag require to be estimated. Second, the susceptibility of the span to vortex induced vibration requires to be considered. In the absence of any quantitative measurements, span assessment requires to be made on the basis of conservative assumptions regarding such unknowns as span length, end fixity, axial tension etc.

Careful analysis of the vibration signature gives a quantitative measurement relating to each of the above, in addition to allowing more direct assessment of the vulnerability of the span to vortex shedding.

Experience has shown that the application for this system can dramatically reduce the rectification work required.

1. Introduction

The occurrence of free spans on subsea pipeline is, in many situations, inevitable as a result of current action and other influences. If the length of a free span becomes excessive, then rectification is required to ensure that the pipeline is not vulnerable to damage. In the past, the length of pipeline free spans has been estimated visually from ROV video records. It is widely acknowledged that this is often a difficult and unreliable estimate.

The purpose of this paper is to outline the capabilities of a quantitative span assessment technique, based on measuring the vibration signature of the free span. The technique has evolved from a concept first proposed by BP personnel. The initial development work was funded by BP and subsequent trials were supported by a consortium including BP, Statoil, Norsk Hydro, Total and Texaco. Following these trials the system has been used successfully in the field throughout the summer of 1988, 1989 and 1990.

In the following, there is a brief discussion of span assessment criteria. This is

Volume 27: Subtech '91, 347–354.

followed by a description of the information which can be gained from the pipespan vibration signature. Finally, the instrumentation required to obtain a free span vibration signature during a pipeline survey is described.

2.   Span Assessment

There are at least two reasons why the length of a pipeline freespan may be critical. First, the sag of the pipeline under its own weight may lead to a failure of the pipeline as a result of excessive bending stresses. Second, if the natural frequency of the span is sufficiently low, it is well known that transverse vibration can occur as a result of the action of vortex shedding induced by current flowing across the pipe. These two failure mechanisms are considered briefly in the following.

2.1   STATIC STRESSES

The bending stresses in a span can in principal be calculated from a knowledge of the geometry of the pipeline, the span length etc. However, there remain two difficulties. First, the nature of the end conditions of the span are unknown. Maximum stress calculations could be based on anything from simply supported to fully fixed conditions. In addition, the touchdown point is often difficult to locate precisely, resulting in a degree of uncertainty in the span length.

As a result of these uncertainties, it is possible for estimates of the stress in a pipeline free span to be in error by a factor of 2 or more. In the absence of better information, conservative assumptions must be made.

2.2   VORTEX SHEDDING

Vortex shedding becomes of importance if the natural frequency of the pipeline free span is matched by the vortex shedding frequency. In this event, large amplitude vibration can occur, potentially leading to pipeline failure as a result of accumulated fatigue damage.

In principle, the natural frequency of a span can be calculated from a pipeline survey based on the estimated span length. As before, however, the difficulties in assessing span length and end conditions introduce large uncertainties.

To summarise, there are two separate calculations required to determine whether a particular free span is critical. Both rely on input data which in the past was either difficult to obtain or unavailable. As a result, conservative assumptions must be made and it appears that resources are frequently wasted in rectifying spans unnecessarily.

3.   Vibration Based Span Assessment

If the natural frequencies of a pipeline free span can be measured directly, a more accurate assessment of the span condition can be made.

With regard to vortex shedding the situation is straightforward. The natural frequency of the span in both the transverse and in-line direction can be measured.

This information can then be used directly to determine the critical current velocity at which the vortex shedding frequency will match the span natural frequency. The critical current velocity can be compared with that estimated for the area resulting in a clear statement of status of the span with regard to vortex shedding.

Natural frequencies can also be used to improve estimates of pipeline static stresses. This can be carried out at various levels of sophistication.

First, the span length can be calculated from the natural frequency and this calculated length used to determine the peak bending stress. Since both natural frequency and peak stresses depend on end conditions in a similar way, uncertainties relating to end conditions are reduced. Moreover there is no longer any uncertainty associated with visual observation of video records and the span assessment is totally quantitative.

Analysis of the free span natural frequencies also provides a means of quantifying the axial tension in the span.

The axial tension in a free span can significantly affect both the peak stress and natural frequency. Axial tensions are produced in a pipeline during laying. During the life of the line, these tensions may relax but are certainly influenced by:

internal pressure
product temperature
sag of free span

The axial tension in a freespan can be calculated from measurements of its natural frequency. The method is based on recording differences in the ratio of the natural frequencies. For example the first three natural frequencies of a simply supported span lie in the ratio 1:4:9. If tension dominates over bending, the corresponding ratio is 1:2:3. In practice, if the first three natural frequencies are measured, the tension in the span can be calculated.

To summarise, the use of span vibration monitoring gives:

(i)   Direct measurement of the span natural frequency giving a clear indication of the vulnerability of the span to vortex shedding.

(ii) A quantitative measure of span length and end conditions for calculating peak stresses.

(iii) Means of estimating the residual tension in the span giving an improved estimate of peak stresses.

Free span vibration monitoring thus gives a cost effective means of quantifying span condition, highlighting problem areas and avoiding unnecessary rectification work.

4.   Case Studies

In order to illustrate the ideas presented in the previous section, several case studies taken from work carried out on offshore pipelines are described in the following.

Case Study (1) - Comparison of Rectification Criteria - 10" Gas Line

On this particular line, the danger of vortex shedding was regarded as a major span rectification criterion. A total of eight spans were regarded as marginal and therefore investigated by frequency analysis.

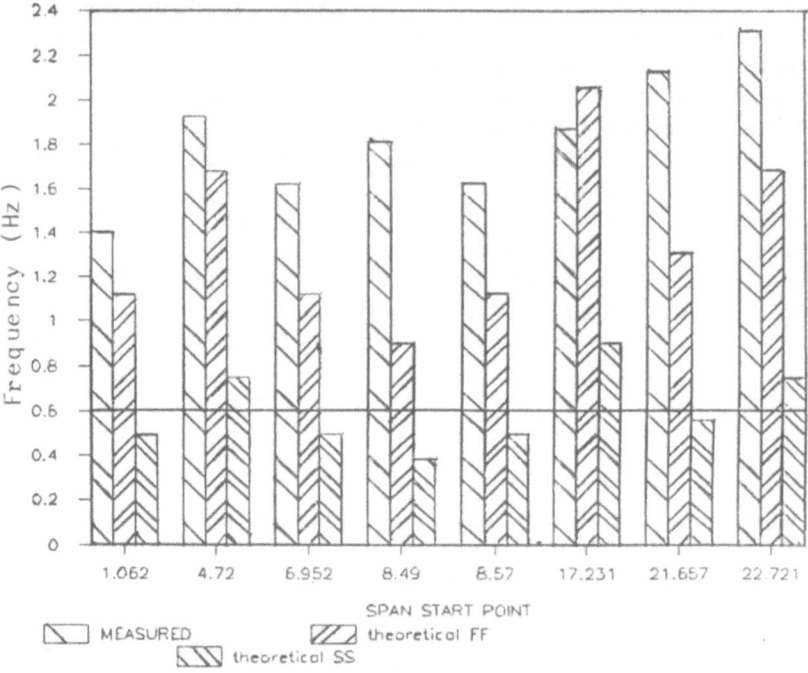

Figure 1. Comparison of Rectification Criteria

Figure 1 above illustrates the results. Natural frequencies were calculated from the span length using both simply supported and fully fixed assumptions. These are compared with the **measured** natural frequency in the vertical plane. It can be seen that generally the measured natural frequency was higher than fully fixed as a result of the curvature (sag) and tension in the span.

More significantly, a line is drawn at 0.6Hz, corresponding to estimated critical frequency for this line. It can be seen that a simply supported end condition assumption implies that five of the eight spans require rectification to eliminate possible vortex induced vibration. In reality, however, there is a large margin of safety on each span as the actual natural frequency is above that predicted even if fully fixed end conditions are assumed. Consequently no span rectification work is required.

Case Study (2) 32" Gas Line

Around 600 km of this line was surveyed, and vibration analysis carried out on critical or ambiguous spans.

Figure 2 shows data from typical span and the type of analysis resulting. Natural frequencies for the span were identified from measurements at 3 positions, corresponding to 40%, 30% and 10% along the span.

The first step in span assessment was to compare the natural frequency with the predicted vortex shedding frequency. The results were then processed to give axial tension, mid span deflection and peak bending stress.

Figure 2 shows a typical computer display of the raw vibration signature and type of results obtained by processing this data.

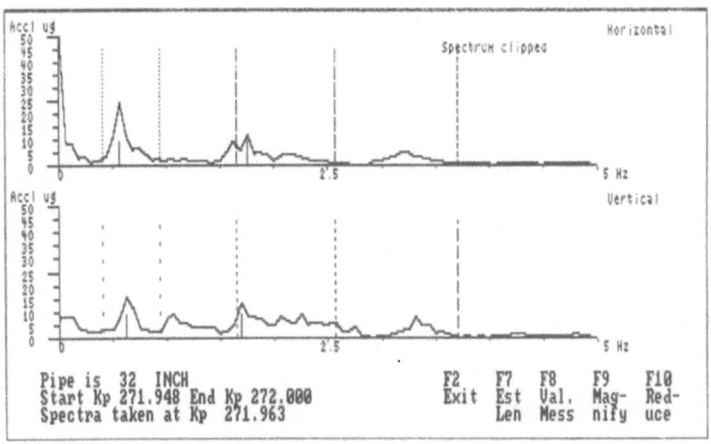

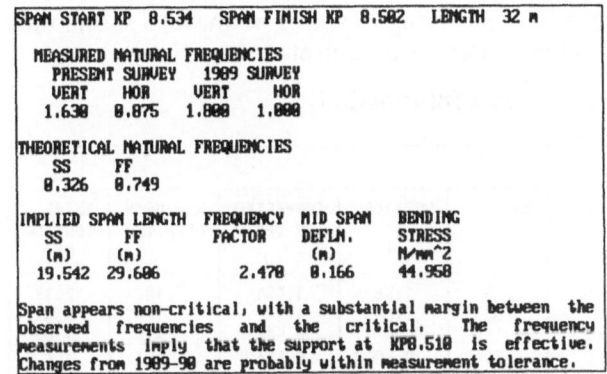

Figure 2. Typical Computer Display and Type of Results

Figure 3 shows a summary of the results for the whole line. The spans shown were identified by the Operators assessment criteria as being critical and were therefore subject to frequency investigation.

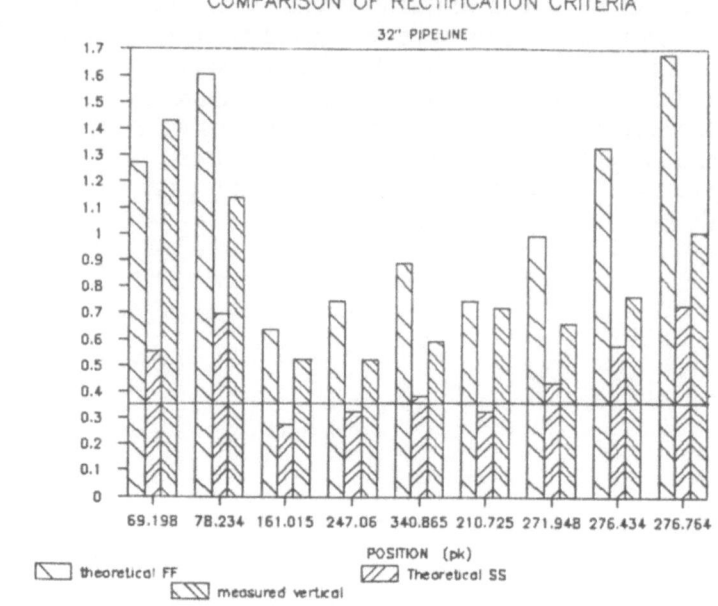

Figure 3.   Summary of Results

Clearly, all spans were well above the critical frequency of .35 Hz, and were
therefore at no risk from vortex shedding. Table 1 below also shows the static stress
information derived from the frequency measurements.

TABLE 1. Static Stress Information

32 INCH						
START OF SPAN	LENGTH	HORIZ. F1	VERTICAL F1	CORRECTED F1 VERT	MID SPAN DEFN.(m)	AXIAL TENSION MN
69.198	46	1.25	1.34	1.431	.18	-2.17
78.234	41	1.06	1.06	1.14	.129	3.0
161.015	65	0.437	0.5	0.524	.215	1.99
247.060	60	0.5000	0.5000	0.5260	.106	3.6
340.865	55	0.5625	0.5625	0.595	.111	0.5
210.725	60	0.5625	0.6875	0.724	.262	2.6
271.948	52	0.5625	0.625	0.663	.202	3.1
276.434	45	0.72	0.7188	0.769	.121	2.7
276.764	40	0.875	0.9375	1.01	.187	1.5

Consideration of these results demonstrated that none of the spans investigated in
fact required rectification.

Case Study (3) - Investigation of Rectified Span

In this case a 74 m span as illustrated in Figure 4 was investigated during an annual survey. Previously, grout bags had been used to provide intermediate support to the span. However, visual inspection indicated that the grout bags were falling away from the line, possibly leaving it unsupported. However, frequency measurements clearly showed that the span natural frequency was around 1.6 Hz. This corresponds to the spacing between the grout bags rather than the overall span length. Moreover, different frequencies were recorded on each subspan. This clearly demonstrated that the grout bags were providing support, so no intervention was required.

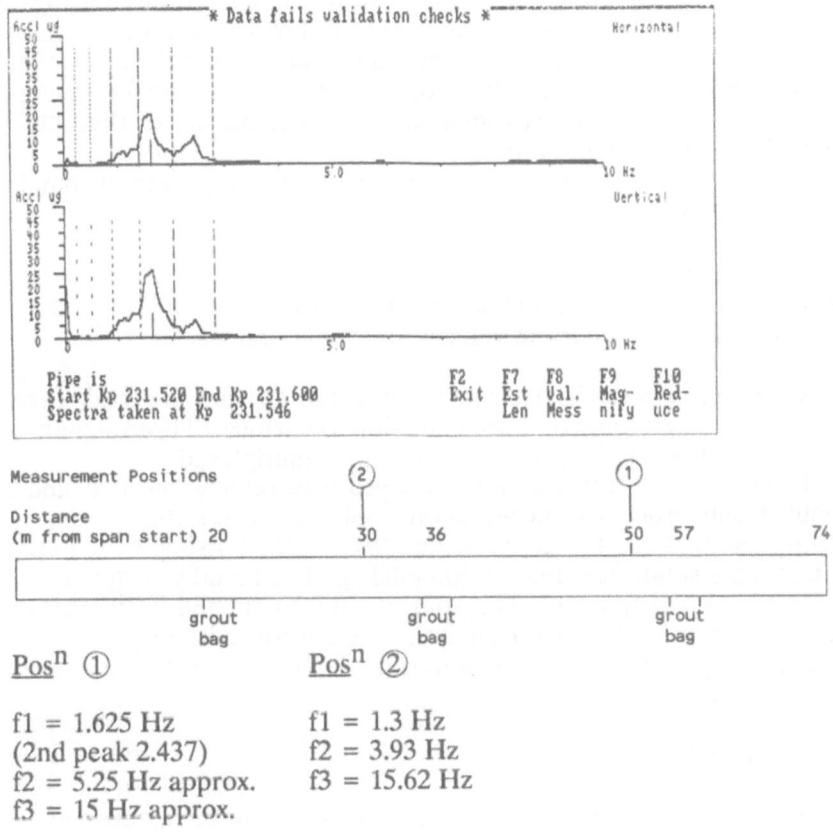

Figure 4. 74 m Span

5. Measurement Techniques

Pipeline vibration spectra can be obtained either during an ROV pipeline survey or, if necessary, by means of diver deployed instrumentation. When span monitoring is integrated into an ROV pipeline survey, it is extremely cost effective since the vibration data is obtained with little interruption to other survey activities, resulting in a negligible increase in vessel and ROV costs.

Vibration monitoring during a pipeline survey is carried out by deploying vibration transducers from the vehicle to record horizontal and vertical motion of the pipeline when a free span is suspected. Vibration of the span can then be excited either as a result of the random turbulence generated by the ROV or, by deflecting the span (increasing vehicle downthrust to maximum then abruptly reducing downthrust). The resulting vibration signal is passed via the ROV umbilical to a compact vibration analysis and recording work station on surface. Here vibration spectra are displayed and archived. The key requirement offshore is the capture of reliable data. For this reason, the software includes validation routines which verify the general pattern of the spectrum. This reliably detects faults ranging from transducer and cable failures to external disturbance of the transducer. At the request of the operator, the computer displays an estimate of span length based on the spectra recorded. This gives a further means of verifying data on site. The span assessment can be made on site or the data archived for detailed investigation onshore.

A support IBM PC software package gives a span report to the pipeline engineer. This report contains:

(1) span length from survey
(2) span vibration spectra and natural frequencies
(3) calculated critical current velocity and static stresses

The vibration monitoring package requires power from the ROV (instrumentation current levels only) and data transmission to surface. Three signals require to be transmitted either by dedicated conductors or multiplexed.

The vibration instrumentation employed is extremely sensitive, and can produce reliable results from pipeline vibration amplitudes of less than 1 mm.

If operational considerations dictate that manned diving is the most appropriate route, then a small instrumentation package is manually strapped to the pipeline span under investigation. The span is then disturbed by the diver, producing adequate vibration levels for analysis. As before, initial span assessment can be made on site with information archived for detailed analysis onshore.

6.   Conclusion

Free span natural frequency measurements can be made rapidly and reliably during an ROV pipeline survey. The information gained in this way gives a precise picture of the condition of the span, allowing

(a) exact calculation of the critical current velocity for vortex shedding
(b) estimate of the end support conditions on the span
(c) estimate of the curvature of the span
(d) estimate of the residual tension in the span